CHEMICAL EQUILIBRIA IN SOILS

WILLARD L. LINDSAY
Centennial Professor
Colorado State University, Fort Collins

Reprint of the 1979 Edition by John Wiley & Sons, Inc.

Copyright © 1979 by Willard L. Lindsay

All rights reserved. No part of this book may be reproduced, stored in a retrieval system, or transmitted by any form or by any means, electronic, mechanical, photocopying, recording, or otherwise, except as may be expressly permitted by the applicable copyright statutes or in writing by the publisher.

Chemical Equilibria in Soils

ISBN: 1-930665-11-3

Library of Congress Control Number: 2001086786

THE BLACKBURN PRESS
P. O. Box 287
Caldwell, New Jersey 07006
U.S.A.
973-228-7077
www.BlackburnPress.com

$$\underline{Al^{3+} > H^+} > Ca^{+2} > Mg^{+2} > K = NH_4 > Na$$

137

CHEMICAL EQUILIBRIA IN SOILS

P. 114

P. 125

ask about Fig. 16.2

P.92 Fig

— if gypsum applied to Sodium affected soil like ND and there is enough SO_4 in the system, then Gypsum will become solid and that won't work in our soil

— in our soil, we have alot of dolomite bc smectite/monmorolitle when it weathers, it gives Mg and that Mg don't go away.

Handwritten note at top:

The difference btw Ion Pair and Complex Ion?

(Ca⁺)(SO₄⁼)
– they still carry their ion charge and can be combined together

Complex Ion
CaH₂PO₄⁺
– the whole complex will have one charge

PREFACE

This book is the outgrowth of approximately 20 years experience on my part in teaching soil chemistry and directing graduate research in soil science. Its objective is to help bridge the gap between soil science and chemistry and to show that most reactions taking place in soils can be understood and predicted from basic chemical relationships.

Emphasis in this text is placed on minerals and solid phases in soils that dissolve and precipitate and, in doing so, control the composition of the soil solution. Solubility relationships are important because they determine the mobility of chemical elements in soils and affect their availability to plants.

This book is designed for students who have had at least one year of inorganic chemistry. The text is intended for soil scientists, plant nutritionists, aquatic chemists, geochemists, sanitary and water engineers, environmentalists, and others who are concerned with the reactions, solubility relationships, and fate of chemical substances in soils.

The book is arranged sequentially; therefore first-time readers should start at the beginning and work diligently to understand each new principle as it is introduced. Problems are given at the end of each chapter to assist the reader in understanding the subject matter and in developing the skills necessary to handle solubility relationships with ease. Without such skills, most readers become discouraged before they appreciate what equilibrium relationships are all about.

Redox reactions are very important in soils because they modify solubility relationships and affect many mineral transformations. In Chapter 2 the reader is introduced to $pe + \text{pH}$ as a new redox parameter that simplifies the handling of many theoretical redox relationships. This parameter partitions the H^+ ions of an overall chemical reaction into those associated with the redox component of the reaction and those associated with the acid-base component. Readers are challenged to make greater use of this convenient parameter.

The solubility relationships used throughout this text are based on selected standard free energies of formation that are summarized and documented in the Appendix. The author was assisted in this exhaustive search and selection by Dr. Muhammad Sadiq. These values have been adjusted, when necessary, to be internally consistent. Hopefully this compilation of thermodynamic data will find many uses beyond the immediate scope of this book. As new solubility data become available, they can be used to expand and modify the developments given herein.

No attempt has been made in this book to apply rigorous kinetic theory to predict chemical reaction rates in soils because the presence of living organisms, catalysts, and unknown constituents in soils are often the controlling factors governing reaction rates. Instead, selected solubility studies in soils have been used to develop practical guidelines for interpreting many of the thermodynamic solubility relationships developed in this book. The solubility relationships that we consider are limited to 25°C. Although temperature changes modify solubility relationships slightly, the added complexity of including them could not be justified.

Undoubtedly some readers will find that their speciality elements are not included in this work. Limits of time and effort have precluded a complete coverage of all elements that are important in soils. Readers are encouraged to apply the principles demonstrated herein to their speciality elements. The rewards for doing so are even more satisfying than reviewing what someone else has already done.

WILLARD L. LINDSAY

Fort Collins, Colorado
May 1979

ACKNOWLEDGMENTS

I am especially grateful to Dr. Muhammad Sadiq who for nearly two years untiringly assisted with the formidable task of collecting and screening the standard free energies of formation documented in the Appendix. This unifying effort has added greatly to the internal consistency of the solubility data used throughout this book.

Appreciation is also expressed to the many graduate students and colleagues, who over the years, through discussion, dialogue, and research have contributed suggestions and insights that are reflected in this work. Special appreciation is extended to Drs. Wendell A. Norvell and Ardell D. Halvorson for their contributions in measuring the activities of metal ions in soils and hydroponic solutions, to Drs. Ahmed A. Abuzkhar and Lee Sommers for their assistance in developing the computer programs for the chelate equilibrium diagrams, to Dr. Dhanpat Rai for his initial work on aluminosilicate equilibria, to Floyd J. Adamsen for his contribution on organic reactions, and to Drs. Paul L. G. Vlek and Jimmy J. Street for their helpful suggestions and encouragement.

Special thanks goes to Lorraine K. Nukaya for typing the preliminary drafts and final manuscript and for always maintaining a cheerful and helpful attitude. Appreciation is extended to David W. Fanning for the line drawings and for taking special pride in the meticulous details of his work.

Appreciation is extended to Dr. Muhammed Sadiq, Dr. Robert D. Heil, Mr. Paul A. Schwab, Mr. Lavoir A. Banks, and Mr. Randal A. Gaseor for reviewing the manuscript and offering many helpful suggestions.

Finally, heartful thanks goes to my wife, Lorna, and our children for their cooperation and understanding, which enabled me to devote so many long hours to what, at times, seemed an endless task.

W. L. L.

CONTENTS

	Symbols	xvii
1	**Introduction**	**1**
	1.1 Dynamic Equilibria in Soils,	2
	1.2 Intensity and Capacity Factors,	4
	1.3 Elemental Composition of Soils,	6
	References,	8
	Problems,	9
2	**Methods of Handling Chemical Equilibria**	**10**
	2.1 Equilibrium Constants,	11
	2.2 Concentration Versus Activity Constants,	12
	2.3 Ionic Strength,	12
	2.4 Activity Coefficients,	13
	2.5 Debye–Hückel Equations,	13
	2.6 Activity Coefficients from Electrical Conductivities,	16
	2.7 Transforming Equilibrium Constants,	18
	2.8 Equilibrium Constants from Thermodynamic Data,	21
	2.9 Redox Relationships,	23
	2.10 $pe + pH$,	23
	2.11 $E°$ versus $\log K°$,	26
	2.12 pe versus Eh,	27

2.13	Redox Measurements in Natural Environments	28
	References,	30
	Problems,	31

3 Aluminum — 34

3.1	Solubility of Aluminum Oxides and Hydroxides,	35
3.2	Solubility of Aluminum Sulfates,	38
3.3	Hydrolysis of Al^{3+}	39
3.4	Fluoride Complexes of Aluminum,	41
3.5	Other Aluminum Complexes,	43
3.6	Estimating Al^{3+} Activity,	45
3.7	Redox Relationships of Aluminum,	46
3.8	Exchangeable Aluminum,	47
	References,	48
	Problems,	49

4 Silica — 50

4.1	Forms of Silica in Soils,	51
4.2	Silicate Species in Solution,	53
	References,	54
	Problems,	55

5 Aluminosilicate Minerals — 56

5.1	Unsubstituted Aluminosilicates,	57
5.2	Sodium Aluminosilicates,	62
5.3	Potassium Aluminosilicates,	65
5.4	Calcium Aluminosilicates,	66
5.5	Magnesium Aluminosilicates,	68
5.6	Summary Stability Diagrams for Aluminosilicates,	71
5.7	Controls of Al^{3+} Activity in Soils,	73
5.8	General Discussion of Aluminosilicates,	75
	References,	76
	Problems,	77

6 Carbonate Equilibria — 78

6.1	The CO_2—H_2O System,	79
6.2	The CO_2—Soil System,	84
	References,	84
	Problems,	84

CONTENTS   xi

7 Calcium   86

7.1 Calcium Silicates and Aluminosilicates,   87
7.2 Other Calcium Minerals,   93
7.3 Complexes of Calcium in Solution,   95
7.4 Redox Relationships of Calcium,   98
7.5 The $CaCO_3-CO_2-H_2O$ System,   98
7.6 The Phase Rule,   101
7.7 The CO_2-H_2O System,   102
7.8 The $CaO-CO_2-H_2O$ System,   102
7.9 The $CaO-CO_2-H_2O-H_2SO_4$ System,   102
     Problems,   103

8 Magnesium   105

8.1 Solubility of Magnesium Silicates,   106
8.2 Magnesium Aluminosilicates,   112
8.3 Oxides, Hydroxides, Carbonates, and Sulfates,   113
8.4 Magnesium Complexes in Solution,   114
8.5 Effect of Redox on Magnesium,   116
     Problems,   116

9 Sodium and Potassium   118

9.1 Solubility of Sodium Minerals,   119
9.2 Solubility of Potassium Minerals,   123
9.3 Complexes of Sodium and Potassium,   125
9.4 Redox Relationships,   126
     Problems,   127

10 Iron   128

10.1 Solubility of Fe(III) Oxides in Soils,   129
10.2 Other Fe(III) Minerals,   133
10.3 Hydrolysis of Fe(III),   134
10.4 Fe(III) Complexes in Soils,   136
10.5 Effect of Redox on Fe(II) Solubility,   139
10.6 Effect of Redox on the Stability of Iron Minerals,   141
10.7 Hydrolysis and Complexes of Fe(II),   146
      References,   148
      Problems,   149

11 Manganese — 150

11.1 Effect of Redox and pH on Manganese Solubility, 151
11.2 Solution Species of Manganese, 157

    References, 160
    Problems, 160

12 Phosphates — 162

12.1 Orthophosphoric Acid, 163
12.2 Aluminum Phosphates, 169
12.3 Iron Phosphates, 173
12.4 Effect of Redox on the Stability of Iron and Aluminum Phosphates, 177
12.5 Solubility of Calcium Phosphates, 180
12.6 Effect of Redox on the Stability of Calcium Phosphates, 185
12.7 Solubility of Magnesium Phosphates, 186
12.8 Manganese Phosphates, 187
12.9 Other Orthophosphates, 189
12.10 Reduced Forms of Phosphorus, 189
12.11 Stability of Polyphosphates in Soils, 190
12.12 Orthophosphate Complexes in Solution, 195
12.13 Reactions of Phosphate Fertilizers with Soils, 197

    References, 204
    Problems, 205

13 Zinc — 210

13.1 Oxidation State of Zinc, 211
13.2 Solubility of Zinc Minerals in Soils, 211
13.3 Zinc Species in Solution, 216

    References, 219
    Problems, 219

14 Copper — 221

14.1 Solubility of Cu(II) Minerals in Soils, 222
14.2 Hydrolysis and Solution Complexes of Cu(II), 228
14.3 Effect of Redox on Copper, 231
14.4 Complexes of Cu(I), 234

    References, 235
    Problems, 236

15 Chelate Equilibria — 238

15.1 Metal Chelates and Their Stability Constants, 239
15.2 Development of Stability-pH Diagrams for Chelates, 244
15.3 Effect of Redox on Metal Chelate Stability, 252
15.4 Chelation in Hydroponics, 256
15.5 Use of Chelates to Estimate Metal Ion Activities in Soils, 259
15.6 Use of Chelating Agents as Soil Tests, 261
15.7 Natural Chelates in Soils, 263

References, 264
Problems, 265

16 Nitrogen — 267

16.1 Oxidation States of Nitrogen, 268
16.2 Equilibrium Between Atmospheric N_2 and O_2, 269
16.3 Effect of Redox on Nitrogen Stability, 272

References, 279
Problems, 280

17 Sulfur — 281

17.1 Effect of Redox on Sulfur Speciation, 282
17.2 Dissociation of Sulfur Acids, 287
17.3 Formation of Elemental Sulfur in Soils, 288
17.4 Formation of Metal Sulfides, 290
17.5 Effect of Sulfides on Metal Solubilities, 295

References, 297
Problems, 297

18 Silver — 299

18.1 Effect of Redox on the Stability of Silver Minerals, 300
18.2 Solubility of Silver Halides and Sulfides, 304
18.3 Stability of Other Silver Minerals, 306
18.4 Stability of Silver Halide Complexes, 308
18.5 Hydrolysis Species and Other Silver Complexes, 310

References, 313
Problems, 313

19 Cadmium — 315

- 19.1 Oxidation States of Cadmium in Soils, — 316
- 19.2 Cadmium Minerals in Soils, — 316
- 19.3 Hydrolysis Species of Cd(II), — 321
- 19.4 Halide and Ammonia Complexes of Cadmium, — 322
- 19.5 Other Cadmium Complexes, — 323
- 19.6 Need for Further Studies, — 326

    References, — 326
    Problems, — 327

20 Lead — 328

- 20.1 Solubility of Lead Minerals, — 329
- 20.2 Hydrolysis Species of Lead, — 338
- 20.3 Halide Complexes of Lead, — 339
- 20.4 Other Complexes of Lead, — 341

    References, — 341
    Problems, — 342

21 Mercury — 343

- 21.1 Stability of Hg(II) Minerals and Complexes, — 344
- 21.2 Stability of Hg(I) Minerals and Complexes, — 353
- 21.3 Stability of Elemental Mercury, — 355
- 21.4 Solubility of Mercury Sulfides in Soils, — 358
- 21.5 Summary Redox Diagram for Mercury, — 359
- 21.6 Organic Mercury Reactions, — 362

    References, — 362
    Problems, — 362

22 Molybdenum — 364

- 22.1 Molybdenum Species in Solution, — 365
- 22.2 Stability of Molybdenum Minerals in Soils, — 367
- 22.3 The Effect of Redox on Molybdenum Solubility, — 369

    References, — 372
    Problems, — 372

23 Organic Transformations — 373

- 23.1 Oxidation States of Carbon, — 374
- 23.2 Products of Glucose Metabolism, — 375

23.3	Reactions of Acetic Acid,	380
23.4	Oxidation to $CO_2(g)$ and Reduction to $CH_4(g)$,	381
23.5	Stability of Graphite,	382
	References,	383
	Problems,	383
Appendix	**Standard Free Energies of Formation**	**385**
Index		**423**

SYMBOLS, CONSTANTS, AND ABBREVIATIONS

A	Constant = 0.509 at 25°C for the Debye-Hückel equation
a_i	Activity of species i
Avog.	Avogadro's number, 6.02252×10^{23} formula units mole^{-1}
amorp	Amorphous solid
atm	Pressure in atmosphere, 1,013,250 dynes cm^{-2}
bar	10^6 dynes cm^{-2}
[ ]	Brackets indicate concentrations in moles liter^{-1}
conc.	Concentration
(c)	Crystalline solid
°C	Temperature in degrees Celsius
d_i	Effective diameter of ions in the Debye-Hückel equation
deg	Degrees of temperature in Kelvin or Celsius
e	Base of the natural logarithm
e^-	Electron
est.	Estimated
ε	Electronic charge in the Boltzman equation
$E°$	Standard electrode potential (volts or millivolts)
EC	Electrical conductivity (millimhos cm^{-1})
Eh	Electrical potential relative to the standard hydrogen electrode (volts or millivolts)

SYMBOLS, CONSTANTS, AND ABBREVIATIONS

F	Faraday constants $= 96{,}487$ coulombs equivalents^{-1} $= 23{,}061$ calories volts^{-1} equivalent^{-1}
γ_i	Activity coefficient of species i
(g)	Gas phase
G	Free energy (kcal)
ΔG_f°	Gibbs standard free energy of formation (kcal mole^{-1})
ΔG_r°	Gibbs standard free energy of reaction (kcal mole^{-1})
ΔH_f°	Standard heat of formation or enthalpy (kcal mole^{-1})
ΔH_r°	Standard heat or enthalpy of reaction (kcal mole^{-1})
(I), (II), etc.	Immediately following a chemical symbol signifies the oxidation state of that element
k	Kilo
κ	Boltzman constant
K_f	Equilibrium formation constant
K_d	Equilibrium dissociation constant
°K	Temperature in degrees Kelvin or absolute
K°	Equilibrium constant expressed in terms of activities
K^c	Equilibrium constant expressed in terms of concentrations
K^m	Equilibrium constant expressed in terms of concentrations except for H$^+$, OH$^-$, and e$^-$, which are expressed in terms of activities
(l)	Liquid phase
ln	Natural logarithm, base $e = 2.71828\ldots$
log	Common logarithm, base 10
ln x	$2.302585 \log x$
M	Concentration in terms of moles liter^{-1}
MF	Mole fraction
n	The number of moles of electrons participating in a reaction
N	Concentration in terms of equivalents liter^{-1}
°	Placed behind a chemical compound indicates an aqueous solution species with no charge, for example, H$_4$SiO$_4$°
( )	Parentheses around chemical species indicate activities (moles liter^{-1})
p	Negative log of base 10.
pe	Negative log of electron activity where electron activity is defined as unity for the standard hydrogen electrode
P_i	Partial pressure of gas i (atm)
Q	Quotient expressing products divided by reactants of a given reaction
R	Universal gas constant $= 1.98717$ cal deg^{-1} mole^{-1} $= 8.3143$ joules deg^{-1} mole^{-1}
r	Correlation coefficient

SYMBOLS, CONSTANTS, AND ABBREVIATIONS

S°	Standard entropy of a substance (cal deg^{-1} mole^{-1})
ΔS_r°	Standard entropy of reaction (cal deg^{-1} mole^{-1})
Σ	summation
T	Temperature in degrees Kelvin (°K)(25°C = 298.16°K)
μ	Ionic strength by itself or micro (10^{-6}) if it precedes another unit of measure
STP	Standard temperature (298.16°K) and pressure (1 atmosphere)
Z	Valency
ψs	Electrical potential on the surface of a clay, Boltzmann equation.

CHEMICAL EQUILIBRIA
IN SOILS

ONE

INTRODUCTION

Soil has been defined as "rock on its way to the ocean." In this respect soil comprises the residual weathering products of rocks and minerals as the more soluble components are leached away.

Many physical, chemical, and biological changes continually take place in soils. Physical processes such as wetting, drying, freezing, thawing, changing temperatures, and leaching modify the surface areas of soil particles. Primary minerals change to secondary minerals as ionic species in solution seek lower free-energy levels. In addition, plants capture energy from the sun and store it in the form of organic compounds. Microorganisms utilize and transform these products through their many biochemical pathways. How can this maze of complex reactions that take place in soil be systematically examined?

During the past century our knowledge of chemistry has advanced tremendously, yet a great gap exists between chemical knowledge and its application to soils where so many parameters are unknown. Such perplexities have caused many soil scientists to resort to empirical investigations of soils without first utilizing the vast knowledge of chemistry that is available.

This book was designed to help bridge the gap between chemistry and soil science and to show how the principles of chemical equilibria can be used to examine many of the chemical reactions that occur in soils.

1.1 DYNAMIC EQUILIBRIA IN SOILS

Soils comprise a multiple-phase system consisting of numerous solid phases (about 50%), a liquid phase (about 25%), and a gas phase (about 25%). The solids include rock consisting of many different primary and secondary minerals. Superimposed on this inorganic matrix is what Truog (1951) described as the "living phase," which includes bacteria, actinomycetes, fungi, algae, protozoa, nematodes, and other forms of life. These living organisms are continually breaking down organic residues and synthesizing many of the products into body tissues while others are released to the surroundings. The manner in which various constituents of the soil interact is depicted diagrammatically in Fig. 1.1.

The soil solution is the focal point in this diagram and is the liquid phase that completely envelops the solid phases. It is the medium from which plants absorb their nutrients (Reaction 1). Small quantities of plant constituents may also be released back into the soil solution (Reaction 2). Ions in the soil solution are buffered by those adsorbed onto soil surfaces or held by exchange sites (Reactions 3 and 4). Removal of ions from the soil solution causes partial desorption of similar ions from the exchange complex.

Soils contain numerous minerals, some of which are crystalline, others are

DYNAMIC EQUILIBRIA IN SOILS

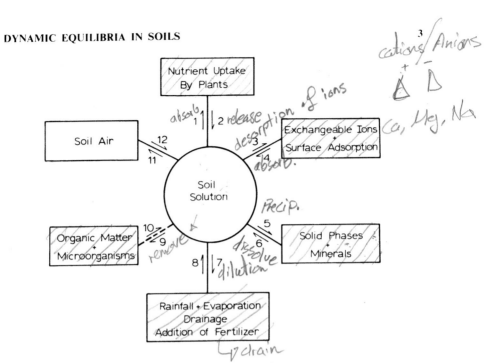

Fig. 1.1 The dynamic equilibria that occur in soils.

amorphous. These minerals impose limits on the chemical composition of the soil solution. If the soil solution becomes supersaturated with respect to any mineral, that mineral can precipitate (Reaction 5) until equilibrium is attained. Similarly, if the soil solution becomes undersaturated with respect to any mineral present in the soil, that mineral can dissolve until equilibrium is attained (Reaction 6).

Reactions 7 and 8 depict several dynamic processes that may occur in soils. For example, rainfall adds water that dilutes the soil solution (Reaction 8). Excess water may drain from the soil profile and carry with it salts and other dissolved constituents (Reaction 7). Fertilizers of various kinds are frequently added to soils. These may dissolve (Reaction 8), form new reaction products (Reaction 5), or be distributed in other ways in soil.

Organic matter and microorganisms also affect the equilibrium relationships in soils. Living organisms remove constituents from the soil solution and incorporate them into their body tissues (Reaction 9). Similarly nutrients are released during the decomposition of organic matter or upon the death of organisms (Reaction 10). These reactions are connected with a broken line to indicate that true equilibrium relationships are generally not achieved but are modified by the metabolic energy relationships of microorganisms that mediate many of these reactions.

Gases in the soil air also tend to attain equilibrium with the soil solution. Gases may either be released to the soil air (Reaction 11) or dissolved in the soil solution (Reaction 12). In soils, plants and microorganisms generally utilize O_2 as an electron acceptor and give off CO_2 from metabolic processes. Diffusion gradients are, therefore, established between the gas phase in the finer pore spaces of soils and the atmosphere above. In waterlogged soils the exchange of $O_2(g)$ and $CO_2(g)$ is greatly restricted because diffusion rates in water are approximately 10^{-4} those in air. As O_2 in the soil is depleted, the soil becomes reduced. Even in unsaturated soils there are often zones of fine-textured materials where reducing conditions may prevail.

The soil solution is affected by all of the reactions depicted in Fig. 1.1, but its composition is ultimately controlled by the mineral phases of the soil. Often the rates of dissolution and precipitation of soil minerals are so slow that true equilibrium is not attained; consequently both kinetic and thermodynamic factors must be considered. Diffusive and convective gradients are both established in soils, and these gradients must also be considered where transport processes are involved.

1.2 INTENSITY AND CAPACITY FACTORS

In soils two very important parameters influence the availability of an element to plants. These are (1) the intensity factor, which is the concentration of an element in the soil solution, and (2) the capacity factor, which is the ability of solid phases in soils to replenish that element as it is depleted from solution. As plants remove ions from solution, the concentration of those ions in the immediate vicinity of roots is reduced and diffusion gradients are established.

The relationship between intensity and capacity factors as they affect nutrient availability is depicted graphically in Fig. 1.2. The concentration of a nutrient in the soil solution, termed the intensity factor, is represented by the vertical axis, and the amount that can potentially come into solution, termed the capacity factor, is represented by the horizontal axis. Depicted at time 0 are three mineral phases: A, B, and C are each capable of maintaining a different level of the same nutrient in the soil solution. Initially mineral A is most soluble and will control the nutrient level in solution. This level causes supersaturation with respect to minerals B and C, which slowly precipitate causing mineral A to dissolve. Once mineral A is completely dissolved, the solution drops from a to b. This drop may not be immediate because of the buffering action of the adsorbed and exchangeable ions of the soil that must attain the new equilibrium level corresponding

INTENSITY AND CAPACITY FACTORS

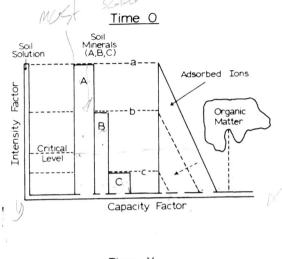

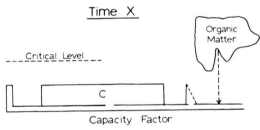

Fig. 1.2 Influence of soil minerals on the solubility and availability of nutrients to plants (adapted from Lindsay, 1972).

to level b in the soil solution. Even after the soil solution attains equilibrium with mineral B, chemical reactions do not cease. The nutrient level at b is still supersaturated with respect to mineral C. Consequently, mineral C slowly precipitates as mineral B dissolves. Eventually mineral B disappears and only mineral C remains.

At time X, when mineral C governs the nutrient level, plants will show a deficiency of this nutrient because its level in solution is below the critical level needed for maximum plant growth. The adsorption mechanism at time X provides very little nutrient reserve. Organic matter can supply some nutrients, but generally it must be broken down by microorganisms before the nutrients are available to plants. The modifying effects of adsorbed ions and the activity of microorganisms are important, but the mineral phases ultimately control the level of nutrients in solution.

1.3 ELEMENTAL COMPOSITION OF SOILS

Soils differ in total chemical composition. These differences are important in equilibrium considerations because they help to determine which elements control the solubility of other elements.

The elemental composition of the lithosphere and of soils is reported in Table 1.1. These data are based on the reports of Clarke and Washington (1924), Swaine (1955), Vinogradov (1959), Jackson (1964), Bowen (1966), Mitchell (1964), and Taylor (1964). The selected average is an arbitrary reference level for soils used in this text. The last column in Table 1.1 gives the maximum concentration of each element in the soil solution if all that element at its average reference level were to dissolve in the water present at 10% of the dry weight of the soil. This parameter is expressed as log M (moles liter^{-1}) and provides a limiting molar concentration for each element in the soil solution.

The molecular ratio of any two elements is useful in relating the stoichiometric combinations of those elements in various minerals. For example, iron may combine with phosphorus to form $FePO_4 \cdot 2H_2O$(strengite) in soils. The molecular ratio of iron to phosphorus in an average soil is $10^{0.83}/10^{-0.71} = 10^{1.54}$ or 34.7-fold. All of the phosphorus may be present as strengite, but all of the iron cannot, because there is not enough phosphorus to combine with it. The conclusion can be drawn that iron may control phosphorus solubility, but phosphorus can not control iron solubility. Such deductions are useful in selecting and eliminating phases that may govern the solubility of different elements in soils where many complex solubility relationships are found.

The average molar concentration of elements in soils given in the last column of Table 1.1 can easily be adjusted to correspond to actual elemental compositions and moisture contents. For example, if a soil treated with sewage sludge contains 500 rather than 50 ppm of zinc, the log of the ratio $(500/50) = 1.00$ can be added to -2.12 (Table 1.1) to give -1.12 M for the maximum concentration of zinc in this soil at 10% moisture. Furthermore, if the moisture content of this soil were 40% instead of 10%, the log of this ratio $(10/40) = -0.60$ can be added to -1.12 to give -1.72 M for the maximum concentration of zinc in this soil containing 500 ppm of zinc at 40% moisture. If a suspension of this soil were prepared in the laboratory to contain 1 g of soil (oven-dry basis) per 100 ml of water, the maximum molar concentration of zinc is obtained as follows:

$$\begin{aligned} \log[Zn^{2+}] &= -2.12 + \log 500/50 + \log 0.1/100 \\ &= -2.12 + 1.00 + (-3.00) \\ &= -4.12 \, M \end{aligned}$$

TABLE 1.1 THE CONTENT OF VARIOUS ELEMENTS IN THE LITHOSPHERE AND IN SOILS

Element	Atomic Weight (g)	Content in Lithosphere (ppm)	Common Range for Soils (ppm)	Selected Average for Soils ppm	Molar Conc. at 10% Moisture log M
Ag	107.87	0.07	0.01–5	0.05	−5.33
Al	26.98	81,000	10,000–300,000	71,000	1.42
As	74.92	5	1–50	5	−3.18
B	10.81	10	2–100	10	−2.03
Ba	137.34	430	100–3,000	430	−1.50
Be	9.01	2.8	0.1–40	6	−2.18
Br	79.91	2.5	1–10	5	−3.20
C	12.01	950		20,000	1.22
Ca	40.08	36,000	7,000–500,000	13,700	0.53
Cd	112.40	0.2	0.01–0.70	0.06	−5.27
Cl	35.45	500	20–900	100	−1.55
Co	58.93	40	1–40	8	−2.87
Cr	52.00	200	1–1,000	100	−1.72
Cs	132.91	3.2	0.3–25	6	−3.35
Cu	63.54	70	2–100	30	−2.33
F	19.00	625	10–4,000	200	−0.98
Fe	55.85	51,000	7,000–550,000	38,000	0.83
Ga	69.72	15	5–70	14	−2.70
Ge	72.59	7	1–50	1	−3.86
Hg	200.59	0.1	0.01–0.3	0.03	−5.83
I	126.90	0.3	0.1–40	5	−3.40
K	39.10	26,000	400–30,000	8,300	0.33
La	138.91	18	1–5,000	30	−2.67
Li	6.94	65	5–200	20	−1.54
Mg	24.31	21,000	600–6,000	5,000	0.31
Mn	54.94	900	20–3,000	600	−0.96
Mo	95.94	2.3	0.2–5	2	−3.68
N	14.01	—	200–4,000	1,400	0.00
Na	22.99	28,000	750–7,500	6,300	0.44
Ni	58.71	100	5–500	40	−2.17
O	16.00	465,000		490,000	2.49
P	30.97	1,200	200–5,000	600	−0.71
Pb	207.19	16	2–200	10	−3.32
Rb	85.47	280	50–500	10	−2.93
S	32.06	600	30–10,000	700	−0.66
Sc	44.96	5	5–50	7	−2.81

(Continued)

TABLE 1.1 (*Continued*)

Element	Atomic Weight (g)	Content in Lithosphere (ppm)	Common Range for Soils (ppm)	Selected Average for Soils ppm	Molar Conc. at 10% Moisture log M
Se	78.96	0.09	0.1–2	0.3	−4.42
Si	28.09	276,000	230,000–350,000	320,000	2.06
Sn	118.69	40	2–200	10	−3.07
Sr	87.62	150	50–1,000	200	−1.64
Ti	47.90	6,000	1,000–10,000	4,000	−0.08
V	50.94	150	20–500	100	−1.71
Y	88.91		25–250	50	−2.25
Zn	65.37	80	10–300	50	−2.12
Zr	91.22	220	60–2,000	300	−1.48

Source: Based on Clarke and Washington (1924), Swaine (1955), Vinogradov (1959), Jackson (1964), Bowen (1966), Mitchell (1964), and Taylor (1964).

Thus the maximum concentration of zinc in this suspension would be $10^{-4.12}$ M.

Data in the last column of Table 1.1 can be adjusted for soils of any elemental content or moisture content. These parameters are useful for predicting maximum solubilities of the various elements in soils and for predicting those minerals that cannot possibly persist in soils.

REFERENCES

Bowen, H. J. M. 1966. Trace Elements in Biochemistry. Academic Press, New York.

Clarke, F. W. and J. S. Washington. 1924. The composition of the earth's crust. U.S. Geol. Sur. Paper 127.

Jackson, M. L. 1964. Chemical composition of soils. *In* F. E. Bear (Ed.), Chemistry of the Soil, 2nd ed. Reinhold, New York, pp. 71–141.

Lindsay, W. L. 1972. Influence of the soil matrix on the availablity of trace elements to plants. Ann. N. Y. Acad. Sci. 199:37–45.

Mitchell, R. L. 1964. Trace elements in soils. *In* F. E. Bear (Ed.), Chemistry of the Soil, 2nd ed. Reinhold, New York, pp. 320–368.

Swaine, D. J. 1955. The trace element content of soils. Commonwealth Bur. Soil Sci. Tech. Comm. No. 48. Herald Printing, York, England.

Taylor, S. R. 1964. The abundance of chemical elements in the continental crust—A new table. Geochim. Cosomochim. Acta 28: 145–146.

Truog, E. 1951. Soil as a medium for plant growth. *In* E. Truog (Ed.), Mineral Nutrition of Plants. The University of Wisconsin Press, Madison, Wisc., pp. 23–55.

Vinogradov, A. P. 1959. The geochemistry of rare and dispersed chemical elements in soils, 2nd ed. Translation from Russian. Consultants Bureau, New York.

PROBLEMS

1.1 Explain how the equilibria depicted in Fig. 1.1 are affected when:
 a. A plant root begins to absorb K^+.
 b. Soluble fertilizers are added to soils.
 c. Soil is subjected to an annual rainfall of 60 inches.
 d. A heavy straw residue is plowed under.
 e. A rice field is flooded during most of the growing season.

1.2 Of what value are equilibrium relationships when it is generally recognized that soils never attain complete equilibrium?

1.3 How might the equilibrium depicted in Fig. 1.1 be affected by diffusive and convective gradients near plant roots?

1.4 Why is it not possible for compounds A, B, and C in Fig. 1.2 to coexist permanently in soils?

1.5 What changes might cause compound C in Fig. 1.2 to become unstable so that compound B could again form? Give an example.

1.6 Explain how a mineral could maintain different levels of a nutrient in different soils.

1.7 From the average elemental contents of soils given in Table 1.1 discuss the following:
 a. The possibility that $AlPO_4 \cdot 2H_2O$ (variscite) could control the solubility of aluminum in soils.
 b. The possibility that $PbMoO_4$ (wulfenite) could control the solubility of lead in soils.
 c. The possibility that $Mn_3(PO_4)_2$ could control the solubility of phosphorus in soils.
 d. The parts per million of manganese in a soil containing 600 ppm of phosphorus below which $Mn_3(PO_4)_2$ can control manganese solubility but above which this mineral can control phosphorus solubility.
 e. The maximum concentration of mercury in moles per liter that could result from shaking 5 g of soil having the indicated average mercury content (Table 1.1) with 500 ml of water.

TWO

METHODS OF HANDLING CHEMICAL EQUILIBRIA

When substances are mixed, they often undergo chemical changes. Most chemical reactions do not go to completion, that is, all of the reactants do not become products. Equilibrium is reached when the forward reaction just balances the reverse reaction. The amounts of products and reactants present at equilibrium differ for each chemical reaction.

One problem in dealing with chemical systems is how to decide when equilibrium has been attained. There is a simple test. Allow the reaction to proceed in the forward direction until nothing more happens and measure the composition of the system. Allow the same reaction to proceed in the reverse direction until nothing more happens. If the composition of the system is the same, regardless of the direction from which equilibrium is approached, then true equilibrium exists.

Some chemical reactions in soils proceed with sufficient speed that equilibrium relationships are immediately attained. Other reactions proceed so slowly that final equilibrium is probably never attained. Regardless of the rate at which equilibrium is attained, equilibrium relationships are useful for predicting chemical changes that can and cannot occur. Equilibrium provides a reference point for predicting which chemical reactions can take place regardless of the rate at which they occur.

2.1 EQUILIBRIUM CONSTANTS

Equilibrium for the reaction:

$$A + B \rightleftharpoons AB \qquad (2.1)$$

can be expressed by a formation constant (K_f)

$$K_f = \frac{AB}{A \cdot B} \qquad (2.2)$$

When Reaction 2.1 is written in the reverse order, that is

$$AB \rightleftharpoons A + B \qquad (2.3)$$

the equilibrium constant is called a dissociation constant (K_d)

$$K_d = \frac{A \cdot B}{AB} \qquad (2.4)$$

It is apparent that the formation and dissociation constants for any reaction are reciprocals of each other. That is,

$$K_f = \frac{1}{K_d} \qquad (2.5)$$

Many chemical reactions are more complex than the example given above, yet their equilibrium constants are defined in the same manner. Consider the reaction

$$aA + bB \rightleftharpoons cC + dD \tag{2.6}$$

The equilibrium constant is expressed as

$$K = \frac{C^c \cdot D^d}{A^a \cdot B^b} \tag{2.7}$$

where the reactants and products are repeated as many times as they appear in the reaction. This is done by raising each term in the equilibrium expression to the power of its coefficient in the reaction.

2.2 CONCENTRATION VERSUS ACTIVITY CONSTANTS

So far equilibrium constants have been defined only in general terms. Actually there are different kinds of equilibrium constants depending upon the units in which reactants and products are expressed. If they are expressed as activities, they define activity constants. If they are expressed as concentrations, they define concentration constants. Each has advantages and disadvantages.

Equilibrium constants expressed in terms of activities have a special significance because they can be calculated from thermodynamic data as will be shown shortly. They are true constants that hold for solutions of all ionic strengths. These constants, however, have the disadvantage that many reactants and products consist of specific ionic or molecular species whose activities are difficult or impossible to measure. For this reason, concentration constants are often used. Concentration constants have the disadvantage that they change with ionic strength and must either be used in systems of the same ionic strength in which they were determined, or ionic strength corrections must be applied.

In this book activities are designated by round brackets ( ) and concentrations by square brackets [ ] whenever it is necessary to distinguish between the two parameters.

2.3 IONIC STRENGTH

Ionic strength is defined as

$$\mu = \tfrac{1}{2}\Sigma c_i Z_i^2 \tag{2.8}$$

DEBYE–HÜCKEL EQUATIONS

where μ is the ionic strength, c_i is the concentration in moles liter^{-1} of ion i, Z_i is the valency of that ion, and Σ indicates that the product of each ion and its valency squared is summed for all ions in solution.

EXAMPLE CALCULATIONS

What is the ionic strength of a 0.01 M NaCl solution? From Eq. 2.8

$$\mu = \tfrac{1}{2}[0.01 \times 1^2 + 0.01 \times (-1)^2]$$
$$\mu = \underline{0.01}$$

What is the ionic strength of a 0.01 M CaCl$_2$ solution?

$$\mu = \tfrac{1}{2}[0.01 \times 2^2 + 0.02 \times (-1)^2]$$
$$\mu = \underline{0.03}$$

2.4 ACTIVITY COEFFICIENTS

Only in infinitely dilute solutions are activities and concentrations equal. The ratio of the activity of an ion, a_i, to its concentration, c_i, is called the activity coefficient, γ_i:

$$\gamma_i = \frac{a_i}{c_i} \qquad (2.9)$$

Knowing γ_i, we are able to convert from concentrations to activities, and vice versa. At infinitely dilute solution $a_i = c_i$, and $\gamma_i = 1$. Generally as ionic strength increases, ions of opposite charge interact in such a way that their "effective" concentration or activity decreases.

2.5 DEBYE–HÜCKEL EQUATIONS

The Debye–Hückel theory of estimating activity coefficients is based on laws of electrostatics and thermodynamics. In essence, it assumes that ions behave like point charges in a continuous medium with a dielectric constant equal to that of the solvent. The resulting equation for calculating activity coefficients of simple ions in aqueous solutions is

$$\log \gamma_i = -AZ_i^2 \mu^{1/2} \qquad (2.10)$$

where $A = 0.509$ for water at 25°C. Comparisons of calculated activity coefficients using Eq. 2.10 with experimentally measured values show a close correspondence up to ionic strengths of about 0.001 M. At higher concentrations the calculated activity coefficients are generally smaller than those measured experimentally.

By extending the Debye–Hückel theory to account for the effective size of hydrated ions, a more precise equation is obtained, that is,

$$\log \gamma_i = -AZ_i^2 \frac{\mu^{1/2}}{1 + Bd_i\mu^{1/2}} \tag{2.11}$$

where $B = 0.328 \times 10^8$ for water at 25°C and d_i is the effective distance of closest approach measured in centimeters and corresponds roughly to the effective size of the hydrated ion. Values of d_i for several selected ions as calculated by Kielland (1937) are reported in Table 2.1. The values of activity coefficients at various ionic strengths calculated from Eq. 2.11 are tabulated in Table 2.2. The constants A and B in this equation are temperature-dependent, but the d_i values do not change appreciably with temperature.

In general the extended Debye–Hückel equation holds fairly well in solutions of ionic strength up to 0.2 M. At higher concentrations ionic interactions are difficult to predict and many activity coefficients become larger than unity due to the repulsion of ions. Butler (1964) has summarized the findings of several workers on the use of various empirical modifications of the Debye–Hückel equation to give the best-fit method of calculating activity coefficients. For estimating unknown activity coefficients, Guntelberg suggested a value of 3×10^{-8} be used for d_i so that Bd_i in Eq. 2.11 is unity, giving

$$\log \gamma_i = -AZ_i^2 \frac{\mu^{1/2}}{1 + \mu^{1/2}} \tag{2.12}$$

This equation gives values of γ_i that are too small for many electrolytes. Guggenheim suggested that a linear term $b\mu$ be included in Eq. 2.12 and that the b factor be determined by best fit of the data. After examining the values of b for a number of 1:1 and 1:2 electrolytes, Davies (1962) proposed the following equation:

$$\log \gamma_i = -AZ_i^2 \left(\frac{\mu^{1/2}}{1 + \mu^{1/2}} - 0.3\mu \right) \tag{2.13}$$

The latter equation is often used in preference to the extended Debye–Hückel equation because the single variable is more adapted to simplify calculations. In Table 2.3 are summarized the γ_i and $\log \gamma_i$ values for several ions of different valence using Eq. 2.13.

At ionic strengths much above 0.1 M, a more accurate representation of

TABLE 2.1 VALUES FOR THE PARAMETER d_i USED IN THE EXTENDED DEBYE-HÜCKEL EQUATION

$10^8 d_i$

Inorganic Ions: Charge 1

9	H^+
6	Li^+
4-4.5	$Na^+, CdCl^+, ClO_2^-, IO_3^-, HCO_3^-, H_2PO_4^-, HSO_3^-, H_2AsO_4^-$
3.5	$OH^-, F^-, NCO^-, HS^-, ClO_3^-, ClO_4^-, BrO_3^-, IO_4^-, MnO_4^-, NCS^-$
3	$K^+, Cl^-, Br^-, I^-, CN^-, NO_2^-, NO_3^-$
2.5	$Rb^+, Cs^+, NH_4^+, Tl^+, Ag^+$

Inorganic Ions: Charge 2

8	Mg^{2+}, Be^{2+}
6	$Ca^{2+}, Cu^{2+}, Zn^{2+}, Sn^{2+}, Mn^{2+}, Fe^{2+}, Ni^{2+}, Co^{2+}$
5	$Sr^{2+}, Ba^{2+}, Ra^{2+}, Cd^{2+}, Hg^{2+}, S^{2-}, S_2O_4^{2-}, WO_4^{2-}$
4.5	$Pb^{2+}, CO_3^{2-}, SO_3^{2-}, MoO_4^{2-}, Co(NH_3)Cl_5^{2+}, Fe(CN)_5NO^{2-}$
4	$Hg_2^{2+}, SO_4^{2-}, S_2O_3^{2-}, S_2O_8^{2-}, SeO_4^{2-}, CrO_4^{2-}, S_2O_6^{2-}, HPO_4^{2-}$

Inorganic Ions: Charge 3

9	$Al^{3+}, Fe^{3+}, Cr^{3+}, Se^{3+}, Y^{3+}, La^{3+}, In^{3+}, Ce^{3+}, Pr^{3+}, Nd^{3+}, Sm^{3+}$
4	$PO_4^{3-}, Fe(CN)_6^{3-}, Cr(NH_3)_6^{3+}, Co(NH_3)_6^{3+}, Co(NH_3)_5 \cdot H_2O^{3+}$

Inorganic Ions: Charge 4

11	$Th^{4+}, Zr^{4+}, Ce^{4+}, Sn^{4+}$
6	$Co(S_2O_3)(CN)_5^{4-}$
5	$Fe(CN)_6^{4-}$

Inorganic Ions: Charge 5

9	$Co(SO_3)_2(CN)_4^{5-}$

Organic Ions: Charge 1

8	$(C_6H_5)_2CHCOO^-, (C_3H_7)_4N^+$
7	$[OC_6H_2(NO_2)_3]^-, (C_3H_7)_3NH^+, CH_3OC_6H_4COO^-$
6	$C_6H_5COO^-, C_6H_4OHCOO^-, C_6H_4ClCOO^-, C_6H_5CH_2COO^-,$ $CH_2{=}CHCH_2COO^-, (CH_3)_2CHCH_2COO^-, (C_2H_5)_4N^+, (C_3H_7)_2NH_2^+$
5	$CHCl_2COO^-, CCl_3COO^-, (C_2H_5)_3NH^+, (C_3H_7)NH_3^+$
4.5	$CH_3COO^-, CH_2ClCOO^-, (CH_3)_4N^+, (C_2H_5)_4NH_2^+, NH_2CH_2COO^-$
4	$NH_3^+CH_2COOH, (CH_3)_3NH^+, C_2H_5NH_3^+$
3.5	$HCOO^-, H_2\text{-citrate}^-, CH_3NH_3^+, (CH_3)_2NH_2^+$

(Continued)

TABLE 2.1 (*Continued*)

$10^8 d_i$	
	Organic Ions: Charge 2
7	$OOC(CH_2)_5COO^{2-}$, $OOC(CH_2)_6COO^{2-}$, Congo red anion^{2-}
6	$C_6H_4(COO)_2^{2-}$, $H_2C(CH_2COO)_2^{2-}$, $(CH_2CH_2COO)_2^{2-}$
5	$H_2C(COO)_2^{2-}$, $(CH_2COO)_2^{2-}$, $(CHOHCOO)_2^{2-}$
4.5	$(COO_2)^{2-}$, H-citrate^{2-}
	Organic Ions: Charge 3
5	Citrate^{3-}

Source: Kielland (1937).

experimental data can be obtained if an equation with several adjustable parameters is used:

$$\log \gamma_i = -AZ_i^2 \frac{\mu^{1/2}}{1+\mu^{1/2}} + b\mu + c\mu^2 + d\mu^3 + \cdots \qquad (2.14)$$

The coefficients, $b, c, d, \ldots$, must be determined experimentally to provide the best fit approximation of activity coefficients for the system in which they are used.

2.6 ACTIVITY COEFFICIENTS FROM ELECTRICAL CONDUCTIVITIES

A convenient and direct method of estimating the ionic strength of a solution is to measure its electrical conductivity. Griffin and Jurinak (1973) examined 27 soil extracts and 124 river waters and obtained the following relationship:

$$\mu = 0.013 \, EC \qquad r = 0.996 \qquad (2.15)$$

where μ is the ionic strength based on concentrations expressed in mole liter^{-1} and EC is electrical conductivity expressed in millimhos cm^{-1} at 25°C. An r of 0.996 indicates a very high correlation coefficient between

TABLE 2.2 SINGLE ION ACTIVITY COEFFICIENTS CALCULATED FROM THE EXTENDED DEBYE–HÜCKEL EQ., 2.11 FOR 25°C

	Ionic Strength (μ)							
d_i^*	0.001	0.0025	0.005	0.01	0.025	0.03	0.05	0.1
	Ionic Charge 1							
9	0.967	0.950	0.934	0.914	0.881	0.874	0.854	0.826
8	0.966	0.950	0.933	0.911	0.877	0.870	0.848	0.817
7	0.966	0.949	0.931	0.909	0.873	0.865	0.841	0.807
6	0.966	0.948	0.930	0.907	0.868	0.860	0.834	0.796
5	0.965	0.947	0.928	0.904	0.863	0.854	0.826	0.783
4	0.965	0.946	0.927	0.902	0.858	0.848	0.817	0.770
3	0.965	0.946	0.925	0.899	0.852	0.841	0.807	0.754
	Ionic Charge 2							
8	0.872	0.813	0.756	0.690	0.592	0.572	0.517	0.445
7	0.871	0.810	0.752	0.683	0.581	0.559	0.500	0.424
6	0.870	0.808	0.748	0.676	0.568	0.546	0.483	0.401
5	0.869	0.805	0.743	0.668	0.555	0.531	0.464	0.377
4	0.867	0.803	0.738	0.661	0.541	0.516	0.445	0.351
	Ionic Charge 3							
9	0.737	0.632	0.540	0.443	0.321	0.299	0.242	0.178
6	0.731	0.619	0.520	0.414	0.280	0.256	0.194	0.128
5	0.728	0.614	0.513	0.404	0.266	0.241	0.178	0.111
4	0.726	0.610	0.505	0.394	0.251	0.226	0.161	0.095
	Ionic Charge 4							
11	0.587	0.452	0.348	0.252	0.151	0.135	0.098	0.063
6	0.572	0.426	0.312	0.209	0.104	0.089	0.054	0.026
5	0.569	0.420	0.305	0.200	0.095	0.080	0.047	0.020

* Effective ionic parameter of ion (i).

the μ calculated by Eq. 2.15 and that based on Eq. 2.8. This empirical relationship is particularly useful when the complete composition of a soil solution is unknown or where ion pair formation and other factors preclude measuring the many different ionic species in solution. Equation 2.15 compares favorably with that of Ponnamperuma et al. (1966) ($\mu = 0.016\,EC$) even though ion pair formation was not included by the latter workers.

TABLE 2.3 THE ACTIVITY COEFFICIENTS (γ_i) FOR IONS OF DIFFERENT VALENCY AND IONIC STRENGTH CALCULATED FROM THE DAVIES' EQ. 2.13

Ionic Strength (μ)	Valency, Z_i				
	1	2	3	4	5
			γ_i		
0.001	0.965	0.867	0.726	0.566	0.411
0.002	0.952	0.820	0.641	0.453	0.290
0.005	0.927	0.739	0.506	0.298	0.151
0.010	0.902	0.662	0.396	0.192	0.076
0.025	0.860	0.546	0.256	0.089	0.023
0.030	0.850	0.522	0.232	0.074	0.017
0.050	0.822	0.455	0.170	0.043	0.007
0.100	0.782	0.373	0.109	0.019	0.002
			$\log \gamma_i$		
0.001	−0.015	−0.062	−0.139	−0.247	−0.386
0.002	−0.021	−0.086	−0.193	−0.344	−0.537
0.005	−0.033	−0.131	−0.296	−0.526	−0.821
0.010	−0.045	−0.179	−0.403	−0.716	−1.119
0.025	−0.066	−0.263	−0.591	−1.051	−1.642
0.030	−0.071	−0.282	−0.635	−1.129	−1.764
0.050	−0.085	−0.342	−0.768	−1.366	−2.135
0.100	−0.107	−0.428	−0.963	−1.712	−2.675

2.7 TRANSFORMING EQUILIBRIUM CONSTANTS

Equilibrium constants are often determined at one ionic strength but are needed for use at another ionic strength. If values for the activity coefficients are available, such transformations can be readily made. Generally three types of constants are used:

K^c concentration constants in which reactants and products all are expressed in terms of concentrations at some specified ionic strength

K^m mixed constants in which all terms are given in concentrations except H^+, OH^-, and e^-, which are given in activities

K^o activity constants in which all terms are expressed in activities

TRANSFORMING EQUILIBRIUM CONSTANTS

The method of transforming from one constant to another is illustrated for the reaction

$$H + L \rightleftharpoons HL \qquad (2.16)$$

where charges on ions are omitted to simplify writing generalized equations

reference point for solubility?

Conversion from K° to K^c

For Reaction 2.16

$$K^\circ = \frac{(HL)}{(H)(L)} \qquad \text{activity} \qquad (2.17)$$

$$K^c = \frac{[HL]}{[H][L]} = \frac{(HL)/\gamma_{HL}}{(H)(L)/\gamma_H \gamma_L} = \frac{(HL)}{(H)(L)} \cdot \frac{\gamma_H \gamma_L}{\gamma_{HL}} \qquad (2.18)$$

$$K^c = K^\circ \frac{\gamma_{\text{reactants}}}{\gamma_{\text{products}}} \qquad (2.19)$$

and

$$K^\circ = K^c \frac{\gamma_{\text{products}}}{\gamma_{\text{reactants}}} \qquad (2.20)$$

Conversion from K° to K^m

For Reaction 2.16

$$K^\circ = \frac{(HL)}{(H)(L)}$$

$$K^m = \frac{[HL]}{(H)[L]} = \frac{(HL)/\gamma_{HL}}{(H)(L)/\gamma_L} = \frac{(HL)}{(H)(L)} \cdot \frac{\gamma_L}{\gamma_{HL}} \qquad (2.21)$$

$$K^m = K^\circ \frac{\gamma_{\text{reactants except } H^+, OH^-, e^-}}{\gamma_{\text{products except } H^+, OH^-, e^-}} \qquad (2.22)$$

be water dissociation with those

and

$$K^\circ = K^m \frac{\gamma_{\text{products except } H^+, OH^-, e^-}}{\gamma_{\text{reactants except } H^+, OH^-, e^-}} \qquad (2.23)$$

Conversion from $K^c_{0.01}$ to $K^c_{0.1}$

Concentration constants can be converted from one ionic strength to another. To convert Reaction 2.16 from 0.01 to 0.1 ionic strength

$$K^c = \frac{[HL]}{[H][L]}$$

$$K^\circ = \frac{[HL]\gamma_{HL}}{[H][L]\gamma_H\gamma_L} = K^c \frac{\gamma_{HL}}{\gamma_H\gamma_L} \qquad (2.24)$$

$$K^\circ = K^c_{0.01} \left|\frac{\gamma_{HL}}{\gamma_H\gamma_L}\right|_{0.01} = K^c_{0.1} \left|\frac{\gamma_{HL}}{\gamma_H\gamma_L}\right|_{0.1} \qquad (2.25)$$

$$K^c_{0.1} = K^c_{0.01} \left|\frac{\gamma_{HL}}{\gamma_H\gamma_L}\right|_{0.01} \cdot \left|\frac{\gamma_H\gamma_L}{\gamma_{HL}}\right|_{0.1} \qquad (2.26)$$

Conversion from $K^c_{0.1}$ to $K^m_{0.1}$

Sometimes it is necessary to convert from concentration constants to mixed constants at the same ionic strength. For reaction 2.16

$$K^m = \frac{[HL]}{(H)[L]}$$

$$K^c = \frac{[HL]}{[H][L]} = \frac{[HL]\gamma_H}{(H)[L]}$$

$$K^m_{0.1} = \frac{K^c_{0.1}}{\gamma_H} \qquad (2.27)$$

Thus equations can be developed to convert various equilibrium constants from one form and ionic strength to another. Such developments can be extended to any reaction.

EXAMPLE CALCULATION

Given the concentration constant at 0.01 ionic strength as log $K^c_{0.01} = 10.17$ for the reaction

$$H^+ + CO_3^{2-} \rightleftharpoons HCO_3^- \qquad (2.28)$$

Change this constant to correspond to an ionic strength of 0.1.

EQUILIBRIUM CONSTANTS FROM THERMODYNAMIC DATA

From Eq. 2.26, activity coefficients in Table 2.2, and d_i values in Table 2.1

$$K^c_{0.1} = 10^{10.17} \times \frac{0.902}{0.914 \times 0.664} \times \frac{0.826 \times 0.364}{0.773}.$$

$$K^c_{0.1} = \underline{10^{9.93}}$$

2.8 EQUILIBRIUM CONSTANTS FROM THERMODYNAMIC DATA

If chemical reactions proceed until the free energy (G) of a system reaches a minimum, then the system is at equilibrium. At equilibrium a system can do no more work on its environment without the input of additional energy. Since the absolute free energy of a substance is not measurable, reference states are selected for each substance and changes in free energy are then measured. The standard free energy change accompanying a chemical reaction (ΔG°_r) is the sum of the free energies of formation (ΔG°_f) of the products in their standard state minus the free energies of formation of the reactants in their standard states. That is,

$$\Delta G^\circ_r = \Sigma \Delta G^\circ_f \text{ products} - \Sigma \Delta G^\circ_f \text{ reactants} \qquad (2.29)$$

The standard free energy change of a reaction is related to the equilibrium constant of that reaction by the relationship:

$$\Delta G^\circ_r = -RT \ln K^\circ \qquad (2.30)$$

where K° is the activity equilibrium constant for the reaction, R is the universal gas constant, and T is the absolute temperature. At 25°C, Eq. 2.30 gives

$$\Delta G^\circ_r(\text{kcal}) = -0.001987 \text{ kcal deg}^{-1} \text{ mole}^{-1} \times 298.15 \text{ deg} \times 2.303 \log K^\circ$$

$$\Delta G^0_r = -1.364 \log K^\circ$$

$$\log K^\circ = -\frac{\Delta G^\circ_r}{1.364} \qquad (2.31)$$

This equation is extremely useful because it permits the calculation of equilibrium constants from thermodynamic data for reactions which are often difficult to measure by conventional methods. Except when noted otherwise, the equilibrium constants used throughout this text were calculated from Eq. 2.29 and 2.31 using the ΔG°_f values documented in the Appendix.

The standard free energy of a reaction (ΔG°_r) can be calculated from changes in the standard enthalpies of reaction (ΔH°_r) and changes in the standard entropies of reaction (ΔS°_r) by the relationship:

$$\Delta G^\circ_r = \Delta H^\circ_r - T\Delta S^\circ_r \qquad (2.32)$$

CH. 2 METHODS OF HANDLING CHEMICAL EQUILIBRIA

where

$$\Delta H_r^\circ = \Sigma \Delta H_f^\circ \text{ products} - \Sigma \Delta H_f^\circ \text{ reactants} \qquad (2.33)$$

and

$$\Delta S_r^\circ = \Sigma S^\circ \text{ products} - \Sigma S^\circ \text{ reactants} \qquad (2.34)$$

Once ΔG_r° is known, the standard free energy of any reactant or product can be determined from Eq. 2.29 providing ΔG_f° values of the other reactants and products are known.

EXAMPLE CALCULATION

Calculate the equilibrium constant for the reaction

$$AgCl(c) \rightleftharpoons Ag^+ + Cl^- \qquad (2.35)$$

using the ΔH_f° and S° values given below from Sadiq and Lindsay (1979):

Species	ΔH_f° (kcal mole^{-1})	S° (cal deg^{-1} mole^{-1})
AgCl(c)	−30.370	23.00
Cl$^-$	−39.933	13.56
Ag$^+$	25.275	17.54
Ag(c)	0	10.17
e$^-$	0	15.603

$\Delta H_r^\circ = 25.275 + (-39.933) - (-30.370) = 15.712$ kcal mole^{-1}
$\Delta S_r^\circ = 17.54 + 13.56 - (23.00) = 8.10$ cal deg^{-1} mole^{-1}

$$\begin{aligned}\Delta G_r^\circ &= \Delta H_r^\circ - T\Delta S_r^\circ \\ &= 15.712 - 298.15(0.00810) \\ &= 13.297 \text{ kcal mole}^{-1}\end{aligned}$$

and from Eq. 2.31

$$\log K^\circ = -\frac{13.297}{1.364} = -9.75$$

pe + pH

2.9 REDOX RELATIONSHIPS (Fe, Manganese)

Oxidation and reduction reactions are common in soils, yet few scientists make full use of theoretical redox relationships. One reason is that redox equilibria are generally expressed in terms of Eh (the millivolt difference in potential between a platinum electrode and the standard hydrogen electrode), whereas other equilibria are expressed in terms of equilibrium constants based on activities (mole liter^{-1}).

Just as pH is based on mole liter^{-1}, redox potentials can be expressed in terms of pe (−log of electron activity) which is compatible with units of mole liter^{-1}. In this way, electrons can be treated as other reactants and products so that both chemical and electrochemical equilibria can be expressed by a single equilibrium constant. = pe

Sillen and Martell (1964) pointed out that the electron (e$^-$) can be considered as any other reactant or product in chemical reactions. They tabulated equilibrium constants for redox reactions in terms of both $E°$ (standard electrode potentials) and log $K°$ (equilibrium activity constants) and encouraged greater use of the log $K°$ constants. Truesdell (1969) showed several advantages of using pe in redox equilibrium calculations. In spite of these innovations, few soil scientists have used these simplifying concepts for handling redox reactions. Instead, Eh continues to be the commonly accepted method of expressing redox relationships in the soils literature.

Reduction Oxidation
↓ H$_2$ state   ↑ O$_2$ state

2.10 pe + pH

Most soil systems consist of aqueous environments in which the dissociation of water into H$_2$(g) or O$_2$(g) imposes redox limits on soils. On the reduced side, the redox limit is given by the reactions

Loss of e$^-$ = oxidation
gain of e$^-$ = reduction

$$H_2O + e^- \rightleftharpoons \tfrac{1}{2}H_2(g) + OH^- \quad (2.36)$$

$$H^+ + OH^- \rightleftharpoons H_2O \quad (2.37)$$

$$H^+ + e^- \rightleftharpoons \tfrac{1}{2}H_2(g) \quad \text{reduction} \quad (2.38)$$

The equilibrium expression for the overall Reaction 2.38 is

$$K° = \frac{(H_2(g))^{1/2}}{(H^+)(e^-)} \quad (2.39)$$

or

$$\log K° = \tfrac{1}{2}\log H_2(g) - \log(H^+) - \log(e^-) \quad (2.40)$$

LEO goes GR

the log of 1 = 0

The equilibrium constant ($K°$) for this reaction is defined as unity (log $K°$ = 0) for standard state conditions in which (H^+) activity is 1 mole liter^{-1} and $H_2(g)$ is the partial pressure of H_2 gas at 1 atm. This is equivalent to setting the electron activity (e^-) at unity for the standard hydrogen half cell reaction. Since log $K°$ in Eq. 2.40 is defined as zero, it follows that

$$pe + pH = -\tfrac{1}{2} \log H_2(g) \qquad (2.41)$$

Thus when $H_2(g) = 1$ atm, $pe + pH = 0$. This represents the most reduced equilibrium conditions expected for natural aqueous environments. When attempts are made to increase the electron activity (e^-) and/or proton activity (H^+), Reaction 2.38 proceeds to the right with the evolution of $H_2(g)$.

On the oxidized side, the redox limit of aqueous systems is given by the reaction

$$H^+ + e^- + \tfrac{1}{4}O_2(g) \rightleftharpoons \tfrac{1}{2}H_2O \qquad (2.42)$$

The equilibrium expression for this reaction is

$$K° = \frac{(H_2O)^{1/2}}{(H^+)(e^-)(O_2(g))^{1/4}} \qquad (2.43)$$

The value of $K°$ can be calculated from the standard free energies of formation in the Appendix and is equal to $10^{20.78}$. In dilute aqueous systems the activity of water is very near unity, so the equilibrium expression in log form becomes

$$-\log(H^+) - \log(e^-) - \tfrac{1}{4}\log O_2(g) = 20.78 \qquad (2.44)$$

or

$$pe + pH = 20.78 + \tfrac{1}{4}\log O_2(g) \qquad (2.45)$$

Thus when $O_2(g)$ is 1 atm, $pe + pH = 20.78$. This corresponds to the most oxidized equilibrium conditions expected in natural aqueous environments. When attempts are made to decrease the electron and/or proton activity, Reaction 2.42 shifts to the left with the evolution of $O_2(g)$.

The redox limits of natural aqueous environments defined by Reactions 2.38 and 2.42 are plotted in Fig. 2.1. Both pe and pH are necessary to specify the redox status of aqueous systems. Redox relationships that lie parallel to the $H_2(g)$ and $O_2(g)$ redox limits can be expressed in terms of $pe + pH$ as well as the partial pressures of $O_2(g)$ or $H_2(g)$.

For example, highly oxidized environments in equilibrium with 1 atm of $O_2(g)$ lie on the uppermost line. This line also corresponds to $10^{-41.56}$ atm of $H_2(g)$ or to a $pe + pH$ of 20.78. Equilibrium with the air containing 0.2 atm of $O_2(g)$ is represented by a line slightly lower ($pe + pH = 20.61$), which is obtained from Eq. 2.45 by substituting 0.20 atm for $O_2(g)$. As electron activity increases, the partial pressure of $O_2(g)$ decreases, and that of $H_2(g)$

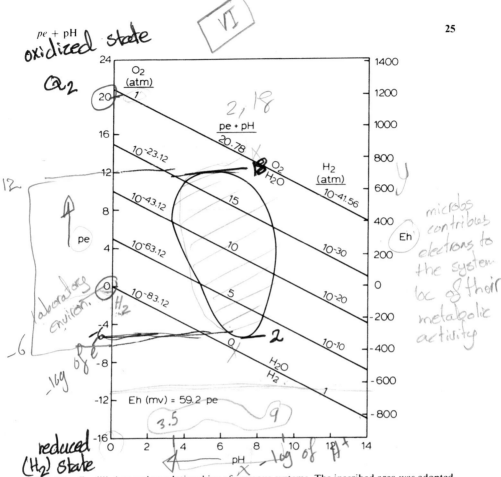

Fig. 2.1 Equilibrium redox relationships of aqueous systems. The inscribed area was adopted from Baas Becking et al. (1960) and is representative of most soils.

increases. The redox limit represented by 1 atm of $H_2(g)$ corresponds to a partial pressure of $O_2(g)$ of $10^{-83.12}$ atm.

— The parameter $pe + pH$ provides a convenient single-term expression for defining the redox status of aqueous systems. The parameter ranges from zero on the reduced side (1 atm H_2) to 20.78 on the oxidized side (1 atm O_2). The partial pressures of $O_2(g)$ and of $H_2(g)$ remain fixed for any given value of $pe + pH$.

Just as pH expresses $-$log of the activity of H^+, so pe denotes $-$log of the activity of electrons. In the case of H^+ this is an absolute activity expressed as mole liter^{-1}. In the case of e^-, the activity is relative to that of the standard hydrogen electrode, which is arbitrarily assigned the value of unity. The actual electron activity of the standard hydrogen half cell is reported to be in the vicinity of 10^{-80} mole liter^{-1} (Hart et al., 1966), but this value need not enter operational calculations.

2.11 $E°$ VERSUS LOG $K°$

The following relationships are obtained from the first and second laws of thermodynamics

$$\Delta G_r = \Delta G_r° + RT \ln Q \qquad (2.46)$$

$$\Delta G_r = -nFE \qquad (2.47)$$

and

$$\Delta G_r° = -nFE° \qquad (2.48)$$

In these expressions, ΔG_r is the Gibbs free energy of reaction, $\Delta G_r°$ is the Gibbs standard free energy of reaction when all products and reactants are in their standard states, R is the gas constant = 1.987 cal deg^{-1} mole^{-1}, T is the temperature in degrees Kelvin (25°C = 298.15°K), Q is the activity quotient of products to reactants for the reaction, n is the number of moles of electrons participating in the reaction, F is the Faraday constant = 23.061 kcal volt^{-1} equivalent^{-1}, E is the cell potential for the reaction, and $E°$ is the standard cell potential when the half cell reactions are written as

$$\text{oxidized} + ne^- \rightleftharpoons \text{reduced} \qquad (2.49)$$

When products and reactants are in their standard states, $Q = K°$ and Eq. 2.46 becomes

$$\Delta G_r = \Delta G_r° + RT \ln K° \qquad (2.50)$$

At equilibrium ΔG_r is zero, so

$$\Delta G_r° = -RT \ln K° \qquad (2.51)$$

Combining Eq. 2.48 and 2.51 gives

$$-nFE° = -RT \ln K° \quad \text{or} \quad E° = \frac{2.303 \, RT \log K°}{nF} \qquad (2.52)$$

Substituting the numerical values for R, T, and F and rearranging gives

$$\log K° = 16.9 \, nE° \text{ (volts)} \qquad (2.53)$$

or

$$E° \text{ (millivolts)} = \frac{59.2}{n} \log K° \qquad (2.54)$$

pe VERSUS *Eh*

LEO goes GR
Loss Electron — Gain Electron
oxidation — reduction

Thus equilibrium relationships involving oxidation or reduction reactions can be expressed either in terms of equilibrium constants ($K°$) or in terms of standard electrode potentials ($E°$). Equations 2.53 and 2.54 can be used to convert from one parameter to the other.

2.12 *pe* VERSUS *Eh*

The redox parameters *pe* and *Eh* are both defined as zero for the standard hydrogen half-cell. Let us consider the equilibrium expression for the general half cell reaction of Eq. 2.49:

$$K°_{2.49} = \frac{(\text{reduced})}{(\text{oxidized})(e^-)^n} \quad (2.55)$$

or

$$\log K°_{2.49} - npe = \log \frac{(\text{reduced})}{(\text{oxidized})} \quad (2.56)$$

On the other hand, the term *Eh* is defined as the potential of the cell

$$\text{Pt, } H_2/H^+ //\text{reduced/oxidized, Pt}$$

where the half cell on the left is the standard hydrogen half cell and that on the right is the general half cell reaction represented by Reaction 2.49. The overall cell reaction becomes

$$\frac{n}{2} H_2(g) \rightleftharpoons nH^+ + ne^-$$

$$\text{oxidized} + ne^- \rightleftharpoons \text{reduced}$$

$$\frac{n}{2} H_2(g) + \text{oxidized} \rightleftharpoons nH^+ + \text{reduced} \quad (2.57)$$

The Nernst equation is obtained by combining Eq. 2.46 through 2.48 which gives

$$-nFE = -nFE° + RT \ln Q \quad (2.58)$$

which reduces to

$$E = E° - \frac{RT}{nF} \ln Q \quad (2.59)$$

Applying this equation to the cell reaction depicted in Eq. 2.57 gives:

$$E = E° - \frac{RT}{nF} \ln \frac{(H^+)^n \text{ (reduced)}}{H_2(g)^{n/2} \text{ (oxidized)}} \qquad (2.60)$$

Since the standard hydrogen electrode is one of the half cells in this reaction, $E° = E°_{2.49}$, $(H^+) = 1$, $H_2(g) = 1$, and E becomes Eh. Thus Equation 2.60 reduces to

$$Eh = E°_{2.49} - \frac{RT}{nF} \ln \frac{\text{(reduced)}}{\text{(oxidized)}} \qquad (2.61)$$

Combining Eq. 2.61 and 2.56 and changing to log base 10 gives

$$Eh = E°_{2.49} - \frac{2.303 \, RT}{nF} (\log K°_{2.49} - npe) \qquad (2.62)$$

It follows from Eq. 2.52. that

$$E°_{2.49} = \frac{2.303 \, RT}{nF} (\log K°_{2.49}).$$

Therefore Eq. 2.62 reduces to

$$Eh = \frac{2.303 \, RT}{F} pe \qquad (2.63)$$

Substituting appropriate numerical values for R, T, and F gives

$$Eh \text{ (millivolts)} = 59.2 \, pe \qquad (2.64)$$

Thus redox relationships can be expressed either in terms of pe or Eh. Both parameters are shown together in Fig. 2.1. Equation 2.64 is useful for converting pe to Eh and vice versa.

2.13 REDOX MEASUREMENTS IN NATURAL ENVIRONMENTS

Baas Becking et al. (1960) have reported numerous redox measurements made in natural environments. Those most frequently measured for soils are shown by the enclosed region in Fig. 2.1. Most soils fall in the pH range of 3.5 to 9 and in the pe range of -6 to $+12$. The $pe + pH$ range of most soils lies between 2 and 18.

Usefulness of the redox parameter $pe + pH$ for describing the redox status of soils is demonstrated in Fig. 2.2 taken from Lindsay and Sadiq (1980). In this study 1:2 suspensions of Weld sandy loam and distilled water were adjusted to different pH values by the addition of either HCl or NaOH.

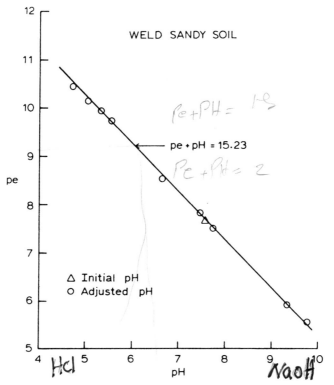

Fig. 2.2 Constancy of *pe* + pH of a Weld sandy loam when pH was adjusted by the addition of either acid or base (from Lindsay and Sadiq, 1980).

The suspensions were left open to the atmosphere while shaking for 24 hours. The *pe* and the pH were then measured with platinum-calomel and glass-calomel electrodes, respectively. The results show that altering pH from 7.57 of the original soil was accompanied by changes in *pe*, while the parameter *pe* + pH remained constant at 15.23, reflecting the characteristic redox status of this soil. Many redox-associated mineral transformations in soils occur at fixed *pe* + pH values (Lindsay and Sadiq, 1980).

The platinum electrode is generally used to measure redox potentials in soils. The platinum electrode apparently does not function well in highly oxidized environments, so redox measurements near or above *pe* + pH of 17 are not too reliable. Further investigations are needed to develop reliable methods for measuring redox in highly oxidized soils. A recent review shows some of the factors involved when preparing functional platinum electrodes (Feltham and Spiro, 1971).

Equilibrium relationships between electrons, $O_2(g)$, and water expressed by Reaction 2.42 are not readily attained, otherwise most soils would reflect a redox level only slightly below the upper line in Fig. 2.1 corresponding to a $pe + $ pH of 20.61. Soils are always more reduced because organisms living therein continuously release electrons to the environment through their respiration processes. This continual release of electrons accounts for the fact that soils and most natural aqueous environments remain reduced. When $O_2(g)$ in soils is depleted, the release of electrons by plant roots and other organisms cause $pe + $ pH to drop and reduction to occur. Submerged or waterlogged soils are examples of natural environments that approach the lower regions of the $pe - $ pH diagram depicted in Fig. 2.1.

The redox parameter $pe + $ pH is used throughout this text to show how redox affects the solubility of various elements and the mineral transformations that occur with changes in redox. This parameter partitions the H^+ ions of an overall chemical reaction into those associated with the redox component of the reaction and those associated with the acid-base component. Use of $pe + $ pH greatly simplifies many theoretical developments of redox relationships, and its use is to be encouraged.

REFERENCES

Baas Becking, L. G. N., I. R. Kaplan, and D. Moore. 1960. Limits of the natural environment in terms of pH and oxidation-reduction potentials. J. Geol. 68:243–284.

Butler, J. N. 1964. Ionic Equilibrium. Addison-Wesley Publishing Co., Inc., Reading, Mass.

Davies, C. W. 1962. Ion Association. Butterworths, London.

Feltham, A. M. and M. Spiro. 1970. Platinized platinum electrodes. Chem. Rev. 71:177–193.

Griffin, R. A. and J. J. Jurinak. 1973. Estimation of activity coefficients from the electrical conductivity of natural aquatic systems and soils extracts. Soil Sci. 116:26–30.

Hart, E. J., G. Sheffield, and E. M. Fielden. 1966. Reaction of the hydrated electron with water. J. Phys. Chem. 70:150–156.

Kielland, J. 1937. Individual activity coefficients of ions in aqueous solutions. J. Am. Chem. Soc. 59:1675–1678.

Lindsay, W. L. and M.Sadiq. 1980. Use of $pe + $ pH as a redox parameter in soils. Unpublished. Am. J. 44 (in press).

Ponnamperuma, F. N., E. M. Tianco, and T. A. Loy. 1966. Ionic strengths of the solutions of flooded soils and other natural aqueous solutions from specific conductance. Soil Sci. 102:408–413.

Sadiq, M. and W. L. Lindsay. 1979. Selection of standard free energies of formation for use in soil chemistry. Colorado State Univ. Exp. Sta. Tech. Bull. 134.

Sillen, L. G. and A. E. Martell. 1964. Stability constants of metal-ion complexes, 2nd ed. Special Publication No. 17. The Chemical Society, London.

Truesdell, A. 1969. The advantage of using pe rather than Eh in redox equilibrium calculations. J. Geol. Ed. 17:17–20.

PROBLEMS

2.1 Write the dissociation constant expression for each of the following compounds or ions, then simplify them by stipulating meaningful standard state conditions:

    a. H_2O. $\quad -vH + OH$ b. $CaCO_3$.
    c. $Ca_5(PO_4)_3(OH)$.   d. HCO_3^-.

2.2 The dissociation constant for $Al(OH)_3$(gibbsite) is log $K° = -33.96$. Write the appropriate equilibrium reaction for the formation of gibbsite and calculate its formation constant.

2.3 Develop an equilibrium expression for the dissociation constant of $Ca_4H(PO_4)_3 \cdot \frac{5}{2}H_2O$(octocalcium phosphate) in terms of $H_2PO_4^-$ rather than PO_4^{3-}.

2.4 Calculate the ionic strength for the following solutions:

    a. 0.01 M KCl.
    b. 0.01 M H_2SO_4.
    c. 0.01 M $AlCl_3$.
    d. A mixture containing 0.006 M $CaCl_2$ and 0.015 M KBr.
    e. A solution having a conductivity of 0.87 mmho cm^{-1}.

2.5 Discuss how you would calculate the ionic strength of a solution having ion pairs and complexes present.

2.6 Prepare a plot of the activity coefficients (γ) versus ionic strength (μ) in the range of 0 to 0.1 ionic strength for the ions: H^+, Ca^{2+}, Al^{3+}, and $H_2PO_4^-$ from the γ_i values given in Tables 2.2 and 2.3. Comment on the effect of ionic charge, effective ionic diameter, and ionic strength on the activity coefficients of ions.

2.7 Calculate the activity coefficient for Ca^{2+} in a solution containing 7.3×10^{-3} M KCl and 2×10^{-3} M $CaCl_2$ using the following:

    a. The Debye–Hückel limited equation where ions are considered as point charges.
    b. The Debye–Hückel extended equation where the effective size of the hydrated ion is considered.
    c. The Debye–Hückel extended equation where the effective diameter of all ions is averaged at 3, that is $Bd_i = 1$.
    d. The Davies modification of the extended Debye–Hückel equation.

2.8

Given	Calculate
a. pH = 6.89	(H^+) =
b. $(H^+) = 1.72 \times 10^{-8}$ M	pH =
c. $(Al^{3+}) = 1.3 \times 10^{-5}$ M	pAl =
d. pH = 0	(OH^-) =
e. pK = 39.3	K =

2.9 Given the equilibrium reaction:

$$H^+ + HPO_4^{2-} \rightleftharpoons H_2PO_4^- \quad \log K° = 7.20$$

derive the necessary equations and calculate the following:
a. $\log K^m_{0.01}$.
b. $\log K^c_{0.03}$.
c. $\log K^c_{0.01}$ starting from $\log K^c_{0.03}$ calculated in part b.

2.10 Using the thermodynamic data for silver given in the example problem of Section 2.8, consider the reaction

$$Ag(c) + Cl^- \rightleftharpoons AgCl(c) + e^-$$

and calculate the following:
a. $\log K°$.
b. $E°$.
c. Value of pe when $(Cl^-) = 1\ M$.
d. Value of Eh when $(Cl^-) = 0.1\ M$.

2.11 Develop the necessary equations and plot the redox relationships for the following aqueous solutions on a pe versus pH diagram:
a. Equilibrium with 1 atm O_2.
b. Equilibrium with air containing 20% $O_2(g)$ at sea level.
c. Equilibrium with 10^{-6} atm O_2.
d. Equilibrium with 10^{-22} atm O_2.
e. Equilibrium with 1 atm H_2.

2.12 From the following thermodynamic data:

Species	$\Delta G°_f$ (kcal)
e^-	0
H^+	0
OH^-	-37.594
Fe^{2+}	-21.80
Fe^{3+}	-4.02
$H_2O(l)$	-56.687
$Fe(OH)_3$(soil)	-170.40
Fe_3O_4(magnetite)	-243.47

Calculate the following:
a. The dissociation constant of water.

b. Log $K°$ for the reaction
$$Fe(OH)_3(soil) \rightleftharpoons Fe^{3+} + 3OH^-$$
c. $E°$ for the reaction
$$Fe(OH)_3(soil) + e^- + 3H^+ \rightleftharpoons Fe^{2+} + 3H_2O$$
d. The pe at which $Fe^{3+}/Fe^{2+} = 1$.
e. The pe + pH at which magnetite and $Fe(OH)_3$(soil) can coexist.

THREE

ALUMINUM

Aluminum is one of the most abundant elements in soils, comprising approximately 7.1% by weight of the earth's crust (Table 1.1). During weathering aluminum is released from primary minerals and is precipitated as secondary minerals, largely as aluminosilicates. Metal ions such as Fe^{2+}, Fe^{3+}, Mg^{2+}, K^+, etc. are often incorporated into many aluminosilicate minerals. As soils weather, silicon is lost more rapidly than aluminum, leaving the latter to precipitate as oxides and hydroxides. In acid soils aluminum is also important as an exchangeable ion. PH<4.5

In this chapter the oxides and hydroxides of aluminum, hydrolysis species, complexes, and exchangeable aluminum are considered. Aluminosilicates will be considered in Chapter 5 and aluminum phosphates in Chapter 12.

3.1 SOLUBILITY OF ALUMINUM OXIDES AND HYDROXIDES

The solubility of aluminum oxides and hydroxides are given by Reactions 1 to 8 of Table 3.1 and are listed in order of decreasing solubility at 25°C. The $\log K°$ values were calculated from the $\Delta G_f°$ values given in the appendix using Eq. 2.29 and 2.31. A plot of $\log Al^{3+}$ versus pH for each of the oxide and hydroxide minerals is given in Fig. 3.1.

The anhydrous aluminum oxides, γ-Al_2O_3(c) and α-Al_2O_3(corundum), are high-temperature minerals that normally do not form in soils. Instead, $Al(OH)_3$(amorp) is precipitated when aluminum salts are added to soils, making the amorphous precipitate the most soluble form of aluminum hydroxide expected in soils. Other minerals in order of decreasing solubility include α-$Al(OH)_3$(bayerite), γ-AlOOH(boehmite), $Al(OH)_3$(norstrandite), γ-$Al(OH)_3$(gibbsite), and α-AlOOH(disapore). The differences in solubility of the latter four minerals are very small (Fig. 3.1). Since gibbsite has been reported as a common mineral in many soils, it is used in this text as a reference mineral that limits the activity of Al^{3+}.

The gibbsite solubility line in Fig. 3.1 was obtained as follows:

$$\gamma\text{-}Al(OH)_3(\text{gibbsite}) + 3H^+ \rightleftharpoons Al^{3+} + 3H_2O \quad \log K° = 8.04$$

$$\frac{(Al^{3+})}{(H^+)^3} = 10^{8.04}$$

$$\log Al^{3+} = 8.04 - 3\text{pH} \qquad (3.1)$$

The other solubility lines in this figure were derived similarly. Another form of Eq. 3.1 that is often used is obtained by dividing Eq. 3.1 by three and rearranging to give:

$$\text{pH} - \tfrac{1}{3}\text{pAl} = 2.68 \qquad (3.2)$$

TABLE 3.1 EQUILIBRIUM REACTIONS OF ALUMINUM MINERALS AND COMPLEXES AT 25°C.

Reaction No.	Equilibrium Reaction	log $K°$
	Oxides and Hydroxides	
1	$0.5\gamma\text{-Al}_2\text{O}_3(c) + 3\text{H}^+ \rightleftharpoons \text{Al}^{3+} + 1.5\text{H}_2\text{O}$	11.49
2	$0.5\alpha\text{-Al}_2\text{O}_3(\text{corundum}) + 3\text{H}^+ \rightleftharpoons \text{Al}^{3+} + 1.5\text{H}_2\text{O}$	9.73
3	$\text{Al(OH)}_3(\text{amorp}) + 3\text{H}^+ \rightleftharpoons \text{Al}^{3+} + 3\text{H}_2\text{O}$	9.66
4	$\alpha\text{-Al(OH)}_3(\text{bayerite}) + 3\text{H}^+ \rightleftharpoons \text{Al}^{3+} + 3\text{H}_2\text{O}$	8.51
5	$\gamma\text{-AlOOH}(\text{boehmite}) + 3\text{H}^+ \rightleftharpoons \text{Al}^{3+} + 2\text{H}_2\text{O}$	8.13
6	$\text{Al(OH)}_3(\text{norstrandite}) + 3\text{H}^+ \rightleftharpoons \text{Al}^{3+} + 3\text{H}_2\text{O}$	8.13
7	$\gamma\text{-Al(OH)}_3(\text{gibbsite}) + 3\text{H}^+ \rightleftharpoons \text{Al}^{3+} + 3\text{H}_2\text{O}$	8.04
8	$\alpha\text{-AlOOH}(\text{diaspore}) + 3\text{H}^+ \rightleftharpoons \text{Al}^{3+} + 2\text{H}_2\text{O}$	7.92
	Sulfates	
9	$\text{Al}_2(\text{SO}_4)_3(c) \rightleftharpoons 2\text{Al}^{3+} + 3\text{SO}_4^{2-}$	20.84
10	$\text{Al}_2(\text{SO}_4)_3 \cdot 6\text{H}_2\text{O}(c) \rightleftharpoons 2\text{Al}^{3+} + 3\text{SO}_4^{2-} + 6\text{H}_2\text{O}$	3.45
11	$\text{KAl}_3(\text{SO}_4)_2(\text{OH})_6(\text{alunite}) + 6\text{H}^+ \rightleftharpoons \text{K}^+ + 3\text{Al}^{3+} + 2\text{SO}_4^{2-} + 6\text{H}_2\text{O}$	3.04
	Hydrolysis	
12	$\text{Al}^{3+} + \text{H}_2\text{O} \rightleftharpoons \text{AlOH}^{2+} + \text{H}^+$	-5.02
13	$\text{Al}^{3+} + 2\text{H}_2\text{O} \rightleftharpoons \text{Al(OH)}_2^+ + 2\text{H}^+$	-9.30
14	$\text{Al}^{3+} + 3\text{H}_2\text{O} \rightleftharpoons \text{Al(OH)}_3^\circ + 3\text{H}^+$	-14.99
15	$\text{Al}^{3+} + 4\text{H}_2\text{O} \rightleftharpoons \text{Al(OH)}_4^- + 4\text{H}^+$	-23.33
16	$\text{Al}^{3+} + 5\text{H}_2\text{O} \rightleftharpoons \text{Al(OH)}_5^{2-} + 5\text{H}^+$	-34.24
17	$2\text{Al}^{3+} + 2\text{H}_2\text{O} \rightleftharpoons \text{Al}_2(\text{OH})_2^{4+} + 2\text{H}^+$	-7.69
	Complexes	
18	$\text{Al}^{3+} + \text{F}^- \rightleftharpoons \text{AlF}^{2+}$	6.98
19	$\text{Al}^{3+} + 2\text{F}^- \rightleftharpoons \text{AlF}_2^+$	12.60
20	$\text{Al}^{3+} + 3\text{F}^- \rightleftharpoons \text{AlF}_3^\circ$	16.65
21	$\text{Al}^{3+} + 4\text{F}^- \rightleftharpoons \text{AlF}_4^-$	19.03
22	$\text{Al}^{3+} + 3\text{NO}_3^- \rightleftharpoons \text{Al(NO}_3)_3^\circ$	0.12
23	$\text{Al}^{3+} + \text{SO}_4^{2-} \rightleftharpoons \text{AlSO}_4^+$	3.20
24	$\text{Al}^{3+} + 2\text{SO}_4^{2-} \rightleftharpoons \text{Al(SO}_4)_2^-$	1.90
25	$2\text{Al}^{3+} + 3\text{SO}_4^{2-} \rightleftharpoons \text{Al}_2(\text{SO}_4)_3^\circ$	-1.88
	Other Equilibria	
26	$\text{CaSO}_4 \cdot 2\text{H}_2\text{O}(\text{gypsum}) \rightleftharpoons \text{Ca}^{2+} + \text{SO}_4^{2-} + 2\text{H}_2\text{O}$	-4.64
27	$\text{CaF}_2(\text{fluorite}) \rightleftharpoons \text{Ca}^{2+} + 2\text{F}^-$	-10.41
28	$\text{Al}^{3+} + 3e^- \rightleftharpoons \text{Al}(c)$	-86.02

SOLUBILITY OF ALUMINUM OXIDES AND HYDROXIDES

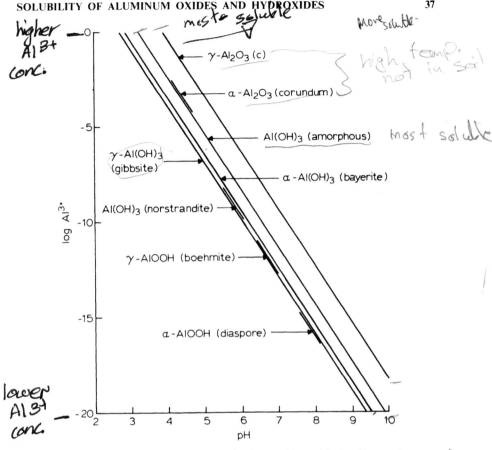

Fig. 3.1 The solubility of various aluminum oxides and hydroxides.

In this text log Al^{3+} + 3pH is used rather than pH − $\frac{1}{3}$ pAl to express aluminum hydroxide solubility relationships because it is believed to be less confusing.

Several of the aluminum oxide and hydroxide minerals in Fig. 3.1 can exist in soils. The activity of Al^{3+} in equilibrium with any of these minerals is pH dependent, decreasing 1000-fold for each unit increase in pH. When Al^{3+} is controlled by $Al(OH)_3$(amorp) rather than gibbsite, the activity of Al^{3+} is $10^{9.66}/10^{8.04} = 10^{1.62}$ or 42 times higher. With time $Al(OH)_3$ (amorp) is expected to transform slowly to one or more of the crystalline forms as it attains more orderly arrangement.

In Chapter 5 it will be shown that the activity of Al^{3+} in soils is often below that of gibbsite due to the presence of various aluminosilicates. Since silicon is removed from soils more rapidly than aluminum, intense weathering causes the eventual disappearance of aluminosilicates. The iron and aluminum that are released generally precipitate as oxides and hydroxides. Gibbsite occurs as an important mineral in many highly weathered soils (Sherman, 1958).

3.2 SOLUBILITY OF ALUMINUM SULFATES

The solubilities of three aluminum sulfate minerals are given by Reactions 9 through 11 of Table 3.1. The minerals $Al_2(SO_4)_3(c)$ and $Al_2(SO_4)_3 \cdot 6H_2O(c)$

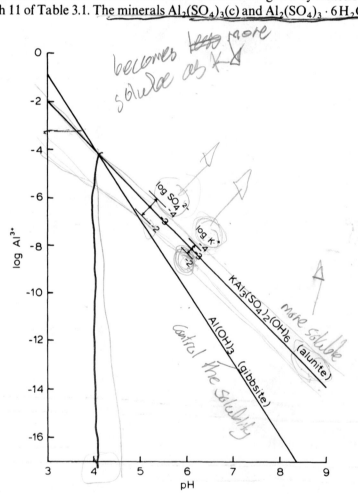

Fig. 3.2 The solubility of alunite compared to gibbsite.

HYDROLYSIS OF Al^{3+}

are too soluble to persist in soils. With SO_4^{2-} in the normal range of 10^{-4} to 10^{-2} M in soils, these minerals would require >1 M Al^{3+} in order to be stable, so they can be discounted as being important in soils.

The mineral $KAl_3(SO_4)_2(OH)_6$ (alunite) is somewhat less soluble and has been reported in acid sulfate soils (Breemen, 1976; Rhodes and Lindsay, 1978). The solubility of alunite is given by Reaction 11 of Table 3.1 and is plotted in Fig. 3.2 along with gibbsite. When SO_4^{2-} and K^+ are 10^{-3} M, alunite is more stable than gibbsite below pH 4.0. With increasing SO_4^{2-} and K^+ activities, alunite remains stable at slightly higher pH values as seen from shifts in the alunite solubility line as these two parameters change.

In acid sulfate soils sulfuric acid is released during the oxidation of sulfides, which lowers pH and increases SO_4^{2-} activity. Under these conditions the precipitation of alunite can be expected. Liming acid sulfate soils causes alunite to become unstable. Breemen (1976) hypothesized the formation of $AlOHSO_4$ to explain the measured log Al^{3+} and pH relationships observed in acid sulfate waters of Thailand. Further investigations are necessary to test this hypothesis and to characterize the solubility of this possible mineral.

3.3 HYDROLYSIS OF Al^{3+}

In aqueous solutions Al^{3+} does not remain as a free ion, but it is surrounded by six molecules of water forming $Al(H_2O)_6^{3+}$. This ion can be represented as follows:

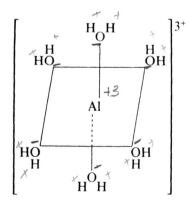

Al hexahydronium ion

As pH increases, protons are removed from the coordinated waters giving a series of hydrolysis products:

$$Al(H_2O)_6^{3+} \rightleftharpoons Al(H_2O)_5(OH)^{2+} + H^+$$
$$\rightleftharpoons Al(H_2O)_4(OH)_2^+ + 2H^+$$
$$\rightleftharpoons Al(H_2O)_3(OH)_3^\circ + 3H^+$$
$$\rightleftharpoons Al(H_2O)_2(OH)_4^- + 4H^+$$
$$\rightleftharpoons Al(H_2O)(OH)_5^{2-} + 5H^+$$

For simplicity these hydrolysis species are generally written without the hydrated water even though the water is present. Equilibrium relationships among the hydrolysis species of Al^{3+} are expressed by the log K° values of Reactions 12 to 17 of Table 3.1. The hydrolysis species are plotted in Fig. 3.3

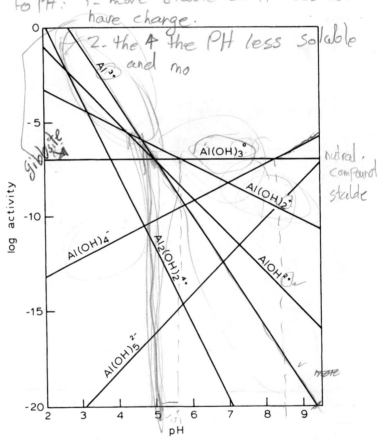

Fig. 3.3 The activity of Al^{3+} and its hydrolysis species in equilibrium with gibbsite.

as a function of pH on the basis that Al^{3+} is in equilibrium with $Al(OH)_3$ (gibbsite). An example of how these solubility relationships were obtained is given for $AlOH^{2+}$. Combining Reactions 12 and 7 of Table 3.1 gives

$$\begin{array}{lr} & \log K° \\ Al^{3+} + H_2O \rightleftharpoons AlOH^{2+} + H^+ & -5.02 \\ Al(OH)_3(\text{gibbsite}) + 3H^+ \rightleftharpoons Al^{3+} + 3H_2O & 8.04 \\ \hline Al(OH)_3(\text{gibbsite}) + 2H^+ \rightleftharpoons AlOH^{2+} + 2H_2O & 3.02 \end{array}$$ (3.3)

$$\frac{(AlOH^{2+})}{(H^+)^2} = 10^{3.02}$$

$$\log AlOH^{2+} = 3.02 - 2pH \tag{3.4}$$

Equation 3.4 is plotted in Fig. 3.3 as the $AlOH^{2+}$ line representing the activity of $AlOH^{2+}$ in equilibrium with gibbsite. The other lines in this figure were obtained similarly.

If solid-phase $Al(OH)_3$(amorp) rather than gibbsite were controlling Al^{3+} activity, then all of the monomeric hydrolysis lines in Fig. 3.3 would rise by 1.62 log units. The $Al_2(OH)_2^{4+}$ species defined by Reaction 17 of Table 3.1 contributes only slightly to total soluble aluminum. A shift in the equilibrium activity of Al^{3+} indicated above results in an upward shift of $2 \times 1.62 = 3.24$ log units for this polymeric species since it contains two aluminum atoms per ion.

There is still considerable controversy about the hydrolysis of Al^{3+} and the stability constants for various monomeric and polymeric species (Frink and Peech, 1963; Dezelic et al., 1971; Frink, 1973; Baes and Mesmer, 1976; Bache and Sharp, 1976). Further investigations are necessary to resolve this controversy.

3.4 FLUORIDE COMPLEXES OF ALUMINUM

The fluoride complexes of aluminum are defined by Reactions 18 through 21 of Table 3.1. These complexes are plotted in Fig. 3.4 on the basis that Al^{3+} activity is fixed at 10^{-10} M. This activity represents equilibrium with gibbsite at pH 6. It is apparent from this plot that fluoride complexes can be extremely important in soils. Only when F^- activity is $<10^{-7}$ M are these complexes less abundant than Al^{3+}. For each log unit increase in F^- activity, AlF^{2+} increases 10-fold, AlF_2^+ 100-fold, $AlF_3°$ 1000-fold, and AlF_4^- 10,000-fold.

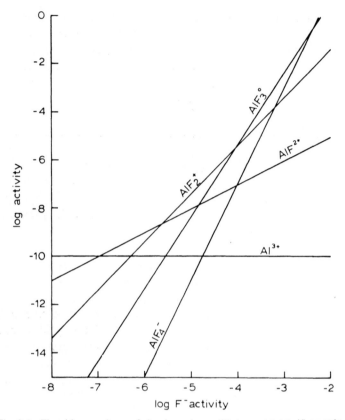

Fig. 3.4 Fluoride complexes of aluminum in equilibrium with 10^{-10} M Al^{3+}.

The activity of F^- in soils normally ranges from 10^{-10} to 10^{-4} M. If CaF_2(fluorite) is present, the F^- activity can be estimated from the reaction

$$CaF_2(\text{fluorite}) \rightleftharpoons Ca^{2+} + 2F^- \qquad \log K^\circ = -10.41$$

Since Ca^{2+} activity in soils is generally near $10^{-2.5}$ M, (Chapter 7), this would permit F^- activity to be $10^{-3.96}$ or approximately 10^{-4} M. With 10^{-10} M Al^{3+} and 10^{-4} M F^- the fluoride complexes consist of $10^{-7.0}$ M AlF^{2+}, $10^{-5.4}$ M AlF_2^+, $10^{-7.0}$ AlF_4^-, and $10^{-5.3}$ M AlF_3°. The ratio of complexed to free Al^{3+} is approximately

$$\frac{AlF_2^+ + AlF_3^\circ}{Al^{3+}} = \frac{10^{-5.4} + 10^{-5.3}}{10^{-10}} = 10^{4.95}$$

This represents a very high degree of complexation. Total aluminum in soil solution is pH dependent and must include hydrolysis species as well (Fig. 3.3). For example, the ratio of fluoride-complexed aluminum to fluoride-free aluminum for a soil at pH 6 in equilibrium with gibbsite is given by

$$\frac{AlF_2^+ + AlF_3^\circ}{Al(OH)_3^\circ + Al(OH)_2^+} = \frac{10^{-5.3} + 10^{-5.4}}{10^{-7} + 10^{-7.3}} = 59.9$$

For this soil fluoride complexes increase total soluble aluminum by approximately 60-fold. Other fluoride minerals such as $KMg_3AlSi_3O_{10}F_2$ (fluorphlogopite) and $Ca_5(PO_4)_3F$ (fluoroapatite) may lower F^- activities in some soils to the range of 10^{-4} to 10^{-10} M (see Chapters 5 and 12). Naturally as fluoride activity is lowered, the fluoride complexes of Al^{3+} contribute less to total soluble aluminum. As pH decreases the ratio $AlF_2^+ + AlF_3^\circ/\Sigma Al(OH)_n$ approaches a value of $10^{4.95}$ when $(F^-) = 10^{-4}$ M.

Although Fig. 3.4 is constructed on the basis that Al^{3+} is 10^{-10} M, its use is not restricted to this Al^{3+} activity. Let us consider a soil of pH 5 in equilibrium with gibbsite. The Al^{3+} in this soil would be 10^{-7} M (Fig. 3.3). The fluoride complexes obtained from Fig. 3.4 at pH 5 are then multiplied by $10^{-7}/10^{-10}$ or 10^3. For a F^- activity of 10^{-5} M in such a soil, AlF_2^+ activity would be $10^{-7.4} \times 10^3 = 10^{-4.4}$ M. In this way it is possible to estimate the activity of aluminum fluoride complexes corresponding to any Al^{3+} or F^- activity.

3.5 OTHER ALUMINUM COMPLEXES

Ions other than OH^- and F^- also complex with Al^{3+}. The importance of such ion complexes can be readily seen from their equilibrium reactions (Table 3.1, Reactions 22 through 25). In the case of $AlSO_4^+$, for example,

$$\frac{(AlSO_4^+)}{(Al^{3+})} = 10^{3.20}(SO_4^{2-}) \tag{3.5}$$

When $(SO_4^{2-}) = 10^{-3.20}$ M the ratio of $(AlSO_4^+)/(Al^{3+}) = 1$. A 10-fold change in (SO_4^{2-}) results in a 10-fold change in this ratio.

Similarly Reaction 24 of Table 3.1 gives:

$$\frac{(Al(SO_4)_2^-)}{(Al^{3+})} = 10^{1.90}(SO_4^{2-})^2 \tag{3.6}$$

When $(SO_4^{2-}) = 10^{-1.90/2}$ or $10^{-0.95}$ M, the molecular ratio of $(Al(SO_4)_2^-/(Al^{3+})$ is unity. Tenfold changes in SO_4^{2-} activity result in 100-fold changes in

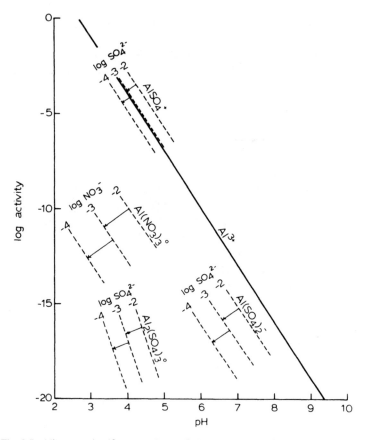

Fig. 3.5 Nitrate and sulfate complexes of aluminum in equilibrium with gibbsite.

this ratio. Similar equations can be developed for $Al(NO_3)_3^°$ and $Al_2(SO_4)_3^°$ from their equilibrium reactions in Table 3.1.

The effect of different activities of NO_3^- and SO_4^{2-} on aluminum complexes in solution in equilibrium with gibbsite is shown in Fig. 3.5. Of the complexes shown here only $AlSO_4^+$ is important. According to Reaction 26 of Table 3.1, the formation of $CaSO_4 \cdot 2H_2O$ (gypsum) limits SO_4^{2-} to less than 10^{-2} M in most soils. Only in sodium-affected soils where pH rises above 8.3 and Ca^{2+} is depressed below 10^{-3} M, can SO_4^{2-} rise sufficiently to permit the $Al(SO_4)_2^-$ complex to be significant (Chapter 7). The $Al(NO_3)_3^°$ complex is not significant in soils because molar concentrations of NO_3^- would be required.

3.6 ESTIMATING Al^{3+} ACTIVITY

The activity of Al^{3+} in soils and water can be estimated from pH, total Al, and activities of F^- and SO_4^{2-}. The major contributing species for most soils (Figs. 3.3 through 3.5) include: *more important than the lower ones*

$$[\text{Total Al}] = [Al^{3+}] + [AlOH^{2+}] + [Al(OH)_2^+] + [Al(OH)_3^\circ]$$
$$+ [Al(OH)_4^-] + [AlF^{2+}] + [AlF_2^+] + [AlF_3^\circ]$$
$$+ [AlSO_4^+] \qquad (3.7)$$

Expressing the ions in terms of activities gives:

$$[\text{Total Al}] = \frac{(Al^{3+})}{\gamma Al^{3+}} + \frac{(AlOH^{2+})}{\gamma AlOH^{2+}} + \frac{(Al(OH)_2^+)}{\gamma Al(OH)_2^+} + \frac{(Al(OH)_3^\circ)}{\gamma Al(OH)_3^\circ}$$
$$+ \frac{(Al(OH)_4^-)}{\gamma Al(OH)_4^-} + \frac{(AlF^{2+})}{\gamma AlF^{2+}} + \frac{(AlF_2^+)}{\gamma AlF_2^+} + \frac{(AlF_3^\circ)}{\gamma AlF_3^\circ} + \frac{(AlSO_4^+)}{\gamma AlSO_4^+}$$

Equilibrium reactions in Table 3.1 can be used to express the activity of each of the hydrolysis species in terms of H^+, Al^{3+}, F^-, and SO_4^{2-} activities giving

$$[\text{Total Al}] = \frac{(Al^{3+})}{\gamma Al^{3+}} + \frac{10^{-5.02}(Al^{3+})}{\gamma AlOH^{2+}(H^+)} + \frac{10^{-9.30}(Al^{3+})}{\gamma Al(OH)_2^+(H^+)^2}$$
$$+ \frac{10^{-14.99}(Al^{3+})}{\gamma Al(OH)_3^\circ(H^+)^3} + \frac{10^{-23.33}(Al^{3+})}{\gamma Al(OH)_4^-(H^+)^4} + \frac{10^{6.98}(Al^{3+})(F^-)}{\gamma AlF^{2+}}$$
$$+ \frac{10^{12.60}(Al^{3+})(F^-)^2}{\gamma AlF_2^+} + \frac{10^{16.65}(Al^{3+})(F^-)^3}{\gamma AlF_3^\circ}$$
$$+ \frac{10^{3.20}(Al^{3+})(SO_4^{2-})}{\gamma AlSO_4^+} \qquad (3.9)$$

Rearranging gives: $(Al^{3+}) = \dfrac{[\text{Total Al}]}{\text{REM}}$, where $\qquad (3.10)$

$$\text{REM} = \frac{1}{\gamma Al^{3+}} + \frac{10^{-5.02}}{\gamma AlOH^{2+}(H^+)} + \frac{10^{-9.30}}{\gamma Al(OH)_2^+(H^+)^2}$$
$$+ \frac{10^{-14.99}}{\gamma Al(OH)_3^\circ(H^+)^3} + \frac{10^{-23.33}}{\gamma Al(OH)_4^-(H^+)^4} + \frac{10^{6.98}(F^-)}{\gamma AlF^{2+}}$$
$$+ \frac{10^{12.60}(F^-)^2}{\gamma AlF_2^+} + \frac{10^{16.65}(F^-)^3}{\gamma AlF_3^\circ} + \frac{10^{3.20}(SO_4^{2-})}{\gamma AlSO_4^+}$$

Equation 3.10 can be used to estimate the activity of Al^{3+} in soils from pH, total Al, ionic strength, and the activities of F^- and SO_4^{2-}.

EXAMPLE PROBLEM

Estimate the Al^{3+} activity of a soil from a 0.01 M $CaCl_2$ extract (50 g soil to 100 ml of solution) which has a pH of 6.3, a F^- activity of 10^{-5} M, a SO_4^{2-} activity of 10^{-3} M, and contains 0.05 ppm of total Al in solution. Discuss whether or not this soil is in equilibrium with various aluminum oxides and hydroxides.

Solution. Equation 3.10 can be used to estimate Al^{3+} activity. In this equation, total Al is 0.05 ppm/27,000 = $10^{-5.73}$ M, H^+ activity is $10^{-6.3}$ M, F^- activity is 10^{-5} M, SO_4^{2-} activity is 10^{-3} M, and the activity coefficients for the various ions can be estimated from the Davies' equation (Table 2.3) at an ionic strength of 0.03 based on the 0.01 M $CaCl_2$ used for extraction. Making these substitutions into Eq. 3.10 gives $(Al^{3+}) = 10^{-9.78}$ M. This value corresponds to a log Al^{3+} + 3pH = 9.12, a value which is intermediate between $Al(OH)_3$(amorp) (9.66) and $Al(OH)_3$(gibbsite) (8.04). This soil is supersaturated with respect to bayerite, boehmite, norstrandite, gibbsite, and diaspore, all of which may form.

When aluminum complexes in addition to those included in Eq. 3.10 are present, they too must be included.

3.7 REDOX RELATIONSHIPS OF ALUMINUM

The only oxidation state of aluminum that is important in soils is Al^{3+}. From Reaction 28 of Table 3.1

$$Al^{3+} + 3e^- \rightleftharpoons Al(c) \qquad \log K° = -86.02$$

comes the relationship

$$\frac{1}{(Al^{3+})(e^-)^3} = 10^{-86.02}$$

or

$$pe = -28.67 + \tfrac{1}{3} \log Al^{3+} \qquad (3.11)$$

Thus Al(c) is highly unstable in the redox range of soils (Fig. 2.1) and would readily oxidize to Al^{3+}.

3.8 EXCHANGEABLE ALUMINUM

As pointed out in Chapter 1, ions in the soil solution equilibrate with adsorbed or exchangeable ions. The Boltzmann relationship can be used to describe the relationship between exchangeable Al^{3+} and H^+ (Lindsay et al., 1959).

$$(M^{Z+})_{exch} = (M^{Z+})_{soln} \exp \frac{-Z\varepsilon\psi s}{\kappa T} \qquad (3.12)$$

where $(M^{Z+})_{exch}$ and $(M^{Z+})_{soln}$ are the cationic activities on exchange sites and in the bulk soil solution outside the influence of the charged clay particle, Z is the valency of the cation, ε is the electronic charge, ψs is the electrical potential on the surface of the clay particle, κ is the Boltzmann constant and T is the absolute temperature (Bolt and Peech, 1953). Applying Eq. 3.12 when M^{Z+} is Al^{3+} and again when M^{Z+} is H^+ and taking their ratio gives:

$$\frac{(Al^{3+})^{1/3}_{exch}}{(H^+)_{exch}} = \frac{(Al^{3+})^{1/3}_{soln}}{(H^+)_{soln}} \qquad (3.13)$$

The log of the right-hand side of this equation can be expressed as

$$\log \frac{(Al^{3+})^{1/3}}{(H^+)} = pH - \tfrac{1}{3}pAl \qquad (3.14)$$

or

$$\log \frac{(Al^{3+})}{(H^+)^3} = \log Al^{3+} + 3pH \qquad (3.15)$$

This means that a soil would tend to maintain a fixed pH $- \tfrac{1}{3}$pAl or log Al^{3+} + 3pH in the soil solution so long as the activities of Al^{3+} and H^+ on the surface of the clay do not change appreciably. These activities tend to remain constant since there are far more ions on the exchange than in solution.

The relationship depicted by Eq. 3.14 or 3.15 was tested for a Mardin silt loam initially at pH 4.2 (Lindsay et al., 1959). This soil was equilibrated with $CaCl_2$ solutions varying in concentration from 10^{-3} to 10^{-1} M. Increasing the concentration of $CaCl_2$ (Table 3.2) shows that both H^+ and Al^{3+} were displaced from the exchange by Ca^{2+}, yet the parameter pH $- \tfrac{1}{3}$pAl remained constant at 2.52. The corresponding value for log Al^{3+} + 3pH is 7.55. This experiment confirms the tendency of soils to maintain a constant log Al^{3+} + 3pH as predicted by the Boltzmann equation independent of solid phases that may be present. Equilibrium with gibbsite gives log Al^{3+} + 3pH of 8.04, whereas equilibrium with $Al(OH)_3$(amorp) gives a value of

TABLE 3.2 THE EFFECT OF INCREASING THE $CaCl_2$ IN SOIL EXTRACTS ON THE VALUE OF $pH - \frac{1}{3} pAl$ OR $\log Al^{3+} + 3pH$ OF A MARDIN SILT LOAM INITIALLY AT pH 4.2

Concentration of $CaCl_2$ (mM)	pH	Total Al (mM)	pH-$\frac{1}{3}$pAl	$\log Al + 3pH$
1	3.94	0.092	2.52	7.55
5	3.79	0.39	2.52	7.55
10	3.75	0.68	2.52	7.55
50	3.71	1.77	2.52	7.55
100	3.70	2.37	2.51	7.51

Source: Lindsay et al. (1959).

9.66 (Reactions 7 and 2 of Table 3.1). Thus the Mardin silt loam represented in Table 3.2 is undersaturated with respect to any of the minerals shown in Fig. 3.1, yet the $\log Al^{3+} + 3pH$ of this soil tends to remain constant over a wide range of displaced Al^{3+}.

The fact that $\log Al^{3+} + 3pH$ for Mardin silt loam depicted in Table 3.2 is lower than that for gibbsite suggests that aluminosilicates may be controlling the Al^{3+} activity. This hypothesis will be examined further in Chapter 5.

REFERENCES

Bache, B. W. and G. W. Sharp. 1976. Soluble polymeric hydroxy-aluminum ions in acid soils. J. Soil Sci. 27:167–174.

Baes, C. F., Jr. and R. E. Mesmer. 1976. The hydrolysis of cations. Wiley-Interscience, New York.

Bolt, G. H. and M. Peech. 1953. The application of the Gouy theory to soil-water systems. Soil Sci. Soc. Am. Proc. 17:210–213.

Breemen, N. van. 1976. Genesis and solution chemistry of acid sulfate soils in Thailand. Agric. Res. Rep. (Versl. landbouwk. Onderz.) 848. Wegeningen, The Netherlands.

Dezelic, N., H. Bilinski, and R. H. H. Wolf. 1971. Precipitation and hydrolysis of metallic ions. IV. Studies on the solubility of aluminum hydroxide in aqueous solution. J. Inorg. Nucl. Chem. 33:791–798.

Frink, C.R. and M. Peech. 1963. Hydrolysis of the aluminum ion in dilute aqueous solutions. Inorg. Chem. 2:473–478.

Frink, C. R. 1973. Aluminum chemistry in acid sulfate soils. In H. Dost (ed.), Acid Sulphate Soils, Proceedings of the International Symposium, Wageningen. ILRI Publ. 18 Vol. 1, pp. 131–168.

Lindsay, W. L., M. Peech, and J. S. Clark. 1959. Determination of aluminum ion activity in soil extracts. Soil Sci. Soc. Am. Proc. 23:266–269.

Rhodes, E. R. and W. L. Lindsay. 1978. Solubility of aluminum in some soils of the humid tropics. J. Soil Sci. 29:324–330.

Sherman, G. D. 1958. Gibbsite-rich soils of the Hawaiian Islands. Hawaii Agr. Exp. Sta. Bull. 116.

PROBLEMS

3.1 Develop the solubility equation for Al^{3+} in equilibrium with $Al(OH)_3$ (amorp) and plot it on a graph similar to Fig. 3.1.

3.2 Develop the equations for $Al(OH)_2^+$ and $Al_2(OH)_2^{4+}$ in equilibrium with $Al(OH)_3$(amorp) and plot them in the graph prepared for Problem 3.1 above. How are these lines different from those shown in Fig. 3.3?

3.3 Develop the equation and calculate the pH at which alunite and gibbsite can coexist when the activity of K^+ is $10^{-3.2}$ M and that of SO_4^{2-} is $10^{-2.2}$ M.

3.4 Using reactions from Table 3.1 and considering activity coefficients as unity, calculate the Total Al in solution at pH 5 and pH 8 corresponding to Fig. 3.3

3.5 Develop the necessary equation to plot the $AlSO_4^+$ complex ion in Fig. 3.3 when the activity of Ca^{2+} is $10^{-2.5}$ M and $CaSO_4 \cdot 2H_2O$ (gypsum) is present.

3.6 A 0.01 M $CaCl_2$ equilibrium extract of a soil has a pH of 4.11, a total Al of 2.48×10^{-3} M, and an activity of SO_4^{2-} of $10^{-2.5}$ M. Calculate the log Al^{3+} + 3pH for this soil and discuss its status with respect to gibbsite:
   a. when fluoride can be ignored
   b. when F^- activity is $10^{-5.5}$ M

3.7 Explain how a soil can maintain a fairly constant log Al^{3+} + 3pH even though the soil is far undersaturated with respect to $Al(OH)_3$ (gibbsite).

FOUR

SILICA

Silicon is the second most abundant element of the earth's crust, and soil contains approximately 32% silicon by weight (Table 1.1). This element is present in primary silicate minerals, secondary aluminosilicates, and various forms of SiO_2. Cations such as Fe^{3+}, Fe^{2+}, Al^{3+}, Cu^{2+}, Zn^{2+}, Mg^{2+}, Ca^{2+}, etc. are sometimes included in silicate minerals. For these reasons it is important to understand the chemical equilibrium relationships of various silicate species.

This chapter is developed to provide a basis for understanding silica equilibrium relationships in soils and soluble silicate species expected in soil solutions. Aluminosilicate minerals are considered in Chapter 5.

4.1 FORMS OF SILICA IN SOILS

In most silicate minerals silicon is surrounded by four oxygen atoms in tetrahedral arrangement. Quartz is generally recognized as the most stable SiO_2 mineral at normal temperatures and pressures. The solubility relationships of the different forms of silica are given by Reactions 1 through 7 of Table 4.1 and are plotted in Fig. 4.1. The solubility of silica minerals in terms of $H_4SiO_4^°$ is expected to range from $10^{-2.74}\ M$ (amorphous silica) to $10^{-4}\ M$ (quartz). The other minerals have intermediate solubilities. The SiO_2(soil) line at $10^{-3.10}\ M$ corresponds to Reaction 4 of Table 4.1 and is

TABLE 4.1 EQUILIBRIUM REACTIONS FOR VARIOUS SILICATE SPECIES AT 25°C

Reaction No.	Equilibrium Reaction	log $K°$
	SiO_2 Minerals	
1	$SiO_2\text{(silica glass)} + 2H_2O \rightleftharpoons H_4SiO_4^°$	−2.71
2	$SiO_2\text{(amorp)} + 2H_2O \rightleftharpoons H_4SiO_4^°$	−2.74
3	$SiO_2\text{(coesite)} + 2H_2O \rightleftharpoons H_4SiO_4^°$	−3.05
4	$SiO_2\text{(soil)} + 2H_2O \rightleftharpoons H_4SiO_4^°$	−3.10
5	$\alpha\text{-}SiO_2\text{(tridymite)} + 2H_2O \rightleftharpoons H_4SiO_4^°$	−3.76
6	$\alpha\text{-}SiO_2\text{(cristobalite)} + 2H_2O \rightleftharpoons H_4SiO_4^°$	−3.94
7	$\alpha\text{-}SiO_2\text{(quartz)} + 2H_2O \rightleftharpoons H_4SiO_4^°$	−4.00
	Silicate Ions	
8	$H_4SiO_4^° \rightleftharpoons H_3SiO_4^- + H^+$	−9.71
9	$H_4SiO_4^° \rightleftharpoons H_2SiO_4^{2-} + 2H^+$	−22.98
10	$H_4SiO_4^° \rightleftharpoons HSiO_4^{3-} + 3H^+$	−32.85
11	$H_4SiO_4^° \rightleftharpoons SiO_4^{4-} + 4H^+$	−45.95
12	$4H_4SiO_4^° \rightleftharpoons H_6Si_4O_{12}^{2-} + 2H^+ + 4H_2O$	−13.32

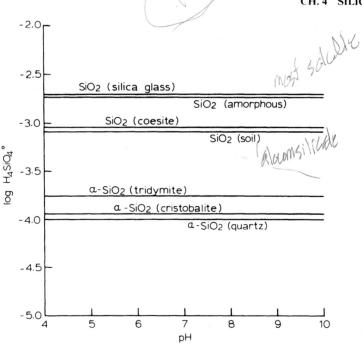

Fig. 4.1 The activity of $H_4SiO_4^0$ maintained by various forms of silica.

based on the experimental results of Elgawhary and Lindsay (1972). They measured the solubility of silica in two soils in which equilibrium was approached from both undersaturation and supersaturation. Other soils may show different levels of soluble silica, nevertheless this value will be used in this text as a reference solubility level of silica in soils. The solubility of SiO_2(soil) is intermediate between quartz and amorphous silica. In highly weathered soils, free SiO_2 may become depleted from soils leaving the sesquioxides of iron and aluminum as the major residual minerals. In such soils the solubility of $H_4SiO_4^0$ drops below that of quartz. The significance of these relationships will be considered further in Chapter 5.

Iller (1955) documented the fact that the chemistry of silica is very complex and that equilibrium relationships are often difficult to attain. Strober (1967) measured the solubility of silicic acid in aqueous suspensions containing different forms of silica including vitreous silica, stishovite, cristobalite, tridymite, quartz, and coesite. He concluded that surface adsorption of silicic acid often prevents equilibrium relationships from being achieved.

Many solubility values of SiO_2(stishovite) and various hydrated silicic acids such as $H_2SiO_3(c)$, $H_4SiO_4(c)$, $H_2Si_2O_5(c)$, and $H_6Si_2O_7(c)$ reported in the literature seem questionable; therefore, no selections of free energies of formation for these species were made.

4.2 SILICATE SPECIES IN SOLUTION

The dissociation of $H_4SiO_4^o$ and the polymerization of silicate species in solution are summarized by Reactions 8 through 12 of Table 4.1. These relationships are also plotted in Fig. 4.2 under the condition that $H_4SiO_4^o$ in soil solution is controlled by SiO_2(soil) discussed above. An example of how the equilibrium relationships in Fig. 4.2 were developed is shown for the $H_3SiO_4^-$ species by combining Reactions 4 and 8 of Table 4.1:

$$
\begin{array}{lr}
& \log K^\circ \\
SiO_2(\text{soil}) + 2H_2O \rightleftharpoons H_4SiO_4^o & -3.10 \\
H_4SiO_4^o \rightleftharpoons H^+ + H_3SiO_4^- & -9.71 \\
\hline
SiO_2(\text{soil}) + 2H_2O \rightleftharpoons H^+ + H_3SiO_4^- & -12.81 \quad (4.1)
\end{array}
$$

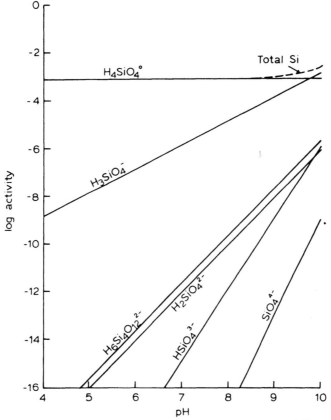

Fig. 4.2 Silicate species in equilibrium with SiO_2(soil) represented by $10^{-3.10}$ M $H_4SiO_4^o$.

for which

$$(H^+)(H_3SiO_4^-) = 10^{-12.81}$$
$$\log H_3SiO_4^- = pH - 12.81 \qquad (4.2)$$

Equation 4.2 is plotted as the $H_3SiO_4^-$ line in Fig. 4.2. Lines for the other solution species in this figure were developed similarly. Thus Fig. 4.2 depicts the major silicate species expected in soil solution. Only at pH values above 8.5 do the ionic silicates contribute significantly to total silica in solution. In the normal pH range of soils, $H_4SiO_4^°$ comprises the major silicate species in solution.

Much controversy has prevailed regarding the solubility of measured silica in soils and how its solubility is affected by pH (McKeague and Cline, 1963). More recently Elgawhary and Lindsay (1972) proposed extracting soils with 0.02 M $CaCl_2$ in order to keep colloidal silica flocculated during extraction and filtration of soils in which silica is measured. When this procedure was used, they found that measured silica in soil solution corresponded more nearly to the levels of $H_4SiO_4^°$ expected from solubility predictions. When $CaCl_2$ was not used, they found that silica solubility varied greatly with time of shaking, pH, and other factors. Hopefully some of the confusion regarding the solubility of silica in soils can be resolved by implementing this technique.

REFERENCES

Elgawhary, S. M. and W. L. Lindsay. 1972. Solubility of silica in soils. Soil Sci. Soc. Am. Proc. 36:439–442.

Iller, R. K. 1955. The Colloid Chemistry of Silica and Silicates. Cornell Univ. Press., Ithaca, N.Y.

McKeague, J. A. and M. G. Cline. 1963. Silica in soils. Adv. Agron. 15:339–396.

Strober, W. 1967. Formation of silicic acid in acqueous suspensions of different silica modifications. *In* R. F. Gould, Ed. Equilibrium Concepts in Natural Water Systems. Adv. Chem. Ser. 67:161–182.

PROBLEMS

4.1 What equations were used to plot the SiO_2(soil) and SiO_2(quartz) lines in Fig. 4.1? What conclusions can be drawn regarding the stability of different forms of silica in soils based on Fig. 4.1?

4.2 Derive the necessary equations and plot the activities of $H_2SiO_4^{2-}$ and $H_6Si_4O_{12}^{2-}$ in equilibrium with SiO_2(quartz). By how many log units do these lines differ from corresponding lines in Fig. 4.2?

4.3 Indicate how you would proceed to determine the equilibrium constant for Reaction 4 of Table 4.1 for a soil of great interest to you.

4.4 From the equilibrium in Table 4.1, calculate an equilibrium constant for the reaction below and interpret its meaning and limitations.

$$SiO_2(quartz) \rightleftharpoons SiO_2(soil)$$

4.5 The ΔG_f° for $CaSiO_3$(wollastonite) is reported as -370.39. Develop the necessary equation and plot log Ca^{2+} versus pH when equilibrium with SiO_2(soil) is established. Knowing that Ca^{2+} in soils in the pH range of 5 to 7.5 is about $10^{-2.5}$ M discuss the potential of $CaSiO_3$(c) as a liming agent. Write a balanced chemical equation for the reaction it would undergo in soils. What is the calcium carbonate equivalent of $CaSiO_3$(c)?

FIVE

ALUMINOSILICATE MINERALS

Aluminosilicates comprise a major portion of the mineral fraction of soils and include both primary and secondary minerals. Most primary minerals were formed at temperatures and pressures much different from those that exist at the earth's surface today. In general primary minerals are unstable in soils and slowly dissolve. Some of the products recombine to form secondary soil minerals that are more stable under the conditions that exist in soils today.

Weathering involves both physical and chemical processes. Physical processes are important because they break up large rocks into smaller particles having greater surface areas. Reactions of solids are largely limited to surface reactions. Thus the greater the surface area, the greater are the rates of chemical reactions. Free energy relationships can be used to predict whether or not specific chemical reactions can take place. This is possible because the free energy of the products of a chemical reaction must always be less than the free energy of reactants. Many weathering processes can be predicted from free energy relationships. Recently studies have shown how these thermodynamic relationships can be used to predict soil mineral transformations (Kittrick, 1977; Marshall, 1977; Rai and Lindsay, 1975).

Thermodynamic stability diagrams are developed and used in this chapter to predict the stability, formation, and weathering of many aluminosilicate minerals found in soils. Equilibrium reactions for the aluminosilicate minerals considered in this chapter are summarized in Table 5.1. The equilibrium constants in this table were calculated from the standard free energy values given in the Appendix.

5.1 UNSUBSTITUTED ALUMINOSILICATES

The solubilities of several unsubstituted aluminosilicates whose solubilities are reasonably well known are given by Reactions 1 thru 7 of Table 5.1. The solubilities of these minerals can be expressed in terms of the activities of Al^{3+}, H^+, and $H_4SiO_4^\circ$. Let us consider kaolinite whose solubility can be expressed as

$$Al_2Si_2O_5(OH)_4(\text{kaolinite}) + 6H^+ \rightleftharpoons 2Al^{3+} + 2H_4SiO_4^\circ + H_2O \quad (5.1)$$

$$\log K^\circ = 5.45$$

$$\frac{(Al^{3+})^2(H_4SiO_4^\circ)^2}{(H^+)^6} = 10^{5.45}$$

$$2\log Al^{3+} + 6pH = 5.45 - 2\log H_4SiO_4^\circ$$
$$\log Al^{3+} + 3pH = 2.73 - \log H_4SiO_4^\circ \quad (5.2)$$

TABLE 5.1 EQUILIBRIUM REACTIONS USED TO CONSTRUCT THE STABILITY DIAGRAMS FOR ALUMINO-SILICATES

Reaction No.	Chemical Reaction	$\log K°$
	Al-Si	
1	$Al_2SiO_5(\text{sillimanite}) + 6H^+ \rightleftharpoons 2Al^{3+} + H_4SiO_4° + H_2O$	15.45
2	$Al_2SiO_5(\text{kyanite}) + 6H^+ \rightleftharpoons 2Al^{3+} + H_4SiO_4° + H_2O$	15.12
3	$Al_2SiO_5(\text{andalusite}) + 6H^+ \rightleftharpoons 2Al^{3+} + H_4SiO_4° + H_2O$	14.48
4	$Al_2Si_2O_5(OH)_4(\text{halloysite}) + 6H^+ \rightleftharpoons 2Al^{3+} + 2H_4SiO_4° + H_2O$	8.72
5	$Al_2Si_2O_5(OH)_4(\text{dickite}) + 6H^+ \rightleftharpoons 2Al^{3+} + 2H_4SiO_4° + H_2O$	5.95
6	$Al_2Si_2O_5(OH)_4(\text{kaolinite}) + 6H^+ \rightleftharpoons 2Al^{3+} + 2H_4SiO_4° + H_2O$	5.45
7	$Al_2Si_4O_{10}(OH)_2(\text{pyrophyllite}) + 6H^+ + 4H_2O \rightleftharpoons 2Al^{3+} + 4H_4SiO_4°$	−1.92
	NaAl-Si	
8	$NaAlSiO_4(\text{nepheline}) + 4H^+ \rightleftharpoons Na^+ + Al^{3+} + H_4SiO_4°$	11.25
9	$NaAlSi_2O_6(\text{jadeite}) + 4H^+ + 2H_2O \rightleftharpoons Na^+ + Al^{3+} + 2H_4SiO_4°$	7.11
10	$NaAlSi_2O_6 \cdot H_2O(\text{analcime}) + 4H^+ + H_2O \rightleftharpoons Na^+ + Al^{3+} + 2H_4SiO_4°$	8.15
11	$NaAlSi_3O_8(\text{Na-glass}) + 4H^+ + 4H_2O \rightleftharpoons Na^+ + Al^{3+} + 3H_4SiO_4°$	10.87
12	$NaAlSi_3O_8(\text{high albite}) + 4H^+ + 4H_2O \rightleftharpoons Na^+ + Al^{3+} + 3H_4SiO_4°$	3.67
13	$NaAlSi_3O_8(\text{low albite}) + 4H^+ + 4H_2O \rightleftharpoons Na^+ + Al^{3+} + 3H_4SiO_4°$	2.74
14	$NaAl_3Si_3O_{10}(OH)_2(\text{paragonite}) + 10H^+ \rightleftharpoons Na^+ + 3Al^{3+} + 3H_4SiO_4°$	17.40
15	$Na_{0.33}Al_{2.33}Si_{3.67}O_{10}(OH)_2(\text{beidellite}) + 7.32H^+ + 2.68H_2O \rightleftharpoons 0.33Na^+ + 2.33Al^{3+} + 3.67H_4SiO_4°$	6.13

KAl-Si

16	$KAlSiO_4(\text{kaliophilite}) + 4H^+ \rightleftharpoons K^+ + Al^{3+} + H_4SiO_4°$	13.05
17	$KAlSi_2O_6(\text{leucite}) + 4H^+ + 2H_2O \rightleftharpoons K^+ + Al^{3+} + 2H_4SiO_4°$	6.72
18	$KAlSi_3O_8(\text{K-glass}) + 4H^+ + 4H_2O \rightleftharpoons K^+ + Al^{3+} + 3H_4SiO_4°$	7.87
19	$KAlSi_3O_8(\text{high sanidine}) + 4H^+ + 4H_2O \rightleftharpoons K^+ + Al^{3+} + 3H_4SiO_4°$	1.40
20	$KAlSi_3O_8(\text{microcline}) + 4H^+ + 4H_2O \rightleftharpoons K^+ + Al^{3+} + 3H_4SiO_4°$	1.00
21	$KAl_2(AlSi_3O_{10})(OH)_2(\text{muscovite}) + 10H^+ \rightleftharpoons K^+ + 3Al^{3+} + 3H_4SiO_4°$	13.44

CaAl-Si

22	$CaAl_2SiO_6(\text{pyroxene}) + 8H^+ \rightleftharpoons Ca^{2+} + 2Al^{3+} + H_4SiO_4° + 2H_2O$	35.25
23	$CaAl_2Si_2O_8(\text{Ca-glass}) + 8H^+ \rightleftharpoons Ca^{2+} + 2Al^{3+} + 2H_4SiO_4°$	33.91
24	$CaAl_2Si_2O_8(\text{hexagonal anorthite}) + 8H^+ \rightleftharpoons Ca^{2+} + 2Al^{3+} + 2H_4SiO_4°$	26.10
25	$CaAl_2Si_2O_8(\text{anorthite}) + 8H^+ \rightleftharpoons Ca^{2+} + 2Al^{3+} + 2H_4SiO_4°$	23.33
26	$CaAl_2Si_2O_8 \cdot 2H_2O(\text{lawsonite}) + 8H^+ \rightleftharpoons Ca^{2+} + 2Al^{3+} + 2H_4SiO_4° + 2H_2O$	17.54
27	$CaAl_2Si_4O_{12} \cdot 2H_2O(\text{wairakite}) + 8H^+ + 2H_2O \rightleftharpoons Ca^{2+} + 2Al^{3+} + 4H_4SiO_4°$	16.05
28	$Ca_2Al_4Si_8O_{24} \cdot 7H_2O(\text{leonhardite}) + 16H^+ + H_2O \rightleftharpoons 2Ca^{2+} + 4Al^{3+} + 8H_4SiO_4°$	17.29

MgAl-Si

29	$Mg_5Al_2Si_3O_{10}(OH)_8(\text{chlorite}) + 16H^+ \rightleftharpoons 5Mg^{2+} + 2Al^{3+} + 3H_4SiO_4° + 6H_2O$	60.30
30	$Mg_2Al_4Si_5O_{18}(\text{Mg-cordierite}) + 16H^+ + 2H_2O \rightleftharpoons 2Mg^{2+} + 4Al^{3+} + 5H_4SiO_4°$	45.46
31	$KMg_3AlSi_3O_{10}F_2(\text{fluorphlogopite}) + 8H^+ + 2H_2O \rightleftharpoons K^+ + 3Mg^{2+} + Al^{3+} + 3H_4SiO_4° + 2F^-$	7.85

(Continued)

TABLE 5.1 (*Continued*)

Reaction No.	Chemical Reaction	log $K°$
	Substituted Al-Si	
32	$(Mg_{2.71}Fe(II)_{0.02}Fe(III)_{0.46}Ca_{0.06}K_{0.1})Si_{2.91}$- $Al_{1.14}O_{10}(OH)_2$(vermiculite) + $10.36H^+ \rightleftharpoons 2.71Mg^{2+} + 0.02Fe^{2+} + 0.46Fe^{3+} + 0.06Ca^{2+}$ $+ 0.1K^+ + 1.14Al^{3+} + 2.91H_4SiO_4° + 0.36H_2O$	38.14
33	$K_{0.6}Mg_{0.25}Al_{2.3}Si_{3.5}O_{10}(OH)_2$(illite) + $8H^+ + 2H_2O \rightleftharpoons 0.6K^+ + 0.25Mg^{2+} + 2.3Al^{3+} + 3.5H_4SiO_4°$	10.35
34	$Mg_{0.2}(Si_{3.81}Al_{1.71}Fe(III)_{0.22}Mg_{0.29})O_{10}(OH)_2$ (Mg-montmorillonite) + $6.76H^+ + 3.24H_2O \rightleftharpoons 0.49Mg^{2+} + 1.71Al^{3+} + 0.22Fe^{3+} + 3.81H_4SiO_4°$	2.68
	Reacting Components	
35	$Fe(OH)_3$(soil) + $3H^+ \rightleftharpoons Fe^{3+} + 3H_2O$	2.70
36	$Fe^{3+} + e^- \rightleftharpoons Fe^{2+}$	13.04
37	α-FeOOH(goethite) + $3H^+ \rightleftharpoons Fe^{3+} + 2H_2O$	−0.02
38	$Al(OH)_3$(amorp) + $3H^+ \rightleftharpoons Al^{3+} + 3H_2O$	9.66
39	$Al(OH)_3$(gibbsite) + $3H^+ \rightleftharpoons Al^{3+} + 3H_2O$	8.04
40	SiO_2(amorp) + $2H_2O \rightleftharpoons H_4SiO_4°$	−2.74
41	SiO_2(soil) + $2H_2O \rightleftharpoons H_4SiO_4°$	−3.10
42	SiO_2(quartz) + $2H_2O \rightleftharpoons H_4SiO_4°$	−4.00
43	CaF_2(fluorite) $\rightleftharpoons Ca^{2+} + 2F^-$	−10.41
44	Ca^{2+}(soil) $\rightleftharpoons Ca^{2+}$	−2.50

CH. 5 UNSUBSTITUTED ALUMINOSILICATES

By selecting different values for log $H_4SiO_4^\circ$, Eq. 5.2 can be solved for log $Al^{3+} + 3pH$ to give:

log $H_4SiO_4^\circ$	log $Al^{3+} + 3pH$
−2.5	5.23
−4.0	6.73
−5.5	8.23

These relationships are plotted in Fig. 5.1 and represent the solubility line for kaolinite labeled 6, which refers to Reaction 6 of Table 5.1. The other solubility lines in this figure were developed similarly.

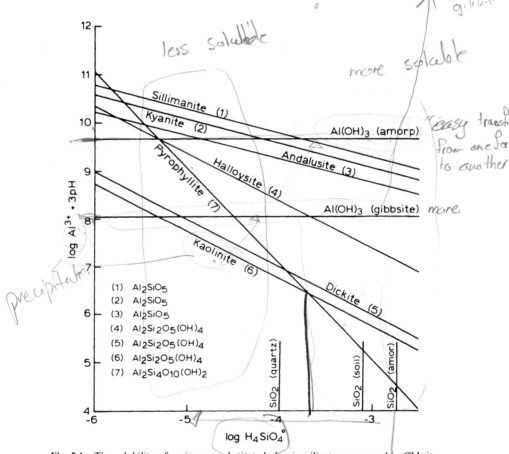

Fig. 5.1 The solubility of various unsubstituted aluminosilicates compared to gibbsite.

From Fig. 5.1 it is easy to see which minerals are most stable and the conditions that affect their stabilities. Solubilities of the $Al_2Si_2O_5$ minerals decrease in the order sillimanite > kyanite > andulusite. Solubilities of the $Al_2Si_2O_5(OH)_4$ minerals decrease in the order halloysite > dickite > kaolinite. Only one mineral having a Si/Al ratio of two is included, namely $Al_2Si_4O_{10}(OH)_2$(pyrophyllite). At $H_4SiO_4^\circ$ activities greater than $10^{-3.7}$ M, pyrophyllite is more stable than kaolinite, whereas below this $H_4SiO_4^\circ$ level, kaolinite is more stable.

The solubility of $Al(OH)_3$(gibbsite) is shown by the horizontal line at log Al^{3+} + 3 pH = 8.04, while that of $Al(OH)_3$(amorp) lies at 9.66. At $H_4SiO_4^\circ$ activities below $10^{-5.31}$ M gibbsite is the most stable mineral represented in Fig. 5.1. The solubility relationships of SiO_2(quartz), SiO_2(soil), and SiO_2(amorp) are included to show the levels of $H_4SiO_4^\circ$ that are imposed by these solid phases, namely $10^{-4.0}$, $10^{-3.10}$, and $10^{-2.74}$ M, respectively. Since primary minerals generally have a higher Si/Al ratio than secondary minerals, silica is released during weathering and tends to maintain $H_4SiO_4^\circ$ somewhere within these limits. In soils developed under more intense weathering environments, silica is leached from soils as first kaolinite and eventually gibbsite become the stable mineral phases. Gibbsite, along with all the other aluminum oxide and hydroxide minerals discussed in Chapter 3, is unstable in soils in which $H_4SiO_4^\circ$ exceeds $10^{-5.31}$ M.

The activity of Al^{3+} in equilibrium with the various aluminosilicate minerals shown in Fig. 5.1 can be readily obtained from the diagram when pH is known. For example, equilibrium with gibbsite is given as

$$\log Al^{3+} + 3\, pH = 8.04$$

which at pH 5 becomes

$$\log Al^{3+} + 3(5) = 8.04$$

or

$$\log Al^{3+} = 8.04 - 15 = -6.96$$

which corresponds to an Al^{3+} activity of $10^{-6.96}$ M. Thus for any given pH, the vertical scale can be converted into Al^{3+} activity. Using combined functions as axes in these plots avoids the necessity of having to prepare a separate plot for each pH.

5.2 SODIUM ALUMINOSILICATES

The solubilities of several sodium aluminosilicates are given by Reactions 8 thru 15 of Table 5.1 and are plotted in Fig. 5.2. An example of how the

SODIUM ALUMINOSILICATES

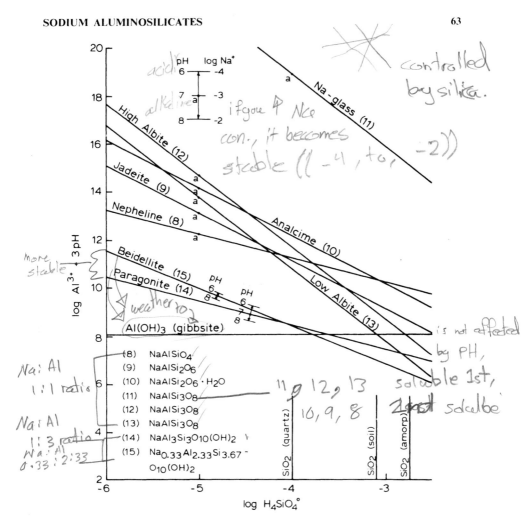

Fig. 5.2 The stability of several sodium aluminosilicates compared to gibbsite, showing the effects of pH and Na$^+$ activity.

solubility lines were developed is given for low albite starting with Reaction 13 of Table 5.1:

$$\text{NaAlSi}_3\text{O}_8(\text{low albite}) + 4\text{H}^+ + 4\text{H}_2\text{O} \rightleftharpoons \text{Na}^+ + \text{Al}^{3+} + 3\text{H}_4\text{SiO}_4^\circ \tag{5.3}$$

$$\log K^\circ = 2.74$$

$$\frac{(\text{Na}^+)(\text{Al}^{3+})(\text{H}_4\text{SiO}_4^\circ)^3}{(\text{H}^+)(\text{H}^+)^3} = 10^{2.74}$$

$$\log \text{Al}^{3+} + 3\text{pH} = 2.74 - \log \text{Na}^+ - \text{pH} - 3 \log \text{H}_4\text{SiO}_4^\circ \tag{5.4}$$

In most well-drained soils, a realistic level of Na$^+$ is approximately 10^{-3} M (see Chapter 9). Substituting this reference level of Na$^+$ into Eq. 5.4 gives

$$\log \text{Al}^{3+} + 3\text{pH} = 5.74 - \text{pH} - 3 \log \text{H}_4\text{SiO}_4^\circ \qquad (5.5)$$

A plot of Eq. 5.5 at pH 7 is shown in Fig. 5.2 and is labeled low albite (13) corresponding to Reaction 13 of Table 5.1. The other lines in this diagram were determined similarly using appropriate reactions from Table 5.1. Although Fig. 5.2 is drawn for pH 7 and 10^{-3} M Na$^+$, it can be used to predict solubility relationships at other pH and Na$^+$ values. The array of short lines and arrows in the upper left of this diagram indicates that all lines marked a shift up or down by one log unit with a one unit change in either pH or log Na$^+$.

The beidellite and paragonite lines shift less than one unit for corresponding changes in pH and Na$^+$. The pH changes are indicated directly on the diagram. A unit change in log Na$^+$ shifts the paragonite line by 0.33 unit and the beidellite line by only 0.14 unit. These changes reflect the Na/Al ratios in these minerals.

Figure 5.2 shows the relative stability of the sodium aluminosilicate minerals. The most soluble mineral is Na-glass. Of the three NaAlSi$_3$O$_8$ minerals, the solubilities decrease in the order of Na-glass > high albite > low albite. Sodium-glass is expected to weather most rapidly in soils. The minerals NaAlSiO$_4$(nepheline), NaAlSi$_2$O$_6$(jadeite), and NaAlSi$_2$O$_6 \cdot$H$_2$O (analcime) are intermediate in solubility and are expected to weather next most rapidly. The most stable sodium aluminosilicates are

$$\text{NaAl}_3\text{Si}_3\text{O}_{10}(\text{OH})_2 \text{(paragonite)}$$

and

$$\text{Na}_{0.33}\text{Al}_{2.33}\text{Si}_{3.67}\text{O}_{10}(\text{OH})_2 \text{(beidellite)}.$$

In neutral soils where H$_4$SiO$_4^\circ$ drops below $10^{-3.8}$ M, beidellite and paragonite become less stable than gibbsite.

All of the sodium aluminosilicate minerals depicted in Fig. 5.2 are unstable in soils and will eventually disappear with weathering because kaolinite and pyrophyllite are more stable (Fig. 5.1). From these relationships it is easy to see why Na$^+$ is released in the weathering of primary minerals and ends up in the oceans, because Na$^+$ is not strongly refixed into secondary minerals. In poorly drained soils, Na$^+$ often accumulates as the major soluble cation.

5.3 POTASSIUM ALUMINOSILICATES

The solubilities of several potassium aluminosilicate minerals are given by Reactions 16 through 21 of Table 5.1 and are plotted in Fig. 5.3. An example of how these solubility lines were developed is shown for muscovite beginning with Reaction 21 of Table 5.1:

$$KAl_2(AlSi_3O_{10})(OH)_2(\text{muscovite}) + 10H^+ \rightleftharpoons$$

$$K^+ + 3Al^{3+} + 3H_4SiO_4^\circ \quad (5.6)$$

$$\log K^\circ = 13.44 \quad \text{↑ higher + log K}$$

$$\frac{(K^+)(Al^{3+})^3(H_4SiO_4^\circ)^3}{(H^+)(H^+)^9} = 10^{13.44}$$

$$3\log Al^{3+} + 9\,pH = 13.44 - \log K^+ - pH - 3\log H_4SiO_4^\circ$$

$$\log Al^{3+} + 3\,pH = 4.48 - 0.33\log K^+ - 0.33\,pH - \log H_4SiO_4^\circ \quad (5.7)$$

When K^+ is 10^{-3} M, Eq. 5.7 becomes

$$\log Al^{3+} + 3\,pH = 5.48 - 0.33\,pH - \log H_4SiO_4^\circ \quad (5.8)$$

A plot of Eq. 5.8 corresponding to pH 7 is given in Fig. 5.3 and is labeled the muscovite line. Substitution of pH 6 or 8 instead of 7 into Eq. 5.8 shifts the muscovite line upward or downward by 0.33 of a unit ($\log Al^{3+} + 3\,pH$). The solubility lines for the other mineral in Fig. 5.3 were obtained similarly using appropriate equations from Table 5.1. Minerals (16) through (20) in Fig. 5.3 all shift upward or downward by one unit for each unit change in pH or $\log K^+$ as shown by the short lines and arrows in the upper left part of this diagram.

Figure 5.3 shows that the solubilities of potassium aluminosilicates decrease in the order of $KAlSi_3O_8$(K-glass) > $KAlSiO_4$(kaliophilite) > $KAlSi_2O_6$(leucite) > $KAlSi_3O_8$(high sanidine) > $KAlSi_3O_8$(microcline) > $KAl_2(AlSi_3O_{10})(OH)_2$(muscovite). Of these minerals K-glass is expected to weather most rapidly. For a soil of pH 7, muscovite becomes more stable than gibbsite when $H_4SiO_4^\circ$ exceeds $10^{-4.8}$ M. The pH dependency of muscovite causes this equilibrium $H_4SiO_4^\circ$ activity to shift slightly. Muscovite is more stable than gibbsite over most of the $H_4SiO_4^\circ$ range expected in soils. Only in the later stages of soil weathering, when silica has been largely removed, is gibbsite stable. It will be shown later (Fig. 5.6) that muscovite is generally metastable to kaolinite.

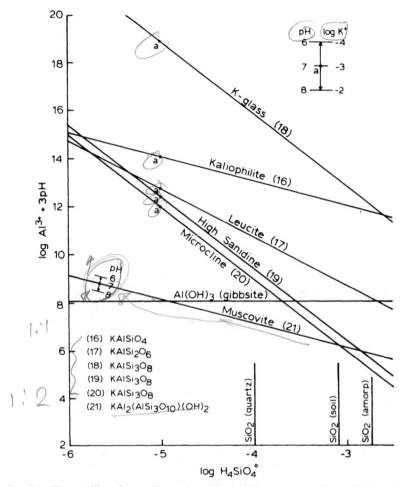

Fig. 5.3 The stability of several potassium aluminosilicates compared to gibbsite, showing the effect of pH and K^+ activity.

5.4 CALCIUM ALUMINOSILICATES

The solubilities of several calcium aluminosilicates are given by Reactions 22 through 28 of Table 5.1 and are plotted in Fig. 5.4. An example of how these stability lines were developed is shown for pyroxene starting with Reaction 22 of Table 5.1.

$$CaAl_2SiO_6(\text{pyroxene}) + 8H^+ \rightleftharpoons Ca^{2+} + 2Al^{3+} + H_4SiO_4^\circ + 2H_2O \tag{5.9}$$

CALCIUM ALUMINOSILICATES

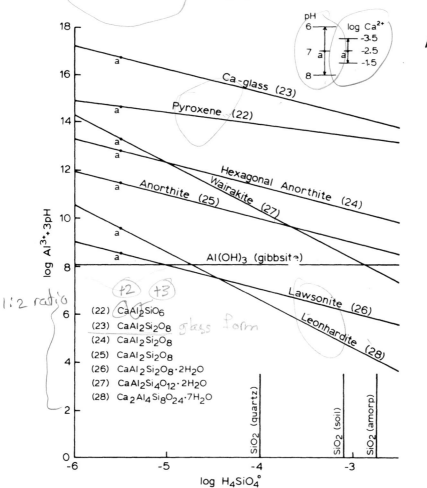

Fig. 5.4 The stability of several calcium aluminosilicates compared to gibbsite with indicated changes for pH and Ca^{2+}.

$$\log K° = 35.25$$

$$\frac{(Ca^{2+})(Al^{3+})^2(H_4SiO_4°)}{(H^+)^2(H^+)^6} = 10^{35.25}$$

$$2 \log Al^{3+} + 6 pH = 35.25 - 2pH - \log Ca^{2+} - \log H_4SiO_4°$$

$$\log Al^{3+} + 3 pH = 17.63 - \tfrac{1}{2} \log Ca^{2+} - pH - \tfrac{1}{2} \log H_4SiO_4° \quad (5.10)$$

If Ca^{2+} is taken as $10^{-2.5}$ M, Eq. 5.10 becomes

$$\log Al^{3+} + 3pH = 18.88 - pH - \tfrac{1}{2} \log H_4SiO_4^\circ \qquad (5.11)$$

A plot of Eq. 5.11 for pH 7 is shown in Fig. 5.4 and is labeled the pyroxene line. The other stability lines in this figure were obtained similarly. Shifts in these lines with changes in pH and log Ca^{2+} are indicated by the short lines and arrows.

The calcium aluminosilicates in Fig. 5.4 decrease in solubility in the order

$CaAl_2SiO_8$(calcium glass) > $CaAl_2SiO_6$(pyroxene)
> $CaAl_2Si_2O_8$(hexagonal anorthite) > $CaAl_2Si_4O_{12} \cdot 2H_2O$(wairakite)
> $CaAl_2Si_2O_8$(anorthite) > $CaAl_2Si_2O_8 \cdot 2H_2O$(lawsonite)
> $Ca_2Al_4Si_8O_{24} \cdot 7H_2O$(leonhardite).

Only the latter two minerals are more stable than gibbsite in the $H_4SiO_4^\circ$ range of most weathering soils. A comparison of Fig. 5.1 and 5.4 reveals that leonhardite and lawsonite maintain similar levels of log Al^{3+} + 3pH as does kaolinite at pH 7. Decreasing pH favors the dissolution of these calcium aluminosilicates with precipitation of kaolinite. Leaching of Ca^{2+} from soils can lower soluble Ca^{2+} below $10^{-2.5}$ M and hasten the dissolution of these minerals.

All of the calcium aluminosilicates in Fig. 5.4 that lie above the gibbsite line can be expected to weather fairly rapidly in soils. Lawsonite and leonhardite are more stable and are expected to weather more slowly. These two minerals are considered further in Chapter 7 as possibly controlling Ca^{2+} near $10^{-2.5}$ M in neutral and near-neutral soils.

5.5 MAGNESIUM ALUMINOSILICATES

The solubilities of several magnesium aluminosilicates and substituted aluminosilicates are given by Reactions 29 through 34 of Table 5.1. All of these minerals except Mg-montmorillonite are plotted in Fig. 5.5. Included are two aluminosilicate minerals whose stabilities are highly pH dependent, namely chlorite and vermiculite. From Reaction 29 of Table 5.1 can be developed the equilibrium reaction for chlorite:

$$\log Al^{3+} + 3pH = 30.15 - 2.5 \log Mg^{2+} - 5pH - 1.5 \log H_4SiO_4^\circ \qquad (5.12)$$

When Mg^{2+} is 10^{-3} M, this equation reduces to

$$\log Al^{3+} + 3pH = 37.65 - 5pH - 1.50 \log H_4SiO_4^\circ \qquad (5.13)$$

Equation 5.13 is plotted in Fig. 5.5 and is labeled the chlorite line. Its stability

MAGNESIUM ALUMINOSILICATES

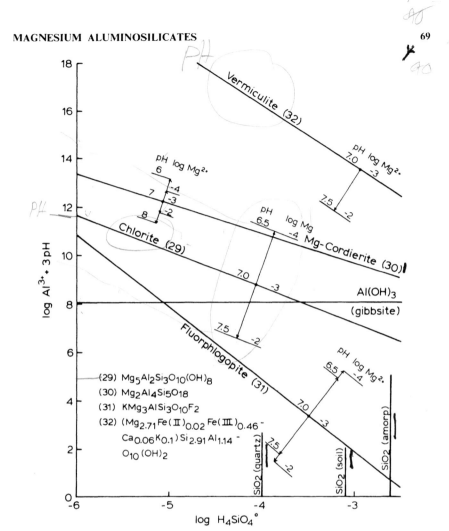

Fig. 5.5. The stability of several magnesium aluminosilicates in equilibrium with 10^{-3} M K^+, 10^{-4} M F^-, and soil-Fe at $pe + pH = 17.0$ with indicated changes for pH and Mg^{2+}.

(29) $Mg_5Al_2Si_3O_{10}(OH)_8$
(30) $Mg_2Al_4Si_5O_{18}$
(31) $KMg_3AlSi_3O_{10}F_2$
(32) $(Mg_{2.71}Fe(II)_{0.02}Fe(III)_{0.46}Ca_{0.06}K_{0.1})Si_{2.91}Al_{1.14}O_{10}(OH)_2$

is highly pH dependent, changing 5 log units for each unit change in pH. As shown here, chlorite is most stable in alkaline environments and may be expected as a stable mineral above pH 7.5. Isomorphous substitution of various cations into the chlorite lattice may possibly shift the stability relationships slightly from those shown here. There is sufficient uncertainty regarding the solubility value of chlorite used here (Sadiq and Lindsay, 1979) that caution should be used in attempting to draw too detailed conclusions.

The equilibrium expression for vermiculite obtained from Reaction 32 of Table 5.1 is

$$\log \text{Al}^{3+} + 3\text{pH} = 33.46 - 6.09\text{pH} - 2.38 \log \text{Mg}^{2+}$$
$$- 0.02 \log \text{Fe}^{2+} - 0.40 \log \text{Fe}^{3+}$$
$$- 0.05 \log \text{Ca}^{2+} - 0.09 \log \text{K}^{+}$$
$$- 2.55 \log \text{H}_4\text{SiO}_4^\circ \quad (5.14)$$

When $\text{Mg}^{2+} = \text{K}^{+} = 10^{-3} M$, $\text{Ca}^{2+} = 10^{-2.5} M$, $\log \text{Fe}^{3+} = 2.70 - 3\text{pH}$, and $\log \text{Fe}^{2+} = 15.74 - 2\text{pH} - (pe + \text{pH})$, Eq. 5.14 becomes

$$\log \text{Al}^{3+} + 3\text{pH} = 39.60 - 4.85\text{pH} + 0.02(pe + \text{pH}) - 2.55 \log \text{H}_4\text{SiO}_4^\circ \quad (5.15)$$

A plot of Eq. 5.15 is shown as the vermiculite line in Fig. 5.5 at $pe + \text{pH} = 17.0$. This mineral is fairly unstable in soils. Increasing pH one unit lowers the line by 4.85 log units (Eq. 5.15). Changes in redox ($pe + \text{pH}$) have only a slight effect on the stability of vermiculite since it contains very little reduced Fe^{2+}. For example, a change in $pe + \text{pH}$ from 17 to 10 only lowers the vermiculite line by 0.14 log unit (Eq. 5.15). Minerals containing more Fe^{2+} would be affected greater by changes in redox. Changes in the activity of Mg^{2+} from the $10^{-3} M$ reference level shifts the vermiculite line by 2.38 log units (Eq. 5.15) for each unit change in log Mg^{2+}.

The mineral $\text{Mg}_2\text{Al}_4\text{Si}_5\text{O}_{18}$(Mg-cordierite) is intermediate in solubility between vermiculite and chlorite. Its solubility is much less pH dependent. Since this mineral lies above that of gibbsite, it is expected to weather fairly rapidly in soils.

The mineral $\text{KMg}_3\text{AlSi}_3\text{O}_{10}\text{F}_2$(fluorphlogopite) is highly stable in soils as shown from Fig. 5.5. The reference F^- activity used to establish the solubility line for this mineral was $10^{-4} M$, the level in equilibrium with CaF_2(fluorite) and $10^{-2.5} M$ Ca^{2+} (Reactions 43 and 44 of Table 5.1). Since the solubility line for fluorphlogopite lies below that of kaolinite (Fig. 5.1) at pH values above 7, this mineral is more stable than fluorite. Since soils generally contain less fluorine than potassium, magnesium, aluminum, and silica (Table 1.1), fluorine can be expected to precipitate as fluorphlogopite rather than fluorite. This being the case, fluorphlogopite would fix F^- activity below $10^{-4} M$ allowing the fluorphlogopite line to move up so that fluorphlogopite and kaolinite can both coexist. The possibility that F^- activity in soils may be controlled by fluorphlogopite will be discussed further in Chapter 12 in connection with $\text{Ca}_5(\text{PO}_4)_3\text{F}$(fluorapatite).

5.6 SUMMARY STABILITY DIAGRAM FOR ALUMINOSILICATES

The aluminosilicate minerals discussed so far have been plotted on several different stability diagrams. The objective of this section is to bring together the most stable of these minerals into a single stability diagram in order to compare their stabilities. The results are presented in Fig. 5.6.

The stability line for illite in this figure was developed from Reaction 33 of Table 5.1 which gives the equation:

$$2.3 \log Al^{3+} + 8\,pH = 10.35 - 0.6 \log K^+ - 0.25 \log Mg^{2+} - 3.5 \log H_4SiO_4^\circ$$

$$\log Al^{3+} + 3\,pH = 4.50 - 0.26 \log K^+ - 0.11 \log Mg^{2+} - 0.48\,pH - 1.52 \log H_4SiO_4^\circ \qquad (5.16)$$

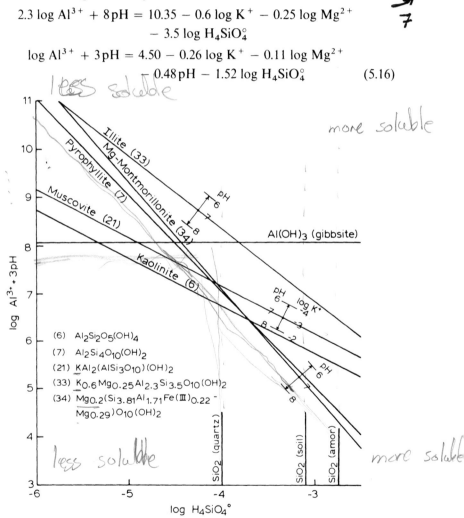

Fig. 5.6 A summary diagram of several mixed aluminosilicates in equilibrium with $10^{-3}\,M\,K^+$, $10^{-3}\,M\,Mg^{2+}$, and soil-Fe with indicated changes for pH and K^+.

When $K^+ = Mg^{2+} = 10^{-3}\,M$, this equation reduces to

$$\log Al^{3+} + 3pH = 5.61 - 0.48\,pH - 1.52 \log H_4SiO_4^\circ \qquad (5.17)$$

Equation 5.17 is plotted in Fig. 5.6 as the illite line. This line was drawn for pH 7, and the accompanying short lines and arrows show that it shifts upward or downward by 0.48 units for each unit change in pH.

The kaolinite line in Fig. 5.6 was obtained from Reaction 7 of Table 5.1 which gives

$$\log Al^{3+} + 3pH = 2.73 - \log H_4SiO_4^\circ \qquad (5.18)$$

The muscovite line was obtained from Reaction 21 of Table 5.1 which gives

$$\log Al^{3+} + 3pH = 4.48 - \tfrac{1}{3} \log K^+ - \tfrac{1}{3}pH - \log H_4SiO_4^\circ \qquad (5.19)$$

When $K^+ = 10^{-3}\,M$ this equation reduces to

$$\log Al^{3+} + 3pH = 5.48 - \tfrac{1}{3}pH - \log H_4SiO_4^\circ \qquad (5.20)$$

The Mg-montmorillonite line was obtained from Reaction 34 of Table 5.1 which gives

$$\log Al^{3+} + 3pH = 1.56 - 0.29 \log Mg^{2+} - 0.13 \log Fe^{3+}$$
$$- 0.95\,pH - 2.23 \log H_4SiO_4^\circ \qquad (5.21)$$

When Mg^{2+} is $10^{-3}\,M$ and Fe^{3+} is governed by $Fe(OH)_3$(soil-Fe), that is, $\log Fe^{3+} = 2.70 - 3pH$ from Reaction 35 of Table 5.1, Eq. 5.21 becomes

$$\log Al^{3+} + 3pH = 2.08 - 0.56\,pH - 2.23 \log H_4SiO_4^\circ \qquad (5.22)$$

Several important stability relationships of the secondary clay minerals in soils are represented in Fig. 5.6. Of the minerals shown Mg-montmorillonite is most stable under high $H_4SiO_4^\circ$ as would occur during the weathering of silica-rich primary minerals. Increasing pH increases the stability of this mineral. The mineral $Al_2Si_4O_{10}(OH)_2$(pyrophyllite) is only slightly less stable. At pH 7.0 if $H_4SiO_4^\circ$ drops below $10^{-3.7}\,M$, $Al_2Si_2O_5(OH)_4$(kaolinite) becomes more stable than Mg-montmorillonite. When $H_4SiO_4^\circ$ drops below $10^{-5.31}\,M$, $Al(OH)_3$(gibbsite) becomes the stable mineral. In the advanced stages of soil weathering soluble silica is largely depleted. Under these conditions the aluminosilicates disappear leaving mainly gibbsite and iron oxides (Chapter 10).

Many factors shift the stability lines shown in Fig. 5.6. For example, increasing pH causes Mg-montmorillonite to become more stable relative to kaolinite, whereas decreasing pH decreases its stability relative to kaolinite. Increasing Mg^{2+} above the $10^{-3}\,M$ reference level lowers the Mg-montmorillonite line. From Eq. 5.21 a 10-fold increase in Mg^{2+} shifts this line downward by only 0.29 log unit.

In soils below pH 8.2 $KAl_2(AlSi_3O_{10})(OH)_2$(muscovite) is metastable with respect to kaolinite when K^+ is 10^{-3} M. Increasing pH above 8.2 and K^+ above 10^{-3} M causes muscovite to become more stable than kaolinite. As soils become acidic and K^+ is leached, muscovite disappears.

The minerals included in Fig. 5.6 are fairly stable in soils and are among those often found there. The diagram helps to show how various soil parameters affect solubility and stability of these minerals. As more reliable solubility values become available for additional aluminosilicate minerals, they can be included to expand the development given here.

5.7 CONTROLS OF Al^{3+} ACTIVITY IN SOILS

The Al^{3+} activity maintained by various oxides and hydroxides of aluminum were developed in Chapter 3 (Fig. 3.1). Since gibbsite is among the most stable of these minerals, it is frequently used to estimate Al^{3+} activities in soils, yet many find it disturbing that aluminum solubility, aluminum toxicity to plants, and percent exchangeable aluminum in soils can not always be predicted from pH alone. In this chapter it has been shown that many aluminosilicates can depress the activity of Al^{3+} below that of gibbsite thus modifying the simple gibbsite-pH relationships. Let us examine some of the solubility relationships of Al^{3+}. The activity of Al^{3+} in equilibrium with gibbsite, kaolinite, and Mg-montmorillonite are plotted together in Fig. 5.7. The gibbsite line was obtained from Reaction 39 of Table 5.1 which gives

$$\log Al^{3+} = 8.04 - 3\,pH \tag{5.23}$$

Only pH affects the activity of Al^{3+} in equilibrium with gibbsite.

The kaolinite line in Fig. 5.7 was obtained from Reaction 6 of Table 5.1 which reduces to

$$\log Al^{3+} = 2.73 - \log H_4SiO_4^\circ - 3\,pH \tag{5.24}$$

Not only pH but also $H_4SiO_4^\circ$ affects the activity of Al^{3+} in equilibrium with kaolinite. The kaolinite line in Fig. 5.7 was drawn with $H_4SiO_4^\circ = 10^{-4.00}$ M representing equilibrium with SiO_2(quartz). The short lines and arrows show how this line shifts downward when $H_4SiO_4^\circ$ is controlled by SiO_2(soil) or SiO_2(amorp). As long as the kaolinite line is below the gibbsite line, gibbsite is not a stable phase. When $H_4SiO_4^\circ$ drops to $10^{-5.31}$ M, the kaolinite line rises to coincide with that of gibbsite indicating that both minerals can coexist at equilibrium. At still lower $H_4SiO_4^\circ$ activities, the kaolinite line moves above the gibbsite line making gibbsite the stable phase.

The Mg-montmorillonite line in Fig. 5.7 is based on Reaction 34 of Table 5.1 which reduces to

$$\log Al^{3+} = 1.56 - 0.29 \log Mg^{2+} - 0.13 \log Fe^{3+}$$
$$- 2.23 \log H_4SiO_4^\circ - 3.95\,pH \tag{5.25}$$

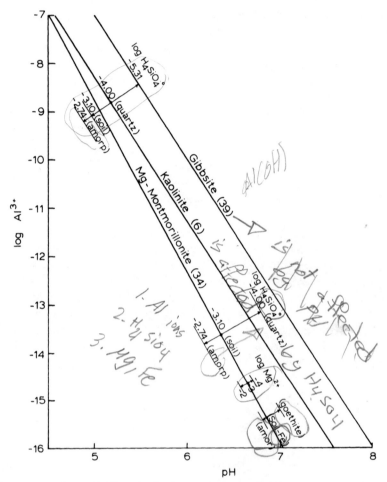

Fig. 5.7 The activity of Al^{3+} maintained by Mg-montmorillonite, kaolinite, and gibbsite as affected by pH, silica, iron oxides, and Mg^{2+}.

The solubility of Al^{3+} in equilibrium with Mg-montmorillonite is affected by Mg^{2+}, Fe^{3+}, $H_4SiO_4^\circ$, and pH. The Mg-montmorillonite line in Fig. 5.7 with $H_4SiO_4^\circ = 10^{-3.10}$ M represents equilibrium with SiO_2(soil). The arrows show an upward shift for equilibrium with quartz and a downward shift for SiO_2(amorp). A 10-fold decrease in $H_4SiO_4^\circ$ increases Al^{3+} by 2.23 log units (Eq. 5.25) whereas the kaolinite line shifts by only one log unit (Eq. 5.24). A 10-fold change in Mg^{2+} shifts the Mg-montmorillonite line upward by 0.29 log unit whereas similar changes in Fe^{3+} cause only 0.13 shift in log Al^{3+}. The Mg-montmorillonite line was drawn for equilibrium with Fe(OH)$_3$(soil). If equilibrium were attained with α-FeOOH(goethite) defined by Eq. 37 of Table 5.1, the line would shift upward by [2.70 − (−0.02)] × 0.13 = 0.35 log unit as shown in Fig. 5.7. The presence of Fe(OH)$_3$(amorp) has only a slight depressing effect.

It is apparent from Fig. 5.7 that Al^{3+} activity in soils is not predictable from pH alone. When the activity of H$_4$SiO$_4^\circ$ > $10^{-5.31}$ this species has the major influence in controlling Al^{3+} activity by forming kaolinite and/or Mg-montmorillonite which depress the activity of Al^{3+} considerably below that of gibbsite.

In this text kaolinite in equilibrium with SiO$_2$(quartz) is frequently used as a reference level of Al^{3+} activity. In making this choice it is recognized that Mg-montmorillonite and possibly other substituted smectites undoubtedly lower Al^{3+} activity in alkaline soils having high H$_4$SiO$_4^\circ$.

In highly weathered soils H$_4$SiO$_4^\circ$ drops below that of SiO$_2$(quartz) allowing Al^{3+} activity to rise until it is limited by gibbsite. Factors that affect Al^{3+} activity can indirectly affect phosphorus solubility through aluminum phosphates. These considerations will be considered further in Chapter 12.

5.8 GENERAL DISCUSSION OF ALUMINOSILICATES

Soils contain many different aluminosilicate minerals (Dixon and Weed, 1977). Isomorphous substitution of ions into the crystal lattices of many of these minerals can be expected to alter slightly the free energy relationships of these minerals. Several methods have been proposed for estimating the free energies of clay minerals based on their stoichiometric composition (Slaughter, 1966; Tardy and Garrels, 1974; Nriagu, 1975; Mattigod and Sposito, 1978).

Minerals capable of accepting Fe^{2+} and Mn^{2+} into their crystal lattices are affected by redox relationships. The more intense the reducing conditions the greater will be the incorporation of these ions into mineral lattices. In soils subjected to reduction, the equilibrium levels of Al^{3+} associated with such minerals are expected to decrease as the minerals become more stable under reducing environments. As oxidizing conditions return, the Al^{3+} activity maintained by the substituted minerals would again rise. Thus redox relationships play an important role in governing the formation and weathering of iron- and manganese-substituted aluminosilicates. The effect of redox on the solubility of many iron and manganese minerals will be considered in Chapters 10 and 11.

The stability diagrams developed in this chapter include only a few of the known aluminosilicate minerals selected on the basis that their solubilities have been fairly well established. Many reported solubility values of aluminosilicates are not reliable, so they must be carefully screened. The selected solubility values included in Table 5.1 and the Appendix were made after careful study of the literature, but it is not claimed to be a final search. As new solubility data and free energy values become available, the developments presented herein can be expanded to include them. In this text Mg-montmorillonite was selected to represent the smectite minerals. Further information on the various isomorphous substitutions would permit a comparison of how various soil factors can affect the solubility and stability relationships of this group of minerals.

Perhaps the greatest value of the present development is its usefulness for predicting the weathering sequences of various minerals and for identifying those reactions that can and cannot occur.

The predictions made herein need to be tested by examining soils in natural environments to see which minerals are dissolving and which are precipitating. Many of the predicted solubility relationships can be easily tested. Hopefully soil scientists will be encouraged to apply the principles of chemical equilibria to soil genesis and mineral weathering. Equilibrium processes are important because they ultimately govern the solubility relationships of all elements in soils.

REFERENCES

Dixon, J. B. and S. B. Weed, Eds. 1977. Minerals in soil environments. Soil Sci. Soc. Am., Madison, Wisconsin.

Kittrick, J. A. 1977. Mineral equilibria and the soil system. In J. B. Dixon and S. B. Weed, Eds. Minerals in Soil Environments. pp. 1-25. Soil Science Society of America, Madison, Wisconsin.

Marshall, C. E. 1977. The Physical Chemistry and Mineralogy of Soils. Vol. 2. Soils in Place. Wiley-Interscience, New York.

Mattigod, S. V. and G. Sposito. 1978. Improved method for estimating the standard free energies of formation ($\Delta G^°_{f, 298.15}$) of smectities and vermiculities. Geochim. Cosmochim. Acta 42:1753-1762.

Nriagu, J. O. 1975. Thermochemical approximations for clay minerals. Am. Mineral. 60:834-839.

Rai, Dhanpat, and W. L. Lindsay. 1975. A thermodynamic model for predicting the formation, stability, and weathering of common soil minerals. Soil Sci. Soc. Am. Proc. 39:991-996.

Sadiq, M. and W. L. Lindsay. 1979. Selection of standard free energies of formation for use in soil chemistry. Colorado State University Experimental Station Tech. Bull 134.

Slaughter, M. 1966. Chemical binding in silicate minerals. Part III. Application of energy calculations to the prediction of silicate mineral stability. Geochim. Cosmochim. Acta 30:323-339.

Tardy, Y. and R. M. Garrels. 1974. A method of estimating the Gibbs energies of formation of layer silicates. Geochim. Cosmochim. Acta 38:1101-1116.

PROBLEMS

5.1 Derive the necessary equations and plot the pyrophyllite line in Fig. 5.1.

5.2 Calculate the following:
   a. The equilibrium level of $H_4SiO_4^\circ$ at which kaolinite and gibbsite can coexist.
   b. The log $Al^{3+} + 3pH$ at which kaolinite and quartz coexist.

5.3 From Fig. 5.2 estimate the activity of Al^{3+} in equilibrium with low albite and quartz at pH 5.5 when $Na^+ = 10^{-4}$ M.

5.4 From Fig. 5.3 estimate the activity of Al^{3+} in equilibrium with microcline and quartz at pH 8 when $K^+ = 10^{-2}$ M. Under these conditions which is most stable microcline, gibbsite, or muscovite?

5.5 Develop the equations and prepare a plot containing chlorite and kaolinite on a log $Al^{3+} + 3pH$ versus log $H_4SiO_4^\circ$ diagram. Discuss the relative stabilities of these minerals as affected by silica, Mg^{2+}, and pH.

5.6 Discuss the weathering conditions under which Mg-montmorillonite, kaolinite, and gibbsite could form and be the stable minerals in soils.

5.7 Using the conditions specified in Fig. 5.6 develop the equation to calculate the pH at which kaolinite, Mg-montmorillonite, and chlorite can coexist. If the Mg^{2+} activity were then lowered, how would this affect the equilibrium pH maintained by these three minerals?

5.8 Using equations from Table 5.1, develop the necessary relationships to answer the following questions based on Fig. 5.7:
   a. How does a change in Mg^{2+} from 10^{-3} to $10^{-4.5}$ M affect Al^{3+} in equilibrium with Mg-montmorillonite?
   b. How does a 100-fold decrease in $H_4SiO_4^\circ$ affect the Al^{3+} in equilibrium with kaolinite? In equilibrium with Mg-montmorillonite?
   c. How would Al^{3+} activity in equilibrium with Mg-montmorillonite change when the following are each decreased by 10-fold: $H_4SiO_4^\circ$, Fe^{3+}, Mg^{2+}, and H^+.

SIX
CARBONATE EQUILIBRIA

Carbonate reactions are among the most important chemical reactions that occur in soils. Since $CO_2(g)$ can escape from soils to the atmosphere and return to precipitate carbonate minerals, soils must be considered as open systems with regard to carbonates. Respiratory processes occurring in plant roots and other organisms in soils continually produce $CO_2(g)$ which lowers pH and modifies the solubility relationships of many plant nutrients. Carbonates are also used as liming agents to raise the pH of acid soils. For these and other reasons, carbonate equilibria merit careful study.

This chapter provides a simple, straightforward development of carbonate equilibria in soils and differs somewhat from previous developments (Garrels and Christ, 1965; Stumm and Morgan, 1970). Only the CO_2–H_2O system is included in this chapter. The solubility relationships of the different carbonate minerals are included in the separate chapters dealing with the individual elements.

6.1 THE CO_2–H_2O SYSTEM

Carbon dioxide dissolves in water to form both dissolved CO_2° and undissociated carbonic acid, $H_2CO_3^\circ$. Kern (1960) reports that the ratio of CO_2° molecules to $H_2CO_3^\circ$ molecules at 25°C is approximately 386. Thus only a small fraction of the total dissolved CO_2° is actually present as $H_2CO_3^\circ$. Despite this fact, it is convenient to treat all dissolved CO_2° as $H_2CO_3^\circ$. For most purposes, there is no problem in doing so because the hydration status of dissolved species need not be stated in thermodynamic consideration. The equilibrium reactions relating the various carbonate species are summarized in Table 6.1.

The solubility of CO_2 in water at 25°C can be represented by

$$CO_2(g) + H_2O \rightleftharpoons H_2CO_3^\circ \qquad \log K^\circ = -1.46 \qquad (6.1)$$

TABLE 6.1 EQUILIBRIUM REACTIONS IN THE CO_2–H_2O SYSTEM AT 25°C

Reaction No.	Equilibrium Reaction	$\log K^\circ$
1	$CO_2(g) + H_2O \rightleftharpoons H_2CO_3^\circ$	-1.46
2	$H_2CO_3^\circ \rightleftharpoons H^+ + HCO_3^-$	-6.36
3	$HCO_3^- \rightleftharpoons H^+ + CO_3^{2-}$	-10.33
4	$CO_2(g) + H_2O \rightleftharpoons H^+ + HCO_3^-$	-7.82
5	$CO_2(g) + H_2O \rightleftharpoons 2H^+ + CO_3^{2-}$	-18.15

giving

$$H_2CO_3^° = 10^{-1.46} CO_2(g) \tag{6.2}$$

$$\log H_2CO_3^° = -1.46 + \log CO_2(g) \tag{6.3}$$

where $H_2CO_3^°$ is expressed as moles per liter and $CO_2(g)$ is expressed in atmospheres.

In solution carbonic acid dissociates to give

$$H_2CO_3^° \rightleftharpoons H^+ + HCO_3^- \quad \log K° = -6.36 \tag{6.4}$$

Thus

$$\frac{(H^+)(HCO_3^-)}{H_2CO_3^°} = 10^{-6.36} \tag{6.5}$$

$$\log \frac{(HCO_3^-)}{H_2CO_3^°} = pH - 6.36 \tag{6.6}$$

At pH 6.36, the molar ratio of (HCO_3^-) to $(H_2CO_3^°)$ is unity. This ratio increases 10-fold for each unit increase in pH and decreases 10-fold for each unit decrease in pH. Reactions 6.1 and 6.4 can be combined to give

		$\log K°$
$CO_2(g) + H_2O \rightleftharpoons H_2CO_3^°$		-1.46
$H_2CO_3^° \rightleftharpoons H^+ + HCO_3^-$		-6.36
$CO_2(g) + H_2O \rightleftharpoons H^+ + HCO_3^-$		$-7.82 \quad (6.7)$

Thus

$$(HCO_3^-) = \frac{10^{-7.82} CO_2(g)}{(H^+)} \tag{6.8}$$

$$\log HCO_3^- = -7.82 + pH + \log CO_2(g) \tag{6.9}$$

The bicarbonate ion also dissociates to give:

$$HCO_3^- \rightleftharpoons H^+ + CO_3^{2-} \quad \log K° = -10.33 \tag{6.10}$$

Thus

$$\frac{(H^+)(CO_3^{2-})}{(HCO_3^-)} = 10^{-10.33} \tag{6.11}$$

$$\log \frac{(CO_3^{2-})}{(HCO_3^-)} = pH - 10.33 \tag{6.12}$$

THE CO$_2$–H$_2$O SYSTEM

At pH 10.33 the molar ratio of (CO$_3^{2-}$) to (HCO$_3^-$) is unity. Each unit increase in pH increases this ratio by 10-fold, and each unit decrease in pH decreases it by 10-fold. Combining Reactions 6.7 and 6.10 gives

$$
\begin{array}{lll}
 & & \log K^\circ \\
\mathrm{CO_2(g) + H_2O} \rightleftharpoons \mathrm{H^+ + HCO_3^-} & & -7.82 \\
\mathrm{HCO_3^-} \rightleftharpoons \mathrm{H^+ + CO_3^{2-}} & & -10.33 \\
\hline
\mathrm{CO_2(g) + H_2O} \rightleftharpoons \mathrm{2H^+ + CO_3^{2-}} & & -18.15 \quad (6.13)
\end{array}
$$

Thus

$$(\mathrm{CO_3^{2-}}) = \frac{10^{-18.15}\mathrm{CO_2(g)}}{(\mathrm{H^+})^2} \tag{6.14}$$

$$\log \mathrm{CO_3^{2-}} = -18.15 + 2\mathrm{pH} + \log \mathrm{CO_2(g)} \tag{6.15}$$

Convenient relationships that follow from Eq. 6.3, 6.9, and 6.15 are given in Table 6.2.

The mole fraction distribution of the various carbonate species in solution is shown in Fig. 6.1 where activity coefficients are taken as unity. Total carbonates in solution consist of [H$_2$CO$_3^\circ$ + HCO$_3^-$ + CO$_3^{2-}$] so the mole fraction (MF) for HCO$_3^-$ is

$$MF_{\mathrm{HCO_3^-}} = \frac{\mathrm{HCO_3^-}}{\mathrm{H_2CO_3^\circ + HCO_3^- + CO_3^{2-}}} \tag{6.16}$$

Using appropriate expressions from Eq. 6.2, 6.8, and 6.14 for each of the terms in Eq. 6.16 gives

$$MF_{\mathrm{HCO_3^-}} = \frac{10^{-7.82}/(\mathrm{H^+})}{10^{-1.46} + 10^{-7.82}/(\mathrm{H^+}) + 10^{-18.15}/(\mathrm{H^+})^2} \tag{6.17}$$

TABLE 6.2 DISTRIBUTION OF CARBONATE SPECIES IN SOLUTION AS A FUNCTION OF CO$_2$(g)

CO$_2$(g), atm	log CO$_2$(g), atm	log H$_2$CO$_3^\circ$, M	log HCO$_3^-$, M	log CO$_3^{2-}$, M
0.0003	−3.52	−4.98	pH − 11.34	2pH − 21.67
0.003	−2.52	−3.98	pH − 10.34	2pH − 20.67
0.01	−2.00	−3.46	pH − 9.82	2pH − 20.15
0.1	−1.00	−2.46	pH − 8.82	2pH − 19.15
1.0	0.00	−1.46	pH − 7.82	2pH − 18.15

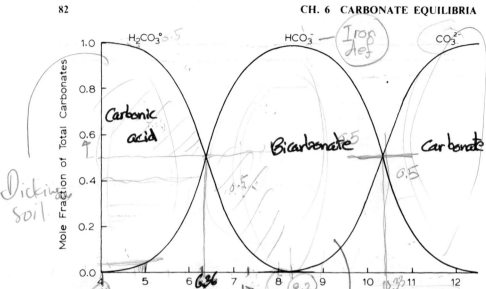

Fig. 6.1 The effect of pH on the distribution of carbonate species in solution.

A plot of Eq. 6.17 is shown as the HCO_3^- curve in Fig. 6.1. The curves for $H_2CO_3^\circ$ and CO_3^{2-} were obtained similarly using appropriate mole fraction expressions.

Since the $CO_2(g)$ term is common to all carbonate species, it can be canceled from the mole fraction expressions (Eq. 6.17). Thus the relationships shown in Fig. 6.1 are independent of the partial pressure of $CO_2(g)$ and, therefore, tell nothing about the total amount of carbonate that may be present in solution. The plot is useful in showing the important effect of pH on the distribution of carbonate species in solution. The intersection of curves at pH 6.36 and 10.33 corresponds to the pK° values for the first and second dissociation constants of carbonic acid (Eq. 6.4 and 6.10).

Another very useful plot showing the actual activities of the various carbonate species in solution as a function of pH in equilibrium with 0.0003 atm $CO_2(g)$ is given in Fig. 6.2. These relationships were plotted from Eq. 6.3, 6.9, and 6.15. In this plot $H_2CO_3^\circ$ is present at $10^{-4.98}$ M and is independent of pH. The activity of HCO_3^- increases 10-fold for each unit increase in pH, whereas that of CO_3^{2-} increases 100-fold. This figure shows the activities of all the carbonate species expected in soils when equilibrium is attained with atmospheric $CO_2(g)$. The pH dependence of total carbonate solubility is readily seen. Again the intersections of the various lines in this figure occur

THE CO_2–H_2O SYSTEM

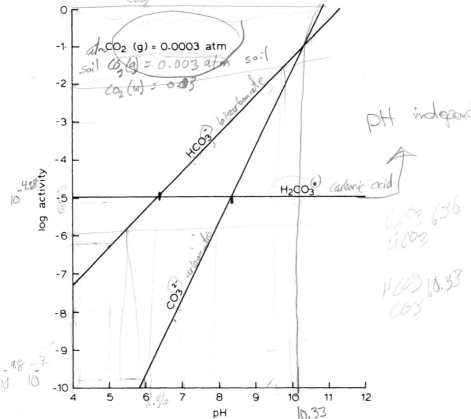

Fig. 6.2 The effect of pH on the activities of carbonate species in equilibrium with 0.0003 atm of $CO_2(g)$.

at pH values corresponding to the $pK°$ for the acid dissociation reactions involved. For example, the dissociation of HCO_3^- to give equimolar CO_3^{2-} and HCO_3^- occurs at pH 10.33 corresponding to the $pK°$ of 10.33 for this reaction (Eq. 6.12).

The relationships depicted in Fig. 6.2 are given for a $CO_2(g)$ partial pressure of 0.0003 atm, but the graph can be used to obtain the carbonate activities for *any* CO_2 level. Increasing $CO_2(g)$ by 10-fold shifts all the lines in this figure upward by one log unit whereas 10-fold decreases in $CO_2(g)$ shifts them down by one log unit. For example, let us say that $CO_2(g)$ increases from 0.0003 to 0.1 atm. The lines on this graph then shift upward by $\log(0.1/0.0003) = 2.52$ log units. Thus at pH 8.0 the HCO_3^- level of $10^{-3.34}$ M at pH 8 would be shifted to $10^{-3.34+2.52}$ or $10^{-0.82}$ M.

6.2 THE CO_2-SOIL SYSTEM

Since soils are open to the atmosphere, there is continual opportunity for loss or gain of $CO_2(g)$. In most soils the partial pressure of $CO_2(g)$ is expected to be slightly higher than that in the air because $CO_2(g)$ is continually being released by the respiration of roots and other organisms in soils. In this book a level of 0.003 atm of $CO_2(g)$ or approximately 10 times the atmospheric level will frequently be used as a reference level for soils. In flooded soils, the $CO_2(g)$ level generally goes much higher because diffusion of gases through water is much slower than through air. The $CO_2(g)$ levels in flooded soils often range from 0.01 to 0.3 atm.

The solubility relationships of carbonates in soils are expected to correspond to those depicted in Fig. 6.2 with the modification that $CO_2(g)$ may vary and cause the solubility lines to move up or down accordingly.

The activity of CO_3^{2-} in acid soils is quite low (Fig. 6.2), so very few metal carbonates can form. However, in alkaline soils CO_3^{2-} reaches levels at which many metal carbonates can form. Consequently, many carbonate minerals become important in alkaline soils and impose limits on the solubilities of many metal ions. These solubility limits will be considered in the separate chapters that follow for the trace elements. The equilibrium relationships developed herein will be used to develop the solubility limits and relate them to the partial pressure of $CO_2(g)$.

REFERENCES

Garrels, R. M. and C. L. Christ. 1965. Solutions, Minerals, and Equilibria. Harper and Row, New York. pp. 74–92.

Kern, D. B. 1960. The hydration of carbon dioxide. J. Chem. Ed. 37:14–23.

Stumm, W. and J. J. Morgan, 1970. Aquatic Chemistry. Wiley-Interscience, New York. pp. 118–160.

PROBLEMS

6.1 Develop the necessary equations and plot the mole fraction of total carbonate in solution present as CO_3^{2-} as a function of pH. How will changes in the partial pressure of $CO_2(g)$ affect this mole fraction distribution? Explain.

6.2 Derive the necessary equations to calculate the pH at which the $H_2CO_3^\circ$ and CO_3^{2-} lines in Fig. 6.2 cross. How is this pH affected by increasing $CO_2(g)$ from 0.0003 to 0.003 atm? Explain.

PROBLEMS

6.3 Calculate the pH of distilled water in equilibrium with
  a. Air containing 0.0003 atm $CO_2(g)$.
  b. Pure $CO_2(g)$ at 1 atm.

6.4 A liter of distilled water has been equilibrated with air containing 0.0003 atm $CO_2(g)$. This solution is placed in a closed system where no exchange of $CO_2(g)$ is permitted and slowly titrated with NaOH. Calculate the milliequivalents of base required to reach
  a. pH 6.36
  b. pH 8.00
  c. pH 10.33.

6.5 A liter of distilled water is equilibrated with 0.003 atm of $CO_2(g)$. Assuming activity coefficients are unity, calculate a hypothetical titration curve:
  a. For a closed system where no additional $CO_2(g)$ is admitted.
  b. For an open system where 0.003 atm of $CO_2(g)$ is maintained and the titration is continued to pH 11.0.

SEVEN

CALCIUM

Calcium comprises approximately 3.6% of the lithosphere while the average content of soils is near 1.37% (Table 1.1). Calcium is somewhat variable in soils, and its content is largely influenced by parent material and rainfall. Soils that develop from calcareous parent materials often have calcium carbonate somewhere in the profile. With advanced weathering and high rainfall, calcium carbonate and most other calcium minerals disappear from soils.

In this chapter the stabilities of various calcium minerals and solution complexes are examined. Reference solubilities of Ca^{2+} for both calcareous and noncalcareous soils are proposed. The solubility relationships of calcium phosphates, sulfides, and molybdates are examined in Chapters 12, 17, and 22, respectively. Since the anions of these minerals are generally less abundant in soils than calcium, these minerals are not expected to control Ca^{2+} solubility.

7.1 CALCIUM SILICATES AND ALUMINOSILICATES

The solubilities of several calcium silicates and aluminosilicates are given by Reactions 1 through 12 of Table 7.1 and are plotted in Fig. 7.1. The reference levels of Al^{3+} and $H_4SiO_4^\circ$ used to develop this diagram are the equilibrium levels with $Al_2Si_2O_5(OH)_4$(kaolinite) and SiO_2(quartz). Furthermore, Mg-montmorillonite can also coexist with kaolinite and quartz if pH is near 7.8, Mg^{2+} is near 10^{-3} M, and Fe^{3+} is controlled by soil-Fe (Fig. 5.6). For these reasons the kaolinite-quartz equilibrium was selected as the logical weathering environment for examining the stability of calcium minerals in soils.

An example of how the solubility lines in Fig. 7.1 were developed is given for wairakite, using the following equilibrium reactions:

$$
\begin{array}{lr}
 & \log K^\circ \\
CaAl_2Si_4O_{12} \cdot 2H_2O(\text{wair}) & \\
\quad + 8H^+ + 2H_2O \rightleftharpoons Ca^{2+} + 2Al^{3+} + 4H_4SiO_4^\circ & 16.05 \\
2Al^{3+} + 2H_4SiO_4^\circ + H_2O \rightleftharpoons Al_2Si_2O_5(OH)_4(\text{kaol}) + 6H^+ & -5.45 \\
2H_4SiO_4^\circ \rightleftharpoons 2SiO_2(\text{quartz}) + 4H_2O & 2(4.00) \\
\hline
CaAl_2Si_4O_{12} \cdot 2H_2O(\text{wair}) \quad Ca^{2+} + Al_2Si_2O_5(OH)_4(\text{kaol}) & \\
\quad + 2H^+ \quad\quad\quad\quad + 2SiO_2(q) + H_2O & 18.60
\end{array}
$$

(7.1)

$$\frac{(Ca^{2+})}{(H^+)^2} = 10^{18.60}$$

$$\log Ca^{2+} = 18.60 - 2pH \tag{7.2}$$

TABLE 7.1 EQUILIBRIUM REACTIONS OF CALCIUM AT 25°C

Reaction No.	Equilibrium Reaction	$\log K°$
	Silicates	
1	$\beta\text{-CaSiO}_3(\text{wollastonite}) + 2\text{H}^+ + \text{H}_2\text{O} \rightleftharpoons \text{Ca}^{2+} + \text{H}_4\text{SiO}_4°$	13.27
2	$\text{CaSiO}_3(\text{pseudowollastonite}) + 2\text{H}^+ + \text{H}_2\text{O} \rightleftharpoons \text{Ca}^{2+} + \text{H}_4\text{SiO}_4°$	14.23
3	$\beta\text{-Ca}_2\text{SiO}_4(\text{larnite}) + 4\text{H}^+ \rightleftharpoons 2\text{Ca}^{2+} + \text{H}_4\text{SiO}_4°$	39.62
4	$\gamma\text{-Ca}_2\text{SiO}_4(\text{Ca olivine}) + 4\text{H}^+ \rightleftharpoons 2\text{Ca}^{2+} + \text{H}_4\text{SiO}_4°$	37.82
	Aluminosilicates	
5	$\text{CaAl}_2\text{SiO}_6(\text{pyroxene}) + 8\text{H}^+ \rightleftharpoons \text{Ca}^{2+} + 2\text{Al}^{3+} + \text{H}_4\text{SiO}_4° + 2\text{H}_2\text{O}$	35.25
6	$\text{CaAl}_2\text{Si}_2\text{O}_8(\text{Ca-glass}) + 8\text{H}^+ \rightleftharpoons \text{Ca}^{2+} + 2\text{Al}^{3+} + 2\text{H}_4\text{SiO}_4°$	33.91
7	$\text{CaAl}_2\text{Si}_2\text{O}_8(\text{hexagonal anorthite}) + 8\text{H}^+ \rightleftharpoons \text{Ca}^{2+} + 2\text{Al}^{3+} + 2\text{H}_4\text{SiO}_4°$	26.10
8	$\text{CaAl}_2\text{Si}_2\text{O}_8(\text{anorthite}) + 8\text{H}^+ \rightleftharpoons \text{Ca}^{2+} + 2\text{Al}^{3+} + 2\text{H}_4\text{SiO}_4°$	23.33
9	$\text{CaAl}_2\text{Si}_2\text{O}_8 \cdot 2\text{H}_2\text{O}(\text{lawsonite}) + 8\text{H}^+ \rightleftharpoons \text{Ca}^{2+} + 2\text{Al}^{3+} + 2\text{H}_4\text{SiO}_4° + 2\text{H}_2\text{O}$	17.54
10	$\text{CaAl}_2\text{Si}_4\text{O}_{12} \cdot 2\text{H}_2\text{O}(\text{wairakite}) + 8\text{H}^+ + 2\text{H}_2\text{O} \rightleftharpoons \text{Ca}^{2+} + 2\text{Al}^{3+} + 4\text{H}_4\text{SiO}_4°$	16.05
11	$\text{Ca}_2\text{Al}_4\text{Si}_8\text{O}_{24} \cdot 7\text{H}_2\text{O}(\text{leonhardite}) + 16\text{H}^+ + \text{H}_2\text{O} \rightleftharpoons 2\text{Ca}^{2+} + 4\text{Al}^{3+} + 8\text{H}_4\text{SiO}_4°$	17.29
12	$\text{CaMg}(\text{SiO}_3)_2(\text{diopside}) + 4\text{H}^+ + 2\text{H}_2\text{O} \rightleftharpoons \text{Ca}^{2+} + \text{Mg}^{2+} + 2\text{H}_4\text{SiO}_4°$	21.16
	Carbonates	
13	$\text{CaCO}_3(\text{calcite}) + 2\text{H}^+ \rightleftharpoons \text{Ca}^{2+} + \text{CO}_2(\text{g}) + \text{H}_2\text{O}$	9.74
14	$\text{CaCO}_3(\text{aragonite}) + 2\text{H}^+ \rightleftharpoons \text{Ca}^{2+} + \text{CO}_2(\text{g}) + \text{H}_2\text{O}$	9.97
15	$\text{CaCO}_3 \cdot 6\text{H}_2\text{O}(\text{ikaite}) + 2\text{H}^+ \rightleftharpoons \text{Ca}^{2+} + \text{CO}_2(\text{g}) + 7\text{H}_2\text{O}$	11.78
16	$\text{CaMg}(\text{CO}_3)_2(\text{dolomite}) + 4\text{H}^+ \rightleftharpoons \text{Ca}^{2+} + \text{Mg}^{2+} + 2\text{CO}_2(\text{g}) + 2\text{H}_2\text{O}$	18.46

Soil, Oxides, Hydroxides, Ferrites

No.	Reaction	log K
17	Soil-Ca $\rightleftharpoons$ Ca^{2+}	-2.50*
18	CaO(lime) + 2H$^+$ $\rightleftharpoons$ Ca^{2+} + H$_2$O	32.95
19	Ca(OH)$_2$(portlandite) + 2H$^+$ $\rightleftharpoons$ Ca^{2+} + 2H$_2$O	22.80
20	CaFe$_2$O$_4$(c) + 8H$^+$ $\rightleftharpoons$ Ca^{2+} + 2Fe^{3+} + 4H$_2$O	21.42

Sulfates

No.	Reaction	log K
21	CaSO$_4$(insoluble) $\rightleftharpoons$ Ca^{2+} + SO$_4^{2-}$	-4.41
22	α-CaSO$_4$(soluble) $\rightleftharpoons$ Ca^{2+} + SO$_4^{2-}$	-2.45
23	β-CaSO$_4$(soluble) $\rightleftharpoons$ Ca^{2+} + SO$_4^{2-}$	-1.75
24	CaSO$_4 \cdot$ 2H$_2$O(gypsum) $\rightleftharpoons$ Ca^{2+} + SO$_4^{2-}$ + 2H$_2$O	-4.64

Fluorides

No.	Reaction	log K
25	CaF$_2$(fluorite) $\rightleftharpoons$ Ca^{2+} + 2F$^-$	-10.41

Solution Complexes

No.	Reaction	log K
26	Ca^{2+} + Cl$^-$ $\rightleftharpoons$ CaCl$^+$	-1.00
27	Ca^{2+} + 2Cl$^-$ $\rightleftharpoons$ CaCl$_2^\circ$	0.00
28	Ca^{2+} + CO$_2$(g) + H$_2$O $\rightleftharpoons$ CaHCO$_3^+$ + H$^+$	-6.70
29	Ca^{2+} + CO$_2$(g) + H$_2$O $\rightleftharpoons$ CaCO$_3^\circ$ + 2H$^+$	-15.01
30	Ca^{2+} + NO$_3^-$ $\rightleftharpoons$ CaNO$_3^+$	-4.80
31	Ca^{2+} + 2NO$_3^-$ $\rightleftharpoons$ Ca(NO$_3$)$_2^\circ$	-4.50
32	Ca^{2+} + H$_2$O $\rightleftharpoons$ CaOH$^+$ + H$^+$	-12.70
33	Ca^{2+} + 2H$_2$O $\rightleftharpoons$ Ca(OH)$_2^\circ$ + 2H$^+$	-27.99
34	Ca^{2+} + H$_2$PO$_4^-$ $\rightleftharpoons$ CaH$_2$PO$_4^+$	1.40
35	Ca^{2+} + H$_2$PO$_4^-$ $\rightleftharpoons$ CaHPO$_4$ + H$^+$	-4.46
36	Ca^{2+} + H$_2$PO$_4^-$ $\rightleftharpoons$ CaPO$_4^-$ + 2H$^+$	-13.09
37	Ca^{2+} + SO$_4^{2-}$ $\rightleftharpoons$ CaSO$_4^\circ$	2.31

(Continued)

TABLE 7.1 (Continued)

Reaction No.	Equilibrium Reaction	log $K°$
	Redox and Other Reactions	
38	$Ca^{2+} + 2e^- \rightleftharpoons Ca(c)$	97.16
39	$H_2PO_4^- \rightleftharpoons HPO_4^{2-} + H^+$	-7.20
40	$HPO_4^{2-} \rightleftharpoons PO_4^{3-} + H^+$	-12.35
41	$H_2PO_4^- \rightleftharpoons PO_4^{3-} + 2H^+$	-19.55

* Arbitrary reference level for Ca^{2+} used for noncalcareous soils.

— if the Ph is @ 8.0 what will happen to Calcite?

— " " " " 7.5 " " " " " ?

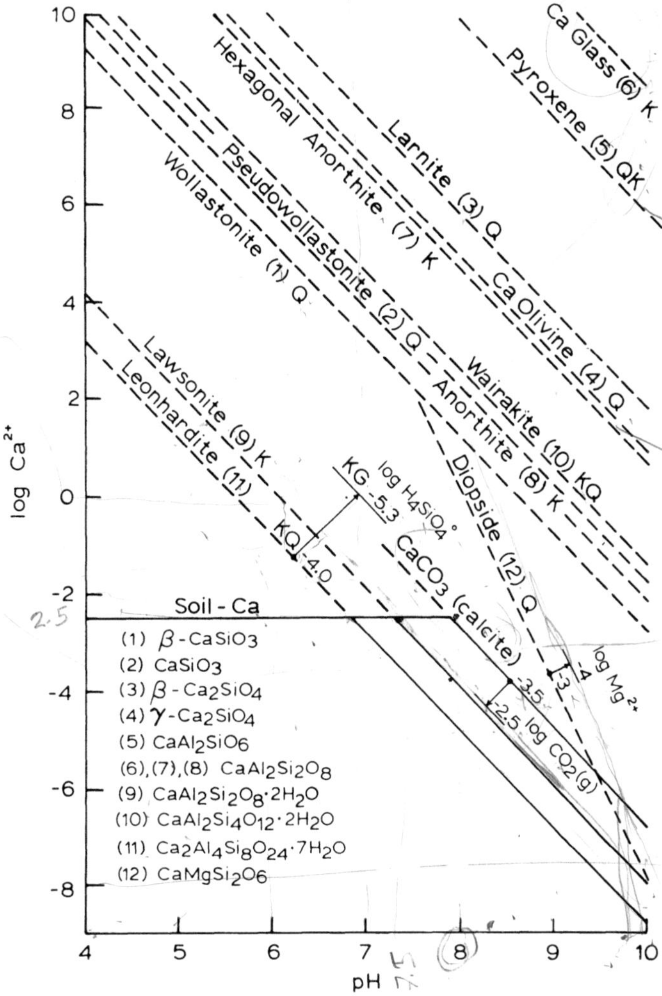

CALCIUM SILICATES AND ALUMINOSILICATES

Fig. 7.1 The solubility of calcium minerals in equilibrium with kaolinite (K), quartz (Q), and/or gibbsite (G) as indicated.

(1) β-CaSiO$_3$
(2) CaSiO$_3$
(3) β-Ca$_2$SiO$_4$
(4) γ-Ca$_2$SiO$_4$
(5) CaAl$_2$SiO$_6$
(6),(7),(8) CaAl$_2$Si$_2$O$_8$
(9) CaAl$_2$Si$_2$O$_8\cdot$2H$_2$O
(10) CaAl$_2$Si$_4$O$_{12}\cdot$2H$_2$O
(11) Ca$_2$Al$_4$Si$_8$O$_{24}\cdot$7H$_2$O
(12) CaMgSi$_2$O$_6$

Equation 7.2 is plotted as the wairakite solubility line in Fig. 7.1. The other lines in this diagram were developed similarly. The number (10) accompanying this mineral identifies its chemical formula in Fig. 7.1 and refers to Reaction 10 in Table 7.1 where the equilibrium constant for the indicated reaction is given. The equilibrium constants in this table were calculated from the ΔG_f° values documented in the Appendix. The KQ associated with wairakite indicates that kaolinite and quartz were used to establish the reference levels for Al^{3+} and $H_4SiO_4^\circ$. By this means, the solubility relationships depicted in Fig. 7.1 can be readily identified and documented.

The calcium minerals included in Fig. 7.1 decrease in solubility in the order:

$CaAl_2Si_2O_8$(Ca-glass) > $CaAl_2SiO_6$(pyroxene) > β-Ca_2SiO_4(larnite)
> γ-Ca_2SiO_4(Ca-olivine) > $CaAl_2Si_2O_8$(hexagonal anorthite)
> $CaAl_2Si_4O_{12} \cdot 2H_2O$(wairakite) > $CaSiO_3$(pseudowollastonite)
> $CaAl_2Si_2O_8$(anorthite) > β-$CaSiO_3$(wollastonite)
> $CaMgSi_2O_6$(diopside) > $CaCO_3$(calcite)
> $CaAl_2Si_2O_8 \cdot 2H_2O$(lawsonite)
> $Ca_2Al_4Si_8O_{24} \cdot 7H_2O$(leonhardite).

All minerals included in Fig. 7.1 are unstable in acid soils and can be expected to dissolve eventually. In alkaline soils calcite, leonhardite, and lawsonite appear as possible stable minerals. The soil-Ca line drawn at $Ca^{2+} = 10^{-2.5}$ M is given as a reference activity of Ca^{2+} in acid and near-neutral soils. In such soils soluble Ca^{2+} is largely buffered by exchangeable calcium. As exchangeable bases are depleted from soils, the pH drops causing H^+ and Al^{3+} to enter the exchange sites displacing Ca^{2+} and other cations. This displacement helps to maintain fairly constant levels of Ca^{2+} over a wide range of pH.

It is a common agricultural practice to lime soils that become highly acidic. The two major benefits of liming are that (1) it neutralizes soil acidity and (2) it replenishes exchangeable Ca^{2+}. Liming reactions can be represented as follows:

$CaCO_3$(c) + $2H^+$-exch $\longrightarrow$ Ca^{2+}-exch + CO_2(g) + H_2O (7.3)

$CaCO_3$(c) + 0.66 Al^{3+}-exch $\quad$ Ca^{2+}-exch + 0.66 $Al(OH)_3$(solid)
$+ 2H_2O$ $\quad$ + CO_2(g) + H_2O (7.4)

The calcium silicates and aluminosilicates included in Fig. 7.1 can all act as liming agents. Their lime equivalent is determined only by the bases present which do not include aluminum. For example, pyroxene reacts as follows:

$CaAl_2SiO_6$(pyroxene) + $2H^+$-exch + $2H_2O$ $\rightleftharpoons$
Ca^{2+}-exch + SiO_2(solid) + $2Al(OH)_3$(solid) (7.5)

Thus 1 mole of pyroxene is equivalent to only 1 mole of $CaCO_3$.

Based on the solubility information used to develop Fig. 7.1, leonhardite and lawsonite would appear to be more stable than calcite. However,

OTHER CALCIUM MINERALS

experimental evidence supporting this conclusion was not found. To the contrary, calcium carbonate generally accumulates in alkaline soils. Possibly the thermodynamic measurements used to calculate the solubilities of lawsonite and leonhardite may not be sufficiently accurate to permit this close an interpretation. Increasing $CO_2(g)$ above that of the atmosphere suppresses the calcite solubility lines in Fig. 7.1. Also the solubility of leonhardite increases as silica is weathered from soils and the activity of $H_4SiO_4^\circ$ drops below 10^{-4} M. For these reasons, calcite at a designated partial pressure of $CO_2(g)$ is used in this text as the solid phase most likely to control the Ca^{2+} solubility in alkaline soils.

7.2 OTHER CALCIUM MINERALS

The solubilities of several other calcium minerals are given by Reactions 13 through 25 of Table 7.1 and several are plotted in Fig. 7.2. Calcite is the least soluble calcium carbonate mineral, and $CaCO_3$(aragonite) is only slightly more soluble. Because of their similar solubilities, both minerals may occur together under certain conditions. The hydrated mineral $CaCO_3 \cdot 6H_2O$ (ikaite) is considerably more soluble and is not expected in soils.

The mineral CaO(lime) is much too soluble to persist in soils and is even too soluble to appear in Fig. 7.2. It readily hydrates to form $Ca(OH)_2$(portlandite), which is considerably more soluble than calcite (Fig. 7.2). The $CO_2(g)$ level at which calcite and portlandite can coexist is calculated as follows:

	$\log K^\circ$
$Ca^{2+} + 2H_2O \rightleftharpoons Ca(OH)_2\text{(portlandite)} + 2H^+$	-22.80
$CaCO_3\text{(calcite)} + 2H^+ \rightleftharpoons Ca^{2+} + CO_2(g) + H_2O$	9.74
$CaCO_3\text{(calcite)} + H_2O \rightleftharpoons Ca(OH)_2\text{(portlandite)} + CO_2(g)$	-13.06

(7.6)

from which

$$CO_2(g) = 10^{-13.06} \tag{7.7}$$

Thus when $CO_2(g)$ drops to $10^{-13.06}$ atm, both calcite and portlandite can coexist. If $CO_2(g)$ drops below this level, calcite becomes metastable to portlandite (unlikely in soils).

Of the four sulfate minerals included in Table 7.1 (Reactions 21 to 24), $CaSO_4 \cdot 2H_2O$(gypsum) is most stable. The solubility line for gypsum in

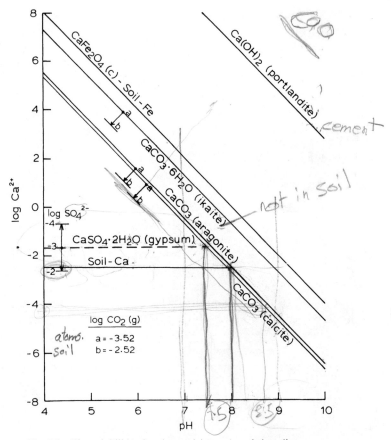

Fig. 7.2 The solubility of various calcium minerals in soils.

Fig. 7.2 is drawn for a sulfate activity of 10^{-3} M showing shifts upward and downward. Gypsum is too soluble to persist in soils unless SO_4^{2-} approaches 10^{-2} M. Generally gypsum is found only in arid soils were very little leaching occurs. It also occurs temporarily in some soils rich in sulfides that oxidize to release large amounts of SO_4^{2-}, for example, some of the polder soils of the Netherlands.

Application of gypsum to sodium-affected soils in the pH range of 8.5 to 10 raises soluble Ca^{2+} above that of calcite (Fig. 7.2) and leads to the precipitation of calcite with the release of protons:

$$Ca^{2+} + CO_2(g) + H_2O \rightleftharpoons CaCO_3(calcite) + 2H^+ \quad (7.8)$$

The pH then drops into the range of 7.5 to 8.0 where gypsum and calcite can coexist. In this way soluble Ca^{2+} is restored to approximately $10^{-2.5}$ M, keeping the soil colloids flocculated and predominantely calcium saturated. Displaced Na^+ can then be leached from the soil as drainage is supplied.

The solubility of CaF_2(fluorite) (Reaction 25 of Table 7.1) in equilibrium with soil-Ca can be represented as

			$\log K°$
CaF_2(fluorite)	$\rightleftharpoons$	$Ca^{2+} + 2F^-$	-10.41
Ca^{2+}	$\rightleftharpoons$	Soil-Ca	2.50
CaF_2(fluorite)	$\rightleftharpoons$	$2F^- +$ Soil-Ca	-7.91 (7.9)

$$(F^-)^2 = 10^{-7.91}$$

which gives

$$\log F^- = -3.96 \qquad (7.10)$$

The presence of fluorite in soils limits F^- activity to approximately 10^{-4} M. Further consideration will be given to $KMg_3AlSi_3O_{10}F_2$(fluorphlogopite) in Chapter 8 and to $Ca_5(PO_4)_3F$(fluorapatite) in Chapter 12 as other possible fluoride-controlling minerals in soils.

7.3 COMPLEXES OF CALCIUM IN SOLUTION

It is sometimes important to know the ionic forms of an element in the soil solution. Several inorganic complexes of Ca^{2+} are given in Table 7.1 (Reactions 26 through 37). The relationships depicted by these equations are plotted in Figs. 7.3 and 7.4. The anion activity ranges used to develop these plots are those commonly found in soils.

Figure 7.3 shows that the chloride and nitrate complexes of calcium contribute very little to total soluble calcium. The activities of Cl^- and NO_3^- affect the stability of these complexes both in the presence of soil-Ca and calcite. In the case of calcite, the extra lines and arrows were omitted to avoid further cluttering of the diagram. The $CaHCO_3^+$ ion is insignificant in acid soils, but it becomes important in submerged soils where pH normally approaches neutral and the partial pressure of $CO_2(g)$ increases. Such changes increase $CaHCO_3^+$ and decrease Ca^{2+}.

In the presence of calcite increasing $CO_2(g)$ by 10-fold depresses log Ca^{2+}, $CaCl^+$, $CaCl_2°$, $CaNO_3^+$, and $Ca(NO_3)_2°$ by one log unit (Fig. 7.3). The

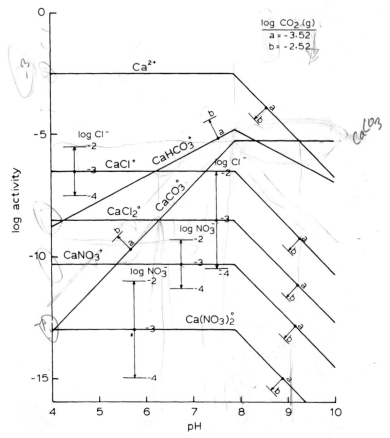

Fig. 7.3 Chloride, nitrate, and bicarbonate complexes of Ca^{2+} in equilibrium with soil-Ca or calcite as affected by $CO_2(g)$, anion activities, and pH.

activity of $CaCO_3^{\circ}$ is constant in the presence of calcite. The exact value can be obtained by combining Reactions 13 and 29 of Table 7.1 to give

$$\begin{array}{lr} & \log K^{\circ} \\ CaCO_3(\text{calcite}) + 2H^+ \rightleftharpoons Ca^{2+} + CO_2(g) + H_2O & 9.74 \\ Ca^{2+} + CO_2(g) + H_2O \rightleftharpoons CaCO_3^{\circ} + 2H^+ & -15.01 \\ \hline CaCO_3(\text{calcite}) \rightleftharpoons CaCO_3^{\circ} & -5.27 \end{array}$$

(7.11)

Thus calcite supports the neutral ion pair, $CaCO_3^{\circ}$, at $10^{-5.27}$ M which is independent of $CO_2(g)$ and pH.

COMPLEXES OF CALCIUM IN SOLUTION

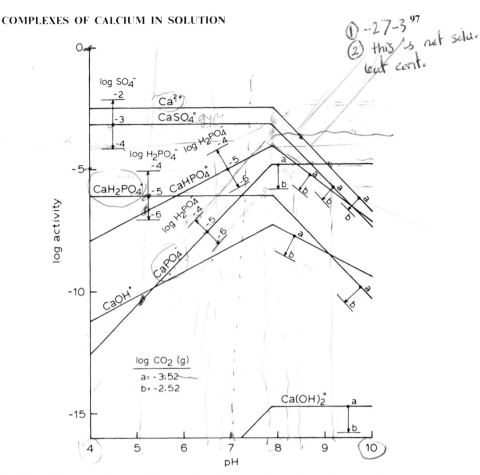

Fig. 7.4 Sulfate, phosphate, and hydroxyl complexes of calcium in equilibrium with soil-Ca or calcite.

Figure 7.4 shows that $CaSO_4^°$ contributes significantly to total calcium in solution if SO_4^{2-} is $> 10^{-4}$ M. The phosphate complexes of Ca^{2+} include $CaH_2PO_4^+$, $CaHPO_4^°$, and $CaPO_4^-$. In acid soils none of these complexes contribute significantly to total soluble calcium. In neutral and calcareous soils, $CaHPO_4^°$ and $CaPO_4^-$ are significant, especially when the activity of $H_2PO_4^-$ is $> 10^{-5}$ M. In the presence of calcite all complexes depicted in Fig. 7.4 decrease one log unit for each 10-fold increase in $CO_2(g)$, but their relative positions remain unchanged. Changes in SO_4^{2-} and $H_2PO_4^-$ affect the stability of calcium complexes in the presence of both calcite and soil-Ca. The extra lines and arrows were deleted in the case of calcite to simplify plotting.

The hydrolysis species $CaOH^+$ and $Ca(OH)_2^0$ are unimportant in the pH range of soils. Speculation that $CaOH^+$ acts as an exchangeable ion that enables more than 100% calcium saturation of soils is highly unlikely since the ratio $CaOH^+/Ca^{2+}$ never exceeds 10^{-4} (Fig. 7.4).

7.4 REDOX RELATIONSHIPS OF CALCIUM

The only important oxidation state of calcium is +2. The redox reaction relating Ca^{2+} to $Ca(c)$ is

$$Ca^{2+} + 2e^- \rightleftharpoons Ca(c) \qquad \log K° = -97.16$$

from which

$$pe = -48.58 + 0.5 \log Ca^{2+} \qquad (7.12)$$

Thus an extremely low pe is required for $Ca(c)$ to be stable. The only way that redox can affect calcium solubility relationships in soils is indirectly as it may affect other cations or anions with which Ca^{2+} combines.

7.5 THE $CaCO_3$–CO_2–H_2O SYSTEM

When calcium carbonate is present in soils, it has a dominating influence on many soil properties. Most calcareous soils fall in the pH range of 7.3 to 8.5, and only in the case of sodium-affected soils does the pH normally rise above 8.5. Because of the importance of calcium carbonate, further consideration will be given to its equilibrium relationships.

The solubility of $CaCO_3$(calcite) can be expressed as follows:

In terms of $CO_2(g)$

$$CaCO_3\text{(calcite)} + 2H^+ \rightleftharpoons Ca^{2+} + CO_2(g) + H_2O \qquad \log K° = 9.74 \qquad (7.13)$$

$$\log Ca^{2+} + 2pH = 9.74 - \log CO_2(g) \qquad (7.14)$$

In terms of $H_2CO_3^0$

$$CaCO_3\text{(calcite)} + 2H^+ \rightleftharpoons Ca^{2+} + CO_2(g) + H_2O \qquad 9.74$$
$$CO_2(g) + H_2O \rightleftharpoons H_2CO_3^0 \qquad -1.46$$

$$\overline{CaCO_3\text{(calcite)} + 2H^+ \rightleftharpoons Ca^{2+} + H_2CO_3^0 \qquad 8.28} \qquad (7.15)$$

THE $CaCO_3-CO_2-H_2O$ SYSTEM

$$\log Ca^{2+} + 2pH = 8.28 - \log H_2CO_3^\circ \qquad (7.16)$$

In terms of HCO_3^-

$$\begin{array}{lr} CaCO_3(\text{calcite}) + 2H^+ \rightleftharpoons Ca^{2+} + CO_2(g) + H_2O & 9.74 \\ CO_2(g) + H_2O \rightleftharpoons H^+ + HCO_3^- & -7.82 \\ \hline CaCO_3(\text{calcite}) + H^+ \rightleftharpoons Ca^{2+} + HCO_3^- & 1.92 \end{array}$$

$$(7.17)$$

$$\log Ca^{2+} + pH \rightleftharpoons 1.92 - \log HCO_3^- \qquad (7.18)$$

In terms of CO_3^{2-}

$$\begin{array}{lr} CaCO_3(\text{calcite}) + 2H^+ \rightleftharpoons Ca^{2+} + CO_2(g) + H_2O & 9.74 \\ CO_2(g) + H_2O \rightleftharpoons 2H^+ + CO_3^{2-} & -18.15 \\ \hline CaCO_3(\text{calcite}) \rightleftharpoons Ca^{2+} + CO_3^{2-} & -8.41 \end{array}$$

$$(7.19)$$

$$\log Ca^{2+} = -8.41 - \log CO_3^{2-} \qquad (7.20)$$

Equations 7.14, 7.16, 7.18, and 7.20 are all equivalent expressions for the solubility of calcite. Rearranging Eq. 7.14 gives

$$pH - 0.5\, pCa = 4.87 - 0.5 \log CO_2(g) \qquad (7.21)$$

which is referred to as the lime potential relationship. In this text the expression $\log Ca^{2+} + 2pH$ is used rather than $pH - 0.5\, pCa$ to express the pH-dependent solubility of calcium minerals.

Convenient solubility relationships of calcite that follow from Eq. 7.13 through 7.21 are summarized below:

$CO_2(g)$, atm	$\log CO_2$, atm	Ca^{2+}, mol/l	$\log Ca^{2+} + 2pH$	$pH - 0.5\, pCa$
0.0003	−3.52	$10^{13.26}(H^+)^2$	13.26	6.63
0.003	−2.52	$10^{12.26}(H^+)^2$	12.26	6.13
0.01	−2.00	$10^{11.74}(H^+)^2$	11.74	5.87
0.1	−1.00	$10^{10.74}(H^+)^2$	10.74	5.37
1.0	0.00	$10^{9.74}(H^+)^2$	9.74	4.87

A useful plot of Eq. 7.14 showing the equilibrium relationships of calcite at various $CO_2(g)$ levels is given in Fig. 7.5. This graph shows the relationships among the variables: pH, $\log Ca^{2+}$, and $CO_2(g)$. The solubility line

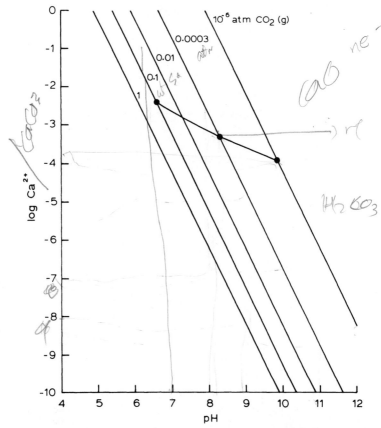

Fig. 7.5 The solubility of $CaCO_3$ (calcite) in terms of pH, $CO_2(g)$, and Ca^{2+} activity. The connecting line and points show the composition of pure $CaO-CO_2-H_2O$ systems at the indicated levels of $CO_2(g)$.

for calcite moves upward as $CO_2(g)$ decreases and downward as it increases. As shown earlier, the calcite line moves up to coincide with that of $Ca(OH)_2$(portlandite) when $CO_2(g)$ drops to $10^{-13.06}$ atm.

The curve that cuts across the calcite lines corresponds to the solution compositions of pure $CaO-CO_2-H_2O$ systems as $CO_2(g)$ changes. Let us derive the composition of the pure calcite-water system corresponding to $CO_2(g)$ at 0.0003 atm. The electroneutrality equation applicable for this system is

$$2[Ca^{2+}] + [CaHCO_3^+] + [H^+] = [HCO_3^-] + 2[CO_3^{2-}] + [OH^-]$$
(7.22)

THE PHASE RULE

and in terms of activities

$$\frac{2(\text{Ca}^{2+})}{\gamma_{\text{Ca}^{2+}}} + \frac{(\text{CaHCO}_3^+)}{\gamma_{\text{CaHCO}_3^+}} + \frac{(\text{H}^+)}{\gamma_{\text{H}^+}} = \frac{(\text{HCO}_3^-)}{\gamma_{\text{HCO}_3^-}} + \frac{2(\text{CO}_3^{2-})}{\gamma_{\text{CO}_3^{2-}}} + \frac{(\text{OH})^-}{\gamma_{\text{OH}^-}} \quad (7.23)$$

The activities of Ca^{2+}, CaHCO_3^+, HCO_3^-, CO_3^{2-}, and OH^- can be expressed in terms of pH and the formation constants for calcite, CaHCO_3^+, carbonic acid, and water. The activity coefficients can be calculated from the Debye-Hückel Eq. 2.11 and ionic strength from Eq. 2.8 using as a first approximation the concentrations of Ca^{2+} and HCO_3^- estimated from Fig. 7.5 and 6.2. Substituting these parameters into Eq. 7.23 gives:

$$10^{13.69}(\text{H}^+)^4 + 10^{0.02}(\text{H}^+)^3 = 10^{-11.31}(\text{H}^+) + 10^{-21.28} \quad (7.24)$$

Several iterations can be used to obtain precise estimates of ionic strength. By substituting various values for (H^+) into Eq. 7.24, a pH value of 8.34 is found to give the correct solution to this equation. Thus a pure $\text{CaO}-\text{CO}_2-\text{H}_2\text{O}$ system in equilibrium with 0.0003 atm $\text{CO}_2(\text{g})$ comes to pH 8.34. The other points along the transecting curve in Fig. 7.5 were obtained similarly using the indicated $\text{CO}_2(\text{g})$ levels.

7.6 THE PHASE RULE

The chemical phase rule states that

$$F = C - P + 2 \quad (7.25)$$

where F is the number of degrees of freedom or the smallest number of independent variables needed to completely define a system (it can be thought of as the maximum number of variables that may be changed without causing the appearance or disappearance of a phase, thereby destroying the system); C is the number of components, that is, the least number of chemically independent species required to describe the composition of every phase in the system; and P is the number of phases present. A phase is a region of uniform chemical and physical properties within a system. The last term in Eq. 7.25, accounts for temperature and pressure as variables. If temperature and pressure are fixed, these two degrees of freedom are removed. In certain cases where electric or magnetic fields are involved, the last term must be increased from 2 to 3 or 4, etc.

7.7 THE CO_2–H_2O SYSTEM

When temperature and pressure are fixed, $F = C - P$. The components in this system can be selected as $CO_2(g)$ and H_2O. The phases consist of $CO_2(g)$ and a solution. Applying the phase rule gives:

$$F = C - P = 2 - 2 = 0$$

This means that when the partial pressure of $CO_2(g)$ is specified, the solution composition is also fixed. The composition of the solution phase in this system can be calculated for any given $CO_2(g)$ level.

7.8 THE CaO–CO_2–H_2O SYSTEM

If temperature and pressure are fixed for the CaO–CO_2–H_2O system, the phase rule states:

$$F = C - P$$

There are now three components: CaO, CO_2, and H_2O. At equilibrium three phases are possible. For example, $CO_2(g)$, $CaCO_3$ (calcite), and solution could exist at equilibrium giving:

$$F = C - P = 3 - 3 = 0$$

This means that $CaCO_3$(calcite) placed in pure water at a fixed $CO_2(g)$ has a unique solution composition. The equilibrium constants and activity coefficients of the species involved make it possible to calculate the composition of the equilibrium solution as demonstrated in Section 7.5.

7.9 THE CaO–CO_2–H_2O–H_2SO_4 SYSTEM

Again if temperature and pressure are fixed

$$F = C - P$$

The components could include CaO, CO_2, H_2O, and H_2SO_4 giving four components. At equilibrium four possible phases may exist such as $CO_2(g)$, $CaCO_3$(calcite), $CaSO_4 \cdot 2H_2O$(gypsum), and the solution phase. Again the phase rule indicates

$$F = C - P = 4 - 4 = 0$$

Thus at a specified $CO_2(g)$ if either $CaCO_3$(calcite) or $CaSO_4 \cdot 2H_2O$ (gypsum) is present but not both, there is 1 degree of freedom and the addition of a component such as H_2SO_4 can alter the composition of the solution phase. If both $CaCO_3$(calcite) and $CaSO_4 \cdot 2H_2O$(gypsum) are present, the equilibrium solution will have a unique composition, and the addition of a component such as H_2SO_4 merely dissolves calcite and precipitates gypsum according to the reaction

$$H_2SO_4 + CaCO_3\text{(calcite)} + H_2O \rightleftharpoons$$
$$CaSO_4 \cdot 2H_2O\text{(gypsum)} + CO_2(g) \quad (7.26)$$

and the composition of the solution phase remains fixed. It is possible to calculate the composition of the solution phase that will have a fixed composition so long as $CO_2(g)$ is fixed and both solid phases are present. Using the electroneutrality equation appropriate for this system gives a unique solution of pH 7.8 when $CO_2(g)$ is 0.0003 atm.

Since soils contain many different chemical components in addition to CaO, CO_2, H_2O, and H_2SO_4, the calcite-gypsum equilibria can occur in soils at pH values slightly displaced from 7.8, but generally the displacement will not be very far. Addition of one or more of the four components to soils, however, will not affect the pH of a soil as demonstrated by Eq. 7.26. For example, addition of H_2SO_4, CaO, $CaCO_3$, $CaSO_4 \cdot 2H_2O$, or other combinations of these four components to a soil containing calcite and gypsum will not affect the pH of that soil. Increasing $CO_2(g)$, however, will lower the pH at which the two minerals can coexist.

PROBLEMS

7.1 Develop the equations and plot the solubility line for leonhardite shown in Fig. 7.1. Also develop the equations and plot this line when gibbsite and kaolinite control Al^{3+} and $H_4SiO_4^\circ$ activities.

7.2 Write the balanced chemical reactions showing the weathering of wairakite:
  a. In acid soils.
  b. In calcareous soils.
  c. Discuss how the pH of the two soils is affected.

7.3 Develop the necessary relationships to show that $CaHCO_3^+$ in a calcareous soil is independent of $CO_2(g)$.

7.4 Calculate the contribution that $CaHCO_3^+$ makes to total soluble calcium in a submerged soil of pH 7.3 when $CO_2(g)$ is 0.05 atm and SO_4^{2-} activity is $10^{-3.5}$ M (assume activity coefficients are unity).

7.5 From Fig. 7.4 estimate the activity of $CaHPO_4^\circ$ in a soil of pH 8.2 having 10^{-4} M HPO_4^{2-} and 0.003 atm $CO_2(g)$.

7.6 Develop the necessary equations and plot the log Ca^{2+} versus pH for calcite in equilibrium with 0.003 atm of $CO_2(g)$.

7.7 Calculate the pH of a $CaO-CO_2-H_2O$ system that attains equilibrium with 0.003 atm $CO_2(g)$. *Hint*: Begin with the electroneutrality equation and use the Debye–Hückel equation to estimate ion activity coefficients through successive approximations of ionic strength. Also calculate the activities of other constituents in the equilibrium solution and plot this solution composition on the diagram developed in Problem 7.6.

7.8 Calculate the pH of the $CaO-CO_2-H_2O-H_2SO_4$ system in equilibrium with 0.003 atm $CO_2(g)$, $CaCO_3$(calcite) and $CaSO_4 \cdot 2H_2O$(gypsum) using the procedure suggested for Problem 7.7 above. Also calculate the activities of the other parameters in the equilibrium solution.

7.9 Discuss the effect of adding the following constituents to the system $CaO-CO_2-H_2O$ system initially in equilibrium with $CaCO_3$ (calcite).
  a. $CaCl_2$.
  b. HCl.
  c. NaOH.
  d. NaCl.
  e. H_2SO_4.
  f. Soil.
  g. Increasing $CO_2(g)$.

7.10 Discuss the effect adding the following constituents to the system $CaO-CO_2-H_2O-H_2SO_4$ system where calcite and gypsum are initially present:
  a. $CaCl_2$.
  b. HCl.
  c. NaOH.
  d. CaO.
  e. H_2SO_4.
  f. Soil.
  g. Increasing $CO_2(g)$.

7.11 Calculate what will happen to a 0.5 M solution of $NaHCO_3$ at pH 8.5 (Olsen's bicarbonate extract for available phosphorus) that is left open to the atmosphere.

EIGHT

MAGNESIUM

The magnesium content of the lithosphere is estimated at 2.1%, while the average content of soils is only 0.5% (Table 1.1). These levels reflect the removal of magnesium from soils during weathering. The solubility relationships of several magnesium minerals are examined in this chapter to determine which minerals are stable in soils and which ones are most likely to control the activity of Mg^{2+} and its complexes in the soil solution. The solubility relationships of magnesium phosphates, sulfides, and molybdates are examined in Chapters 12, 17, and 22, respectively.

8.1 SOLUBILITY OF MAGNESIUM SILICATES

The solubilities of several magnesium silicates are given by Reactions 1 through 9 of Table 8.1 and are plotted in terms of log Mg^{2+} versus pH in Fig. 8.1. In this development $H_4SiO_4^°$ was fixed by SiO_2(soil) representing the early stages of weathering where most primary magnesium silicates are weathering. To plot $Mg_{1.6}Fe(II)_{0.4}SiO_4$(olivine), Fe^{2+} was fixed by $Fe(OH)_3$ (soil). Redox levels are represented at $pe + pH$ of 10 and 17. The magnesium silicates included in Fig. 8.1 decrease in solubility in the order:

$Mg_{1.6}Fe(II)_{0.4}SiO_4$(olivine) > Mg_2SiO_4(foresterite)
> $MgSiO_3$(clinoenstatite) > $Mg_3Si_4O_{10}(OH)_2 \cdot 2H_2O$(vermiculite)
> $MgCa(SiO_3)_2$(diopside)

in equilibrium with calcite

> $Mg_3Si_2O_5(OH)_4$(chrysotolite) > $Mg_2Si_3O_5(OH)_4$(sepiolite)
> $Mg_6Si_4O_{10}(OH)_8$ > (serpentine) > $Mg_3Si_4O_{10}(OH)_2$(talc).

Dolomite, $MgCa(CO_3)_2$, in equilibrium with $CaCO_3$(calcite) included as a reference mineral falls between serpentine and talc when $CO_2(g)$ is $10^{-3.52}$ atm. Soils generally have higher CO_2. Increasing CO_2 by 10-fold causes dolomite to become the most stable mineral shown in Fig. 8.1.

The silicate minerals in Fig. 8.1 shift with changes in $H_4SiO_4^°$ activity. The talc line was established using the following equilibria:

$$\log K°$$

$Mg_3Si_4O_{10}(OH)_2$(talc)
$\quad + 4H_2O + 6H^+ \rightleftharpoons 3Mg^{2+} + 4H_4SiO_4^° \qquad 22.26$

$\quad\quad\quad\quad 4H_4SiO_4^° \rightleftharpoons 4SiO_2\text{(soil)} + 8H_2O \qquad 4(3.10)$

$Mg_3Si_4O_{10}(OH)_2$(talc) + $6H^+ \rightleftharpoons$
$\quad\quad\quad\quad 3Mg^{2+} + 4SiO_2\text{(soil)} + 4H_2O \qquad 34.66$

$$\frac{(Mg^{2+})^3}{(H^+)^6} = 10^{34.66} \qquad (8.1)$$

TABLE 8.1 EQUILIBRIUM REACTIONS OF MAGNESIUM AT 25°C

Reaction No.	Equilibrium Reaction	log K°
	Silicates	
1	$MgSiO_3$(clinoenstatite) + $2H^+$ + $H_2O \rightleftharpoons Mg^{2+}$ + $H_4SiO_4^\circ$	11.42
2	$MgCa(SiO_3)_2$(diopside) + $4H^+$ + $2H_2O \rightleftharpoons Mg^{2+}$ + Ca^{2+} + $2H_4SiO_4^\circ$	21.16
3	Mg_2SiO_4(forsterite) + $4H^+ \rightleftharpoons 2Mg^{2+}$ + $H_4SiO_4^\circ$	28.87
4	$Mg_{1.6}Fe(II)_{0.4}SiO_4$(olivine) + $4H^+ \rightleftharpoons 1.6Mg^{2+}$ + $0.4Fe^{2+}$ + $H_4SiO_4^\circ$	26.18
5	$Mg_3Si_2O_5(OH)_4$(chrysotolite) + $6H^+ \rightleftharpoons 3Mg^{2+}$ + $2H_4SiO_4^\circ$ + H_2O	32.87
6	$Mg_2Si_3O_6(OH)_4$(sepiolite) + $2H_2O$ + $4H^+ \rightleftharpoons 2Mg^{2+}$ + $3H_4SiO_4^\circ$	15.89
7	$Mg_3Si_4O_{10}(OH)_2$(talc) + $4H_2O$ + $6H^+ \rightleftharpoons 3Mg^{2+}$ + $4H_4SiO_4^\circ$	22.26
8	$Mg_3Si_4O_{10}(OH)_2 \cdot 2H_2O$(vermiculite) + $2H_2O$ + $6H^+ \rightleftharpoons 3Mg^{2+}$ + $4H_4SiO_4^\circ$	30.39
9	$Mg_6Si_4O_{10}(OH)_8$(serpentine) + $12H^+ \rightleftharpoons 6Mg^{2+}$ + $4H_4SiO_4^\circ$ + $2H_2O$	61.75
	Aluminosilicates	
10	$Mg_5Al_2Si_3O_{10}(OH)_8$(chlorite) + $16H^+ \rightleftharpoons 5Mg^{2+}$ + $2Al^{3+}$ + $3H_4SiO_4^\circ$ + $6H_2O$	60.12
11	$Mg_2Al_4Si_5O_{18}$(cordierite) + $2H_2O$ + $16H^+ \rightleftharpoons 2Mg^{2+}$ + $4Al^{3+}$ + $5H_4SiO_4^\circ$	45.46
12	$K_{0.6}Mg_{0.25}Al_{2.3}Si_{3.5}O_{10}(OH)_2$(illite) + $8H^+$ + $2H_2O \rightleftharpoons 0.6K^+$ + $0.25Mg^{2+}$ + $2.3Al^{3+}$ + $3.5H_4SiO_4^\circ$	10.34
13	$Mg_{0.2}(Si_{3.81}Al_{1.71}Fe(III)_{0.22}Mg_{0.29})O_{10}(OH)_2$(Mg-montmorillonite) + $6.76H^+$ + $3.24H_2O \rightleftharpoons$ $0.49Mg^{2+}$ + $1.71Al^{3+}$ + $0.22Fe^{3+}$ + $3.81H_4SiO_4^\circ$	2.67
14	$Mg_{2.71}Fe(II)_{0.02}Fe(III)_{0.46}Ca_{0.06}K_{0.31}(Si_{3.91}Al_{1.14})O_{10}(OH)_2$(vermiculite) + $10.36H^+ \rightleftharpoons$ $2.71Mg^{2+}$ + $0.02Fe^{2+}$ + $0.46Fe^{3+}$ + $0.06Ca^{2+}$ + $0.1K^+$ + $1.14Al^{3+}$ + $2.91H_4SiO_4^\circ$ + $0.36H_2O$	38.05

(*Continued*)

TABLE 8.1 (Continued)

Reaction No.	Equilibrium Reaction	log $K°$
	Oxides, Hydroxides, and Carbonates	
15	$MgO(periclase) + 2H^+ \rightleftharpoons Mg^{2+} + H_2O$	21.74
16	$Mg(OH)_2(brucite) + 2H^+ \rightleftharpoons Mg^{2+} + 2H_2O$	16.84
17	$MgCO_3(magnesite) + 2H^+ \rightleftharpoons Mg^{2+} + CO_2(g) + H_2O$	10.69
18	$MgCO_3 \cdot 3H_2O(nesquehonite) + 2H^+ \rightleftharpoons Mg^{2+} + CO_2(g) + 4H_2O$	13.49
19	$MgCO_3 \cdot 5H_2O(lansfordite) + 2H^+ \rightleftharpoons Mg^{2+} + CO_2(g) + 6H_2O$	13.62
20	$MgCa(CO_3)_2(dolomite) + 4H^+ \rightleftharpoons Mg^{2+} + Ca^{2+} + 2CO_2(g) + 2H_2O$	18.46
	Sulfates	
21	$MgSO_4(c) \rightleftharpoons Mg^{2+} + SO_4^{2-}$	8.18
	Solution Species	
22	$Soil\text{-}Mg \rightleftharpoons Mg^{2+}$	−3.00*
23	$Mg^{2+} + 2Cl^- \rightleftharpoons MgCl_2°$	−0.03
24	$Mg^{2+} + CO_2(g) + H_2O \rightleftharpoons MgHCO_3^+ + H^+$	−6.76
25	$Mg^{2+} + CO_2(g) + H_2O \rightleftharpoons MgCO_3° + 2H^+$	−14.92
26	$Mg^{2+} + 2NO_3^- \rightleftharpoons Mg(NO_3)_2°$	−0.01
27	$Mg^{2+} + H_2O \rightleftharpoons MgOH^+ + H^+$	−11.45
28	$Mg^{2+} + 2H_2O \rightleftharpoons Mg(OH)_2° + 2H^+$	−27.99
29	$Mg^{2+} + H_2PO_4^- \rightleftharpoons MgHPO_4° + H^+$	−4.29
30	$Mg^{2+} + SO_4^{2-} \rightleftharpoons MgSO_4°$	2.23

Redox Reaction

31	$Mg^{2+} + 2e^- \rightleftharpoons Mg(c)$	−79.92

Other Reactions

32	$Fe(OH)_3(soil) + 3H^+ \rightleftharpoons Fe^{3+} + 3H_2O$	2.70
33	$Fe(OH)_3(soil) + 3H^+ + e^- \rightleftharpoons Fe^{2+} + 3H_2O$	15.74
34	$Al(OH)_3(gibbsite) + 3H^+ \rightleftharpoons Al^{3+} + 3H_2O$	8.04
35	$Al_2Si_2O_5(OH)_4(kaolinite) + 6H^+ \rightleftharpoons 2Al^{3+} + 2H_4SiO_4^\circ + H_2O$	5.45
36	$CaCO_3(calcite) + 2H^+ \rightleftharpoons Ca^{2+} + CO_2(g) + H_2O$	9.74

* From Eq. 8.6 in text.

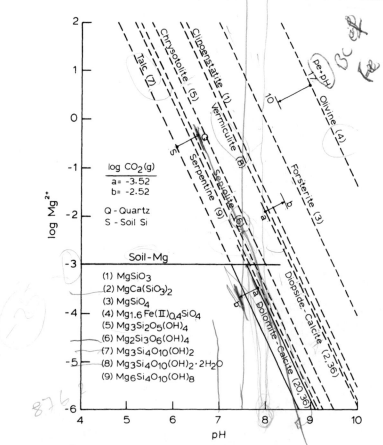

Fig. 8.1 The solubility of several magnesium silicates in equilibrium with soil-Si and soil-Fe with indicated changes for $CO_2(g)$, quartz, and redox.

from which

$$\log Mg^{2+} = 11.55 - 2pH \qquad (8.2)$$

If quartz rather than soil-Si is used to establish $H_4SiO_4^\circ$, the corresponding $\log K^\circ$ for Reaction 8.1 becomes $22.26 + 4(4.00) = 38.26$ which gives a solubility line for talc of

$$\log Mg^{2+} = 12.75 - 2pH \qquad (8.3)$$

This equation reflects an upward shift of 1.20 log units for the talc line shown in Fig. 8.1 by the arrows and short lines going from S(soil-Si) to Q(quartz). Such shifts can be readily obtained from the Si/Mg atomic ratio of the mineral. For example, decreasing $\log H_4SiO_4^\circ$ by one unit shifts the

SOLUBILITY OF MAGNESIUM SILICATES

talc line up by $\frac{4}{3} = 1.33$ log units of Mg^{2+} while that of serpentine moves up only $\frac{4}{6}$ or 0.67 log units. As soils weather, $H_4SiO_4^°$ declines and magnesium silicates become more soluble.

Several magnesium silicates may form in alkaline soils that are high in soluble $H_4SiO_4^°$. These include talc, serpentine, sepolite, and chrysotolite. As $H_4SiO_4^°$ declines or $CO_2(g)$ increases, dolomite-calcite becomes the stable magnesium phase. The dolomite-calcite combination is used in this text as a limit of Mg^{2+} solubility in calcareous soils. The equilibrium reactions describing this limit are as follows:

$$\begin{array}{lr} & \log K^° \\ MgCa(CO_3)_2(\text{dolomite}) + 4H^+ \rightleftharpoons & \\ \quad Mg^{2+} + Ca^{2+} + 2CO_2(g) + H_2O & 18.46 \\ Ca^{2+} + CO_2(g) + H_2O \rightleftharpoons & \\ \quad CaCO_3(\text{calcite}) + 2H^+ & -9.74 \\ \hline MgCa(CO_3)_2(\text{dolomite}) + 2H^+ \rightleftharpoons & \\ \quad Mg^{2+} + CO_2(g) + H_2O + CaCO_3(\text{calcite}) & 8.72 \end{array}$$

(8.4)

from which

$$\log Mg^{2+} = 8.72 - \log CO_2(g) - 2pH \qquad (8.5)$$

Equation 8.5 is plotted in Fig. 8.1 and shows the pH-dependent solubility of Mg^{2+} in equilibrium with dolomite-calcite.

Below pH 7.5 most magnesium minerals are too soluble to persist in soils. A reference solubility of Mg^{2+} in acid soils buffered by exchangeable Mg^{2+} is estimated at 10^{-3} M. This reference level can be expressed as follows:

$$Mg^{2+}(\text{soil}) \rightleftharpoons Mg^{2+} \qquad \log K^° = -3.0 \qquad (8.6)$$

In soils Mg^{2+} may vary slightly from this value depending on the rate of weathering compared to the rate of leaching. If Mg^{2+} in soil solution increases much above $10^{-2.5}$ M), it is removed more rapidly by leaching waters. This reference level for Mg^{2+} was selected to be slightly below that of Ca^{2+} ($10^{-2.5}$ M) to reflect the fact that exchangeable Ca^{2+} is generally higher than exchangeable Mg^{2+} than Ca^{2+}, and in such cases the activity of Mg^{2+} may exceed that of Ca^{2+}. Because of the great instability of magnesium minerals under acid conditons, Mg^{2+} depletion from acid soils can be expected.

8.2 MAGNESIUM ALUMINOSILICATES

The solubilities of several magnesium aluminosilicates are given by Reactions 10 through 14 of Table 8.1 and are plotted in Fig. 8.2. To develop this diagram, Al^{3+} and $H_4SiO_4^0$ were fixed by kaolinite and quartz, Fe^{3+} by soil-Fe, Ca^{2+} at $10^{-2.5}$ M or calcite, K^+ at 10^{-3} M, and $pe + pH$ at 17. Under these conditions the solubilities of illite, cordierite, and vermiculite are much higher than Mg-montmorillonite, chlorite, and dolomite-calcite. The more soluble minerals will disappear, whereas the more stable secondary minerals will remain until they, too, are removed in later stages of weathering as soils become more acid.

The solubilities of aluminosilicate minerals depicted in Fig. 8.2 are affected

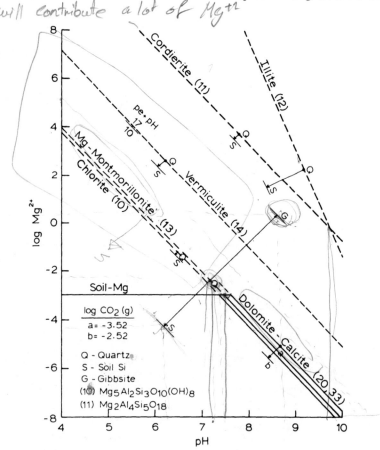

Fig. 8.2 The solubility of magnesium aluminosilicates in equilibrium with kaolinite, quartz, Ca^{2+} at $10^{-2.5}$ M, K^+ at 10^{-3} M, and soil-Fe.

by changes in $H_4SiO_4^°$. The kaolinite-quartz equilibrium was chosen as a chemical environment typical of the intermediate stages of weathering where secondary magnesium aluminosilicates are most likely to be weathering.

Shifts in the equilibrium lines of Fig. 8.2 corresponding to changes in $H_4SiO_4^°$ can be readily obtained from the atomic ratios of silicon, aluminum, and magnesium in a mineral from the equation:

$$\Delta \log Mg^{2+} = \Delta \log H_4SiO_4^° \left(\frac{Al - Si}{Mg} \right) \tag{8.7}$$

With $Mg_5Al_2Si_3O_{10}(OH)_8$(chlorite) as an example, increasing log $H_4SiO_4^°$ by 0.9 units, as would result in going from quartz to soil-Si, shifts the chlorite line by $0.9(2 - 3)/5 = -0.18$ log units of Mg^{2+}. In the case of Mg-montmorillonite the shift would be $0.9(1.71 - 3.81)/0.49 = 3.86$ log units. These shifts are indicated in Fig. 8.2 by the short lines and arrows going from Q(quartz) to S(soil-Si). Only illite and Mg-montmorillonite are affected appreciably by changes in $H_4SiO_4^°$. As $H_4SiO_4^°$ approaches that of soil-Si, Mg-montmorillonite becomes highly stable, even more so than dolomite-calcite.

In the advanced stages of weathering, $H_4SiO_4^°$ is expected to drop to $10^{-5.31}$ M where kaolinite and gibbsite can coexist (Fig. 5.7). Under these conditions, the Mg-montmorillonite line in Fig. 8.2 shifts upward by

$$(4.00 - 5.31)\left(\frac{1.71 - 3.81}{0.49}\right) = 5.61 \text{ log units}$$

reflecting the great instability of Mg-montmorillonite at low $H_4SiO_4^°$.

Other smectic minerals containing various substitutions of ions can be included in these solubility relationships once reliable free energy values are available.

8.3 OXIDES, HYDROXIDES, CARBONATES, AND SULFATES

The solubilities of several oxides, hydroxides, carbonates, and sulfates of magnesium are given by Reactions 15 through 21 of Table 8.1, and are plotted in Fig. 8.3.

Periclase (MgO) is much too soluble to persist in soils. It can hydrolyze to $Mg(OH)_2$(brucite) or it can precipitate as one of the other less soluble minerals. The carbonates of magnesium are also too soluble to persist in soils. They decrease in solubility in the order: $MgCO_3 \cdot 5H_2O$(lansfordite) > $MgCO_3 \cdot 3H_2O$(nesquehonite) > $MgCO_3$(magnesite). Increasing $CO_2(g)$ by a factor of 10 depresses Mg^{2+} activity in equilibrium with each of the

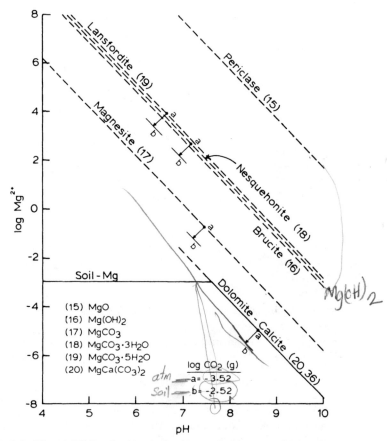

Fig. 8.3 The solubilities of oxides, hydroxides, and carbonates of magnesium.

magnesium carbonate minerals by one log unit. Of the magnesium minerals included in Fig. 8.3 only dolomite-calcite is stable.

The reported solubility for $MgSO_4(c)$ with a dissociation log $K°$ of 8.15 reflects a very high solubility for this mineral. It is much too soluble to form in well-drained soils.

8.4 MAGNESIUM COMPLEXES IN SOLUTION

The stabilities of several magnesium complexes are given by Reactions 22 through 30 of Table 8.1 and are plotted in Fig. 8.4.

MAGNESIUM COMPLEXES IN SOLUTION

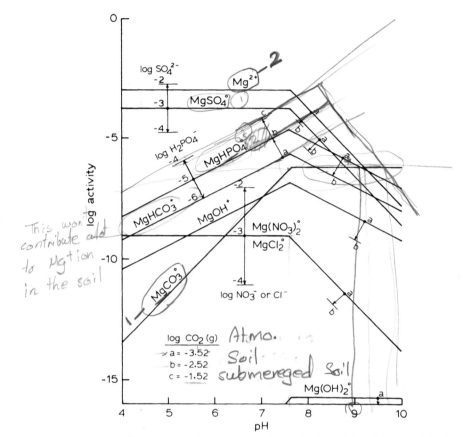

Fig. 8.4 The stability of magnesium complexes in equilibrium with soil-Mg or dolomite-calcite.

The MgSO$_4^\circ$ ion pair contributes significantly to total magnesium in solution as SO$_4^{2-}$ increases above 10^{-4} M. This ion pair increases 10-fold for each 10-fold increase in SO$_4^{2-}$.

The MgHPO$_4^\circ$ ion pair also becomes significant as pH and H$_2$PO$_4^-$ activity increase. For a soil of pH 7.2 with H$_2$PO$_4^-$ at 10^{-5} M, MgHPO$_4^\circ$ contributes approximately 1% to the total magnesium in solution. Increasing either pH or log H$_2$PO$_4^-$ increases its contribution. This ion pair can also contribute to total phosphorus in solution as will be shown in Chapter 12.

The MgHCO$_3^+$ species is insignificant in acid soils but becomes important in slightly alkaline soils and in submerged soils where CO$_2$(g) is generally high.

In the presence of dolomite and calcite, the $MgCO_3^0$ ion pair is independent of either pH or $CO_2(g)$ and has a fixed value of $10^{-6.2}$ M. In the presence of these minerals, $MgHCO_3^+$ is pH-dependent but is not affected by $CO_2(g)$.

The $Mg(NO_3)_2^0$ and $MgCl_2^0$ complexes have similar stabilities, but they are insignificant at the normal level of NO_3^- and Cl^- found in soils. Also the hydrolysis species $MgOH^+$ and $Mg(OH)_2^0$ are insignificant and contribute very little to total soluble magnesium.

8.5 EFFECT OF REDOX ON MAGNESIUM

For magnesium only the +2 oxidation state is important in soils. From Eq. 31 of Table 8.1 comes the relationship

$$pe = -39.96 + \log Mg^{2+} \tag{8.8}$$

which reflects the great instability of Mg(c) in aqueous environments and soils. The only way changes in redox can affect magnesium in soils is indirectly by affecting other cations or anions with which magnesium is associated. *So Redox doesn't affect Mg directly, but*

PROBLEMS

8.1 Using reactions from Table 8.1, derive the necessary equations and plot the solubility of $Mg_{1.6}Fe(II)_{0.4}SiO_4$(olivine) as shown in Fig. 8.1. Include shifts in this line for changes in $pe + pH$ and for equilibrium with SiO_2(quartz).

8.2 Prepare a plot to show how the solubility line for serpentine in Fig. 8.1 changes when $H_4SiO_4^0$ is in equilibrium with
   a. Quartz.
   b. Kaolinite and gibbsite.

8.3 How much does the Mg-montmorillonite line in Fig. 8.2 shift when equilibrium is attained with tridymite rather than quartz?

8.4 Develop an appropriate expression to calculate shifts in the solubility lines of Fig. 8.2 if equilibrium is attained with dickite rather than kaolinite. From this expression calculate the shift for Mg-montmorillonite as well.

8.5 Justify the selection of dolomite as a solid phase limiting the activity of Mg^{2+} in alkaline soils when it is obvious from Figs. 8.1 and 8.2 that there are minerals that may be less soluble (see also Fig. 5.6).

8.6 Examine the equilibrium conditions in a system prepared by adding

10 g each of dolomite, calcite, and magnesite to 50 ml of water open to the atmosphere. Indicate the following:
a. The solids present.
b. The ratio of Ca^{2+}/Mg^{2+} in solution.

8.7 Calculate the contribution of $MgSO_4^\circ$, $MgHPO_4^\circ$, and $MgHCO_3^+$ to total soluble magnesium in a soil having the following activities: $10^{-7.2}$ M (H^+), $10^{-3.5}$ M (SO_4^{2-}), $10^{-4.6}$ (HPO_4^{2-}) when the electrical conductivity of this soil is 0.77 millimho cm^{-1} and $CO_2(g)$ is 0.02 atm.

NINE

SODIUM AND POTASSIUM

The sodium content of the lithosphere is 2.8%, whereas the average content of soils is estimated at 0.63%. For potassium the lithosphere contains approximately 2.6% with an average of 0.83% for soils (Table 1.1). These contents reflect the removal of sodium and potassium during the weathering of soils. The removal of sodium slightly exceeds that of potassium. Both elements are essential to animals, and potassium is one of the three major fertilizer nutrients required by plants. Sodium and potassium are present as exchangeable cations in soils. In poorly drained soils, sodium salts generally accumulate as the major contributor of salinity. For these and other reasons, it is important to understand the chemical reactions of these two elements in soils.

In this chapter the solubility relationships of sodium and potassium minerals are examined under the chemical weathering matrix of soils to see which minerals are stable and to determine if the complexes of these cations are important in soils.

9.1 SOLUBILITY OF SODIUM MINERALS

Sodium silicates are too soluble to form in soils, but several sodium aluminosilicate minerals are important. The solubility relationships of several Na-aluminosilicates found in soils are given by Reactions 1 through 7 of Table 9.1 and are plotted in Fig. 9.1. These minerals are plotted in equilibrium with kaolinite. For those minerals having a ratio of Si/Al > 1 equilibrium with soil-Si is also used. These reference states were selected because they represent the silica-rich environments in which most sodium minerals generally weather.

An example of how the solubility lines in Fig. 9.1 were developed is given for low albite:

$\log K°$

$NaAlSi_3O_8(\text{low albite}) + 4H^+ + 4H_2O \rightleftharpoons$
$\quad Na^+ + Al^{3+} + 3H_4SiO_4° \qquad 2.74$

$Al^{3+} + H_4SiO_4° + 0.5H_2O \rightleftharpoons$
$\quad 0.5Al_2Si_2O_5(OH)_4(\text{kaol}) + 3H^+ \qquad 0.5(-5.45)$

$2H_4SiO_4° \rightleftharpoons$
$\quad 2SiO_2(\text{soil}) + 4H_2O \qquad 2(3.10)$

$NaAlSi_3O_8(\text{low albite}) + H^+ + 0.2H_2O \rightleftharpoons$
$\quad Na^+ + 0.5Al_2Si_2O_5(OH)_4(\text{kaol}) + 2SiO_2(\text{soil}) \qquad 6.22 \qquad (9.1)$

$$\log Na^+ = 6.22 - pH \qquad (9.2)$$

TABLE 9.1 EQUILIBRIUM REACTIONS OF SODIUM AND POTASSIUM MINERALS AND SOLUTION SPECIES

Reaction No.	Equilibrium Reaction	$\log K°$
	Na Aluminosilicates	
1	$NaAlSiO_4(\text{nepheline}) + 4H^+ \rightleftharpoons Na^+ + Al^{3+} + H_4SiO_4°$	11.25
2	$NaAlSi_2O_6(\text{jadeite}) + 4H^+ + 2H_2O \rightleftharpoons Na^+ + Al^{3+} + 2H_4SiO_4°$	7.11
3	$NaAlSi_2O_6 \cdot H_2O(\text{analcime}) + 4H^+ + H_2O \rightleftharpoons Na^+ + Al^{3+} + 2H_4SiO_4°$	8.15
4	$NaAlSi_3O_8(\text{Na-glass}) + 4H^+ + 4H_2O \rightleftharpoons Na^+ + Al^{3+} + 3H_4SiO_4°$	10.87
5	$NaAlSi_3O_8(\text{high albite}) + 4H^+ + 4H_2O \rightleftharpoons Na^+ + Al^{3+} + 3H_4SiO_4°$	3.67
6	$NaAlSi_3O_8(\text{low albite}) + 4H^+ + 4H_2O \rightleftharpoons Na^+ + Al^{3+} + 3H_4SiO_4°$	2.74
7	$NaAl_2(AlSi_3O_{10})(OH)_2(\text{paragonite}) + 10H^+ \rightleftharpoons Na^+ + 3Al^{3+} + 3H_4SiO_4°$	17.40
8	$Na_{0.33}Al_{2.33}Si_{3.67}O_{10}(OH)_2(\text{beidellite}) + 7.32H^+ + 2.68H_2O \rightleftharpoons 0.33Na^+ + 2.33Al^{3+} + 3.67H_4SiO_4°$	6.13
	K Aluminosilicates	
9	$KAlSiO_4(\text{kaliophilite}) + 4H^+ \rightleftharpoons K^+ + Al^{3+} + H_4SiO_4°$	13.05
10	$KAlSi_2O_6(\text{leucite}) + 4H^+ + 2H_2O \rightleftharpoons K^+ + Al^{3+} + 2H_4SiO_4°$	6.72
11	$KAlSi_3O_8(\text{K-glass}) + 4H^+ + 4H_2O \rightleftharpoons K^+ + Al^{3+} + 3H_4SiO_4°$	7.87
12	$KAlSi_3O_8(\text{high sanidine}) + 4H^+ + 4H_2O \rightleftharpoons K^+ + Al^{3+} + 3H_4SiO_4°$	1.40
13	$KAlSi_3O_8(\text{microcline}) + 4H^+ + 4H_2O \rightleftharpoons K^+ + Al^{3+} + 3H_4SiO_4°$	1.00
14	$KAl_2(AlSi_3O_{10})(OH)_2(\text{muscovite}) + 10H^+ \rightleftharpoons K^+ + 3Al^{3+} + 3H_4SiO_4°$	13.44
15	$K_{0.6}Mg_{0.25}Al_{2.3}Si_{3.5}O_{10}(OH)_2(\text{illite}) + 8H^+ + 2H_2O \rightleftharpoons 0.6K^+ + 0.25Mg^{2+} + 2.3Al^{3+} + 3.5H_4SiO_4°$	10.35

Na Complexes

16	$Na^+ + Cl^- \rightleftharpoons NaCl°$	0.00
17	$Na^+ + CO_2(g) + H_2O \rightleftharpoons NaCO_3^- + 2H^+$	−16.89
18	$2Na^+ + CO_2(g) + H_2O \rightleftharpoons Na_2CO_3° + 2H^+$	−18.14
19	$Na^+ + CO_2(g) + H_2O \rightleftharpoons NaHCO_3° + H^+$	−7.58
20	$Na^+ + H_2O \rightleftharpoons NaOH° + H^+$	−14.20
21	$Na^+ + SO_4^{2-} \rightleftharpoons NaSO_4^-$	0.70

K Complexes

22	$K^+ + Cl^- \rightleftharpoons KCl°$	−0.70
23	$2K^+ + CO_2(g) + H_2O \rightleftharpoons K_2CO_3° + 2H^+$	−18.17
24	$K^+ + H_2O \rightleftharpoons KOH° + H^+$	−14.50
25	$K^+ + SO_4^{2-} \rightleftharpoons KSO_4^-$	0.85

Redox Reactions

26	$Na^+ + e^- \rightleftharpoons Na(c)$	−45.89
27	$K^+ + e^- \rightleftharpoons K(c)$	−49.49

CH. 9 SODIUM AND POTASSIUM

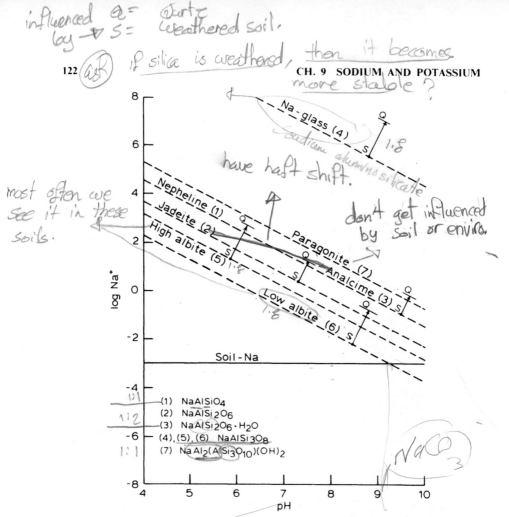

Fig. 9.1 The stability of sodium minerals in equilibrium with kaolinite and, when indicated, soil-Si (S) with shifts to quartz (Q).

Equation 9.2 is plotted as the low albite line in Fig. 9.1 and is marked (S) to indicate equilibrium with soil-Si. The number 6 behind this mineral refers to its chemical formula given in Fig. 9.1 and to its equilibrium reaction given by Reaction 6 of Table 9.1 as calculated from the free energy of formation values in the Appendix. Had SiO_2(quartz) been used rather than SiO_2(soil), the corresponding equilibrium constant for Reaction 9.1 would be increased by $2(0.9) = 1.8$ log units. This shift is indicated in Fig. 9.1 by the arrow and short line labeled Q.

Shifts in each of the lines of Fig. 9.1 for changes in $H_4SiO_4^0$ can be obtained from the relationship

$$\Delta \log Na^+ = \Delta \log H_4SiO_4^0 \left(\frac{Al - Si}{Na} \right) \tag{9.3}$$

SOLUBILITY OF POTASSIUM MINERALS
depends on

where Al, Si, and Na refer to the atomic composition of the mineral. For example, to locate the jadeite line in equilibrium with kaolinite-gibbsite where log $H_4SiO_4^\circ$ drops to -5.31 (Fig. 5.7):

$$\Delta \log Na^+ = (-5.31 - (-3.10))\left(\frac{1-2}{1}\right) = 2.21$$

This means that the line for jadeite in Fig. 9.1 would rise by 2.21 log units of Na^+ making it more soluble than paragonite in highly weathering kaolinitic soils where gibbsite forms.

When kaolinite and soil-Si provide the weathering environment for sodium minerals (Fig. 9.1), their solubilities decrease in the order:

$NaAlSi_3O_8$(Na-glass) $>$ $NaAl_2(AlSi_3O_{10})(OH)_2$(paragonite)
$>$ $NaAlSi_2O_6 \cdot H_2O$(analcime) $=$ $NaAlSiO_4$(nepheline)
$>$ $NaAlSi_2O_6$(jadeite) $>$ $NaAlSi_3O_8$(high albite)
$>$ $NaAlSi_3O_8$(low albite).

Dissolution is favored.

All of these minerals are too soluble to persist in soils because their equilibrium levels of Na^+ are far above the 10^{-3} M Na^+ depicted for soil-Na. Therefore, the sodium minerals depicted in this figure are expected to dissolve. If equilibrium is attained with quartz rather than soil-Si, the minerals having ratios of Si/Al $>$ 1 become more soluble and are expected to weather even more rapidly. The lines for nepheline and paragonite in Fig. 9.1 do not change with changes in $H_4SiO_4^\circ$ because the Si/Al ratio is equal to one and matches that of kaolinite. Beidellite whose solubility is given by Reaction 8 of Table 9.1 is not included in Fig. 9.1. The sodium component of this mineral is exchangeable and not lattice sodium, so the indicated equilibrium relationships would only be valid for sodium-saturated beidellite.

Figure 9.1 demonstrates why sodium is weathered from soils and either accumulates as soluble salts or is carried to the oceans. Sodium minerals are too soluble to precipitate in soils. The 10^{-3} M level of Na^+ is an arbitrary reference level that approximates the level of Na^+ found in well-drained soils and represents a balance between the rate of weathering of sodium minerals and the rate of leaching of Na^+. If higher levels of Na^+ occur, sodium is leached more rapidly, while if lower levels occur, its rate of leaching is diminished. In poorly drained arid soils, the rate of weathering often exceeds the rate of leaching and sodium salts accumulate to become phytotoxic.

9.2 SOLUBILITY OF POTASSIUM MINERALS

The solubilities of several potassium aluminosilicates are given by Reactions 9 through 15 of Table 9.1 and are plotted in Fig. 9.2. When equilibrium is

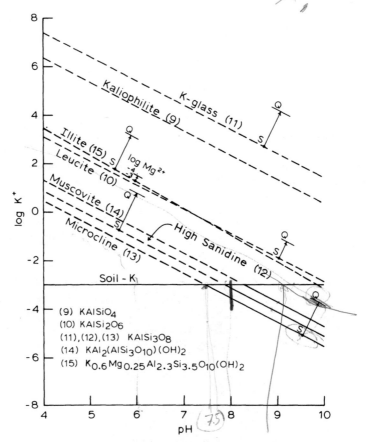

Fig. 9.2 The stability of potassium minerals in equilibrium with kaolinite, and when indicated, soil-Si (S) with shifts to quartz (Q).

maintained with kaolinite and soil-Si, these minerals decrease in the order:

$KAlSi_3O_8$(K-glass) > $KAlSiO_4$(kaliophilite)
> $K_{0.6}Mg_{0.25}Al_{2.3}Si_{3.5}O_{10}(OH)_2$(illite) = $KAlSi_2O_6$(leucite)
> $KAl_2(AlSi_3O_{10})(OH)_2$(muscovite) > $KAlSi_3O_8$(sanidine)
> $KAlSi_3O_8$(microcline).

These minerals are generally more soluble than the reference level of soil-K, which is arbitrarily set at 10^{-3} M K^+. In alkaline soils, microcline, high sanidine, and muscovite are sufficiently stable to keep soluble K^+ from rising above the reference level. Again as $H_4SiO_4^0$ drops to quartz or lower, the lines of minerals having Si/Al > 1 rise, whereas only muscovite and kaliophilite remain fixed.

COMPLEXES OF SODIUM AND POTASSIUM

Potassium is known to be refixed in secondary clay minerals such as illite, yet the solubility value for illite used here (see the appendix) appears to be too soluble to permit its formation in soils (Fig. 9.2). Further details on the refixation of potassium by clay minerals must await more accurate determination of the free energy values of these minerals and the exchange reactions they undergo in soils. Such data are vitally needed to develop meaningful equilibrium relationships of these important minerals in soils.

Comparing Figs. 9.1 and 9.2 shows that potassium minerals in general are less soluble than sodium minerals. The rates of weathering of these minerals also depend upon particle size since chemical reactions are largely limited to surface reactions.

9.3 COMPLEXES OF SODIUM AND POTASSIUM

Several complexes of sodium and potassium in solution are given by Reactions 16 through 25 of Table 9.1 and are plotted in Figs. 9.3 and 9.4.

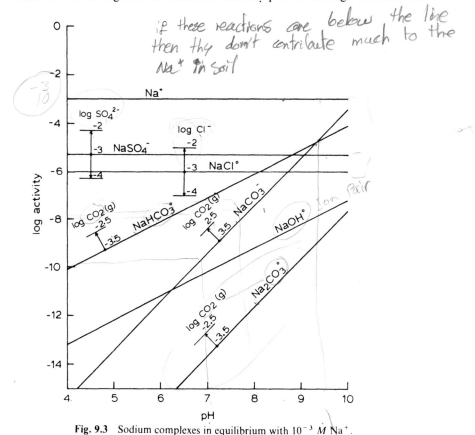

Fig. 9.3 Sodium complexes in equilibrium with 10^{-3} M Na^+.

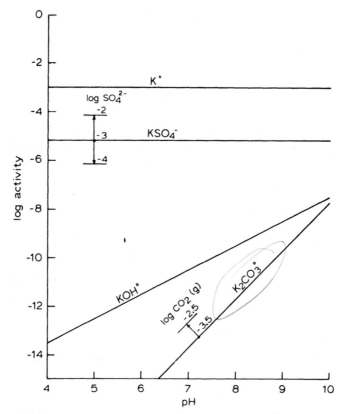

Fig. 9.4 Potassium complexes in equilibrium with 10^{-3} M K^+.

These plots were prepared on the basis of equilibrium with the reference levels of 10^{-3} M Na^+ or K^+. All lines on these diagrams except $Na_2CO_3^\circ$ and $K_2CO_3^\circ$ change by one log unit with corresponding changes of K^+ or Na^+. Since two Na^+ or K^+ ions are included in the latter complexes, they change by two log units.

In general the complexes of Na^+ and K^+ in well-drained soils are of only minor importance and can be fairly well ignored. These include sodium and potassium complexes of SO_4^{2-}, Cl^-, HCO_3^-, OH^-, and CO_3^{2-}.

9.4 REDOX RELATIONSHIPS

The redox relationships of sodium and potassium given by Reactions 26 and 27 of Table 9.1 are

$$Na^+ + e^- \rightleftharpoons Na(c) \quad \log K^\circ = -45.89 \quad (9.4)$$

$$pe = -45.89 + \log \text{Na}^+ \qquad (9.5)$$

and

$$\text{K}^+ + e^- \rightleftharpoons \text{K(c)} \qquad \log K° = -49.49 \qquad (9.6)$$

$$pe = -49.49 + \log \text{K}^+ \qquad (9.7)$$

These redox relationships show that only the oxidation state of +1 is important for these elements in aqueous or soil environments. Changes in redox can not affect sodium and potassium directly, only indirectly as redox affects other constituents with which they react.

PROBLEMS

9.1 Develop equations and plot the solubility lines for high albite and paragonite in Fig. 9.1. How do these lines shift if equilibrium is maintained with SiO_2(amorp) rather than soil-Si?

9.2 Develop equations and plot the solubility line for microcline in terms of log K^+ verus pH for equilibrium with
   a. Kaolinite and soil-Si.
   b. Kaolinite and quartz.
   c. Kaolinite and amorp-Si.
   d. Mg-montmorillonite, soil-Si, soil-Fe, and 10^{-3} M Mg^{2+}.

Discuss these relationships in terms of the weathering of microcline in soils at different stages of weathering.

9.3 Under what conditions, if any, can beidellite be expected to form in soils.

9.4 Develop the relationships expressed in Eq. 9.3.

TEN

IRON

Iron is a major constituent of the lithosphere, comprising approximately 5.1%; its average content in soils is estimated at 3.8% (Table 1.1). In primary minerals iron occurs largely as ferromagnesium minerals. During weathering these minerals dissolve (see olivine Fig. 8.1), and the released iron precipitates as ferric oxides and hydroxides. In this chapter it will be shown that the solubility of iron in soils is largely governed by Fe(III) oxides while hydrolysis, complexation, and redox are important modifying factors. The solubility relationships of iron oxides are examined in great detail, whereas those of phosphates, iron chelates, and iron sulfides will be given in Chapters 12, 15, and 17, respectively.

10.1 SOLUBILITY OF Fe(III) OXIDES IN SOILS

The solubilities of Fe(III) oxides commonly found in soils are given by Reactions 1 through 6 of Table 10.1. These equilibria are written as acid dissociation reactions, but they can easily be converted to basic dissociation reactions. Taking $Fe(OH)_3$(soil), for example:

$$\begin{array}{lll}
 & & \log K° \\
Fe(OH)_3\text{(soil)} + 3H^+ \rightleftharpoons Fe^{3+} + 3H_2O & & 2.70 \\
3H_2O \rightleftharpoons 3H^+ + 3OH^- & & 3(-14.00) \\
\hline
Fe(OH)_3\text{(soil)} \rightleftharpoons Fe^{3+} + 3OH^- & & -39.30 \quad (10.1)
\end{array}$$

Solubilities of the Fe(III) oxides are plotted in Fig. 10.1. These oxides decrease in solubility in the order:

$Fe(OH)_3$(amorp) > $Fe(OH)_3$(soil) > γ-Fe_2O_3(maghemite)
> γ-FeOOH(lepidocrocite) > α-Fe_2O_3(hematite)
> α-FeOOH(goethite).

The activity of Fe^{3+} maintained by each of these oxides decreases 1000-fold for each unit increase in pH. The solubility of Fe^{3+} maintained by $Fe(OH)_3$ (amorp) is $10^{3.54}/10^{-0.02} = 10^{3.56}$ or 3631 times greater than that maintained by α-FeOOH(goethite). When soluble Fe(III) salts are added to soils, Fe^{3+} readily precipitates. Within a few hours its solubility approaches that of by $Fe(OH)_3$(amorp). To determine where in this wide range of solubilities Fe^{3+} in soils might be, Norvell and Lindsay (1981) used the chelation method to determine the activity of Fe^{3+} in soils (see Chapter 15). Their findings indicate that soils generally maintain an Fe^{3+} activity slightly below that of $Fe(OH)_3$

TABLE 10.1   EQUILIBRIUM REACTIONS OF IRON MINERALS AND COMPLEXES AT 25°C

Reaction No.	Equilibrium Reaction	log $K°$
	Fe(III) Oxides and Hydroxides	
1	$Fe(OH)_3(amorp) + 3H^+ \rightleftharpoons Fe^{3+} + 3H_2O$	3.54
2	$Fe(OH)_3(soil) + 3H^+ \rightleftharpoons Fe^{3+} + 3H_2O$	2.70
3	$\frac{1}{2}\gamma\text{-}Fe_2O_3(maghemite) + 3H^+ \rightleftharpoons Fe^{3+} + \frac{3}{2}H_2O$	1.59
4	$\gamma\text{-}FeOOH(lepidocrocite) + 3H^+ \rightleftharpoons Fe^{3+} + 2H_2O$	1.39
5	$\frac{1}{2}\alpha\text{-}Fe_2O_3(hematite) + 3H^+ \rightleftharpoons Fe^{3+} + \frac{3}{2}H_2O$	0.09
6	$\alpha\text{-}FeOOH(goethite) + 3H^+ \rightleftharpoons Fe^{3+} + 2H_2O$	-0.02
	Other Fe(III) Minerals	
7	$FeCl_3(molysite) \rightleftharpoons Fe^{3+} + 3Cl^-$	13.25
8	$Fe_2(SO_4)_3(c) \rightleftharpoons 2Fe^{3+} + 3SO_4^{2-}$	2.89
9	$KFe_3(SO_4)_2(OH)_6(jarosite) + 6H^+ \rightleftharpoons K^+ + 3Fe^{3+} + 2SO_4^{2-} + 6H_2O$	-12.51
	Fe(III) Hydrolysis	
10	$Fe^{3+} + H_2O \rightleftharpoons FeOH^{2+} + H^+$	-2.19
11	$Fe^{3+} + 2H_2O \rightleftharpoons Fe(OH)_2^+ + 2H^+$	-5.69
12	$Fe^{3+} + 3H_2O \rightleftharpoons Fe(OH)_3° + 3H^+$	-13.09
13	$Fe^{3+} + 4H_2O \rightleftharpoons Fe(OH)_4^- + 4H^+$	-21.59
14	$2Fe^{3+} + 2H_2O \rightleftharpoons Fe_2(OH)_2^{4+} + 2H^+$	-2.90
	Fe(III) Complexes	
15	$Fe^{3+} + Cl^- \rightleftharpoons FeCl^{2+}$	1.48
16	$Fe^{3+} + 2Cl^- \rightleftharpoons FeCl_2^+$	2.13
17	$Fe^{3+} + 3Cl^- \rightleftharpoons FeCl_3°$	0.77
18	$Fe^{3+} + Br^- \rightleftharpoons FeBr^{2+}$	-0.60
19	$Fe^{3+} + 3Br^- \rightleftharpoons FeBr_3°$	0.04
20	$Fe^{3+} + F^- \rightleftharpoons FeF^{2+}$	6.00
21	$Fe^{3+} + 2F^- \rightleftharpoons FeF_2^+$	9.20
22	$Fe^{3+} + 3F^- \rightleftharpoons FeF_3°$	11.70
23	$Fe^{3+} + NO_3^- \rightleftharpoons FeNO_3^{2+}$	1.00
24	$Fe^{3+} + SO_4^{2-} \rightleftharpoons FeSO_4^+$	4.15
25	$Fe^{3+} + 2SO_4^{2-} \rightleftharpoons Fe(SO_4)_2^-$	5.38
26	$Fe^{3+} + H_2PO_4^- \rightleftharpoons FeH_2PO_4^{2+}$	5.43
27	$Fe^{3+} + H_2PO_4^- \rightleftharpoons FeHPO_4^+ + H^+$	3.71

TABLE 10.1 (*Continued*)

Reaction No.	Equilibrium Reaction	log $K°$
	Redox Reactions	
28	$Fe(c) \rightleftharpoons Fe^{2+} + 2e^-$	15.98
29	$Fe^{3+} + e^- \rightleftharpoons Fe^{2+}$	13.04
30	$Fe_3O_4(\text{magnetite}) + 8H^+ + 2e^- \rightleftharpoons 3Fe^{2+} + 4H_2O$	35.69
31	$Fe_3O_4(\text{magnetite}) + 8H^+ \rightleftharpoons 3Fe^{3+} + e^- + 4H_2O$	−3.42
32	$Fe_3(OH)_8(\text{ferrosic oxide}) + 8H^+ + 2e^- \rightleftharpoons 3Fe^{2+} + 8H_2O$	43.75
33	$Fe_{0.95}O(\text{wustite}) + 0.10e^- + 2H^+ \rightleftharpoons 0.95Fe^{2+} + H_2O$	12.42
	Fe(II) Minerals	
34	$FeO(c) + 2H^+ \rightleftharpoons Fe^{2+} + H_2O$	13.48
35	$Fe(OH)_2(c) + 2H^+ \rightleftharpoons Fe^{2+} + 2H_2O$	12.90
36	$FeCl_2(\text{lawrencite}) \rightleftharpoons Fe^{2+} + 2Cl^-$	9.00
37	$FeCO_3(\text{siderite}) + 2H^+ \rightleftharpoons Fe^{2+} + CO_2(g) + H_2O$	7.92
38	$FeSO_4(c) \rightleftharpoons Fe^{2+} + SO_4^{2-}$	2.65
39	$FeSO_4 \cdot 7H_2O \rightleftharpoons Fe^{2+} + SO_4^{2-} + 7H_2O$	−2.46
40	$FeSiO_3(c) + 2H^+ + H_2O \rightleftharpoons Fe^{2+} + H_4SiO_4°$	14.79
41	$Fe_2SiO_4(\text{fayalite}) + 4H^+ \rightleftharpoons 2Fe^{2+} + H_4SiO_4°$	19.76
	Fe(II) Hydrolysis	
42	$Fe^{2+} + H_2O \rightleftharpoons FeOH^+ + H^+$	−6.74
43	$Fe^{2+} + 2H_2O \rightleftharpoons Fe(OH)_2° + 2H^+$	−16.04
44	$Fe^{2+} + 3H_2O \rightleftharpoons Fe(OH)_3^- + 3H^+$	−31.99
45	$Fe^{2+} + 4H_2O \rightleftharpoons Fe(OH)_4^{2-} + 4H^+$	−46.38
46	$3Fe^{2+} + 4H_2O \rightleftharpoons Fe_3(OH)_4^{2+} + 4H^+$	−45.39
	Fe(II) Complexes	
47	$Fe^{2+} + 2Br^- \rightleftharpoons FeBr_2°$	0.00
48	$Fe^{2+} + 2Cl^- \rightleftharpoons FeCl_2°$	−0.07
49	$Fe^{2+} + 2F^- \rightleftharpoons FeF_2°$	0.03
50	$Fe^{2+} + H_2PO_4^- \rightleftharpoons FeH_2PO_4^+$	2.70
51	$Fe^{2+} + H_2PO_4^- \rightleftharpoons FeHPO_4° + H^+$	−3.60
52	$Fe^{2+} + SO_4^{2-} \rightleftharpoons FeSO_4°$	2.20

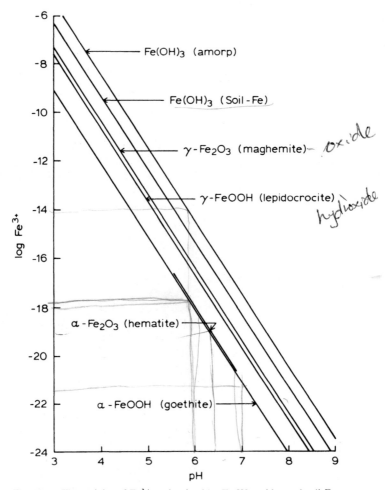

Fig. 10.1 The activity of Fe^{3+} maintained by Fe(III) oxides and soil-Fe.

(amorp) as represented by $Fe(OH)_3$(soil) (Reaction 2, Table 10.1). When they added soluble $FeCl_3$ to their soils, the activity of Fe^{3+} temporarily increased, but over a period of several weeks, it slowly approached the solubility of $Fe(OH)_3$(soil). Throughout this text $Fe(OH)_3$(soil) or soil-Fe is used as a reference solid phase controlling the solubility for Fe^{3+} in soils. Rather than think of soil-Fe as a discrete crystalline phase, it can be thought of as an amorphous phase having a greater degree of structural order than freshly precipitated $Fe(OH)_3$(amorp), hence it has a slightly more negative ΔG_f° (see the appendix).

OTHER Fe(III) MINERALS

Since the crystalline iron oxides in Fig. 10.1 fall below soil-Fe, they are more stable and can be expected to crystallize slowly. The activity of Fe^{3+} maintained by these iron minerals is very low; consequently at high pH transformations from one mineral to another and attainment of final equilibrium is an extremely slow process. Schwertmann and Taylor (1977) have discussed some of the conditions believed to affect the formation of various forms of Fe(III) oxides in soils.

Maghemite (γ-Fe_2O_3) is a magnetic mineral and can be separated from soils with a magnet. The percent composition of various iron oxides in a soil may have little bearing on the Fe^{3+} activity maintained by that soil. Soils generally contain some of several different iron oxides. The solubility of Fe^{3+} is usually controlled by the most soluble oxide present. For this reason soil-Fe generally controls the activity of Fe^{3+} in most soils. Only in well-drained, highly weathered soils not subject to frequent reductions are hematite and goethite expected to lower the solubility of Fe^{3+} toward their equilibrium levels. The solubilities of hematite and goethite are nearly identical, but goethite is generally considered the ultimate weathering product of iron in soils.

From Fig. 10.1 it is easy to see why iron deficiencies are more prevalent on alkaline soils than on acid soils, and why highly weathered soils are lower in available iron than less weathered soils. "Limonite," once considered a common form of iron oxide in soils, is now recognized as finely divided goethite with adsorbed water.

10.2 OTHER Fe(III) MINERALS

The solubilities of other Fe(III) minerals are given by Reactions 7 through 9 of Table 10.1. The solubilities of $FeCl_3$(molysite) and $Fe_3(SO_4)_2$(c) are much too high to permit their formation in soils. The mineral $KFe_3(SO_4)_2(OH)_6$ (jarosite) is often found in acid sulfate soils (Breemen, 1976). The conditions necessary for its precipitation are depicted in Fig. 10.2. Only in soils below pH 4.0, or slightly higher depending on SO_4^{2-} and K^+ activities, can jarosite form. Such conditions are found in acid sulfate soils containing FeS_2(pyrite), which oxidizes to give sulfuric acid and precipitated Fe(III) oxides. Figure 10.2 indicates that in the acid pH range of soils jarosite can form from $Fe(OH)_3$(amorp) and $Fe(OH)_3$(soil) but not from goethite. Release of Fe^{2+} from pyrite and its subsequent oxidation to Fe^{3+} favor jarosite formation. Potassium ions are also needed to precipitate jarosite, but they are relatively minor constituents that can be replaced to some extent by Na^+, H_3O^+, or other cations (Breemen, 1976).

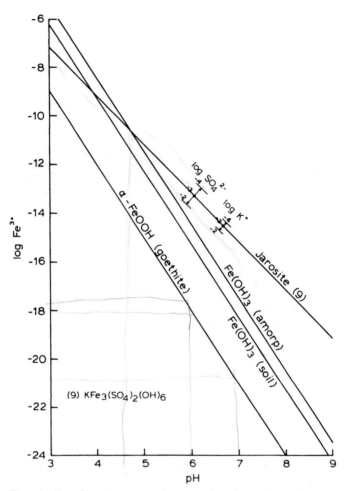

Fig. 10.2 The solubility of jarosite compared to amorphous iron oxide, soil-Fe, and goethite.

10.3 HYDROLYSIS OF Fe(III)

In aqueous solution Fe^{3+} is surrounded by six molecules of water, that is, $Fe(H_2O)_6^{3+}$, similar to the hexahydrated Al^{3+} ion (Section 3.3). Increasing pH removes H^+ from the coordinated water and gives rise to various hydrolysis products defined by Reactions 10 through 14 of Table 10.1. For simplicity the water molecules are usually omitted in representing the hydrolysis species which include $FeOH^{2+}$, $Fe(OH)_2^+$, $Fe(OH)_3^\circ$, $Fe(OH)_4^-$, and the polymer $Fe_2(OH)_2^{4-}$.

A plot of the various Fe^{3+} hydrolysis species in soils as affected by pH is

HYDROLYSIS OF Fe(III)

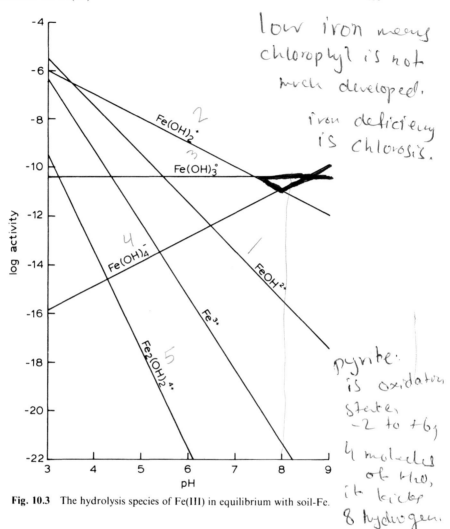

Fig. 10.3 The hydrolysis species of Fe(III) in equilibrium with soil-Fe.

given in Fig. 10.3. These relationships were developed on the basis that equilibrium is maintained between the hydrolysis species and soil-Fe. The manner in which the various lines in this figure were obtained is shown for $Fe(OH)_3^\circ$ by combining Reactions 2 and 12 of Table 10.1.

		log $K°$
$Fe(OH)_3(soil) + 3H^+ \rightleftharpoons Fe^{3+} + 3H_2O$		2.70
$Fe^{3+} + 3H_2O \rightleftharpoons Fe(OH)_3^\circ + 3H^+$		−13.09
$Fe(OH)_3(soil) \rightleftharpoons Fe(OH)_3^\circ$		−10.39

(10.2)

Equation 10.2 shows that soil-Fe maintains a fixed activity of $10^{-10.39}$ M Fe(OH)$_3^{\circ}$ as represented by the horizontal line in Fig. 10.3. The other lines in this figure were obtained similarly.

In the pH range of soils, the hydrolysis species are more abundant than free Fe^{3+}. If the activity of Fe^{3+} rises or falls because of equilibrium with the various other solid phase Fe(III) oxides (Fig. 10.1), so do the activities of the various hydrolysis species. For example, if Fe^{3+} were controlled by goethite rather than soil-Fe, all of the lines in Fig. 10.3 representing monomeric iron species would shift downward $(-0.02) - 2.70 = -2.72$ log units, while the dimeric species, Fe$_2$(OH)$_2^{4+}$, would be lowered by $2(-2.72) = -5.44$ log units.

The hydrolysis species are important in soils because they increase total iron in solution. In the case of insoluble nutrients like iron, transport from soil to plant roots is generally the rate-limiting step in nutrient uptake. Since diffusional and mass flow transport processes depend upon total solubility, the hydrolysis species are very important in plant nutritional considerations. Iron has a minimum solubility in the pH range of 7.4 to 8.5, which is the pH range of soils in which iron deficiencies are most common.

A soil of pH 8 in equilibrium with soil-Fe has only $10^{-21.3}$ M Fe^{3+}, whereas total soluble iron consists of $10^{-10.4}$ M Fe(OH)$_3^{\circ}$, $10^{-10.9}$ M Fe(OH)$_4^-$, and $10^{-11.0}$ M Fe(OH)$_2^+$ giving a total soluble iron of $10^{-10.20}$ M. Yet this level of iron is still below that needed to supply most plants with adequate iron as can be demonstrated by attempting to grow plants in hydroponic solutions supplied with only inorganic iron. Most plants develop severe iron deficiency even at pH 5.0 unless chelating agents are present to increase iron solubility (Chapter 15).

The hydrolysis species of Fe^{3+} in aqueous solutions are always present and cannot be eliminated. Knowing pH and any one of the hydrolysis species permits the calculation of all other species. It makes little difference whether a plant absorbs one ion or another inasmuch as the other ions will immediately dissociate to restore equilibrium relationships among all species.

10.4 Fe(III) COMPLEXES IN SOILS

In addition to the hydrolysis species, Fe^{3+} combines with various anions to form complexes and ion pairs. Several such complexes are given by Reactions 15 through 27 of Table 10.1. The complexes with Cl$^-$, Br$^-$, and F$^-$ are depicted graphically in Fig. 10.4 for various log activity levels of anions in equilibrium with soil-Fe.

None of the complexes included in Fig. 10.4 are very significant in soils.

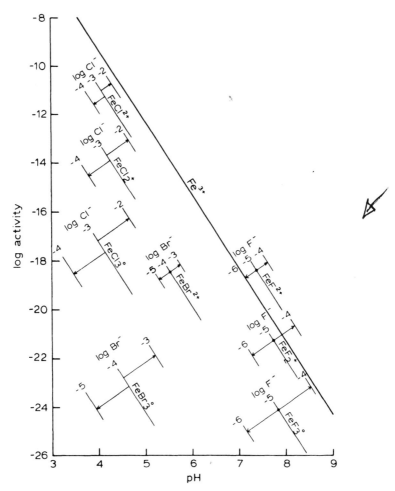

Fig. 10.4 Halide complexes of Fe(III) in equilibrium with soil-Fe.

Only the fluoride complexes are more abundant than Fe^{3+} at the anionic levels represented in this figure.

The stability lines for $FeCl^{2+}$, $FeCl_2^+$, and $FeCl_3^\circ$ increase 1, 2, and 3 log units with 10-fold changes in anion activity. For example, the $Fe(Cl)_3^\circ$ line moves up three log units for a 10-fold increase in Cl^- whereas the $FeCl^{2+}$ line moves up only one log unit.

The Fe(III) complexes of nitrate, sulfate, and phosphate in equilibrium with soil-Fe are shown in Fig. 10.5. Some of these complexes are highly significant. For example, when $H_2PO_4^-$ in soils is 10^{-5} M, the ratio of

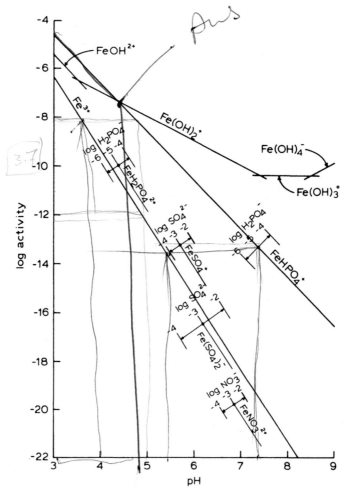

Fig. 10.5 Phosphate, sulfate, and nitrate complexes of Fe(III) in equilibrium with soil-Fe compared to the hydrolysis species of Fe^{3+}.

$FeHPO_4^+/Fe^{3+}$ at pH 5 is $10^{3.7}$ or approximately 5000-fold. At pH 8 this ratio is $10^{6.7}$ or more than 50 million. The complexes $FeH_2PO_4^{2+}$, $FeSO_4^+$, and $Fe(SO_4)_2^-$ are significant relative to Fe^{3+}, but they do not contribute significantly to total iron in solution due to the great preponderance of the hydrolysis species. Only under acid conditions does $FeHPO_4^+$ increase total iron in solution.

If Fe^{3+} in Fig. 10.4 and 10.5 were governed by one of the crystalline iron oxides instead of soil-Fe, the complex species in solution will shift downward

EFFECT OF REDOX ON Fe(II) SOLUBILITY

the same number of log units as Fe^{3+}, but the relative positions of the lines would remain unchanged. These figures are useful for examining the Fe(III) complexes and determining which ions may be significant in soils.

10.5 EFFECT OF REDOX ON Fe(II) SOLUBILITY

The electron activity in soils controls the ratio of Fe^{3+} to Fe^{2+} in solution according to the reaction

$$Fe^{3+} + e^- \rightleftharpoons Fe^{2+} \qquad \log K° = 13.04 \qquad (10.3)$$

for which

$$\log \frac{Fe^{2+}}{Fe^{3+}} = 13.04 - pe \qquad (10.4)$$

$\log \frac{Fe^{2+}}{Fe^{3+}} = 13.04 - pe$

When $pe = 13.04$, the ratio of Fe^{2+}/Fe^{3+} is unity. Changing pe by one unit changes the ratio of Fe^{2+}/Fe^{3+} by 10-fold. Thus the ratio of Fe^{2+}/Fe^{3+} in any aqueous media can be readily calculated from pe using Eq. 10.4.

Combining Reaction 10.3 with the solubility expression for soil-Fe gives

$$
\begin{array}{lr}
 & \log K° \\
Fe^{3+} + e^- \rightleftharpoons Fe^{2+} & 13.04 \\
Fe(OH)_3(soil) + 3H^+ \rightleftharpoons Fe^{3+} + 3H_2O & 2.70 \\
\hline
Fe(OH)_3(soil) + 3H^+ + e^- \rightleftharpoons Fe^{2+} + 3H_2O & 15.74 \quad (10.5)
\end{array}
$$

$$\frac{(Fe^{2+})}{(H^+)^3(e^-)} = 10^{15.74} \qquad (10.6)$$

from which

$$\log Fe^{2+} = 15.74 - (pe + pH) - 2pH \qquad (10.7)$$

Equation 10.7 is plotted in Fig. 10.6 and shows the activity of Fe^{2+} in equilibrium with soil-Fe as a function of redox and pH. The redox parameter $(pe + pH)$ is used because it partitions the protons associated with the redox component of the reactions from those associated with the pH dependency of the reduced species. In this case one proton is associated with the reduction of Fe^{3+} to Fe^{2+}, whereas the other two protons are associated with the 100-fold change in Fe^{2+} with pH.

Lowering redox (Fig. 10.6) increases the activity of Fe^{2+} in equilibrium with soil-Fe 10-fold for each unit decrease in $pe + pH$. The lowest dashed line in this figure represents Fe^{2+} at $pe + pH$ of 20.61, the equilibrium redox

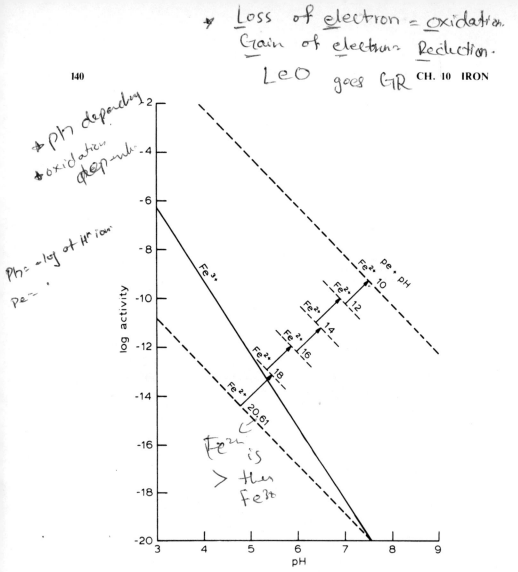

Fig. 10.6 The effect of redox and pH on the Fe^{2+} in equilibrium with soil-Fe.

corresponding to 0.2 atm $O_2(g)$. At this redox, the activities of Fe^{2+} and Fe^{3+} become equal at pH 7.57. Since the $pe + pH$ of soils is generally lower than 20.61 (Section 2.10), the activity of Fe^{2+} in soils is most often greater than that of Fe^{3+}.

If one of the crystalline Fe(III) oxides rather than soil-Fe controls Fe^{3+} activity, the lines in Fig. 10.6 would shift downward. For example, if goethite controls Fe^{3+} activity, all lines shift downward by 2.72 log units (Reactions 2 and 6 of Table 10.1). Thus Fig. 10.6 can be used to obtain the Fe^{2+} activity in equilibrium with any of the Fe(III) oxides providing $pe + pH$ is known.

EFFECT OF REDOX ON THE STABILITY OF IRON MINERALS

10.6 EFFECT OF REDOX ON THE STABILITY OF IRON MINERALS

As soils are reduced the activity of Fe^{2+} increases until the solubility of a more reduced iron mineral is exceeded. The reduced mineral then becomes the stable phase.

Let us consider the equilibrium relationships between soil-Fe and Fe_3O_4 (magnetite). The latter mineral contains two Fe(III) and one Fe(II) ions per mole. Combining Reaction 30 of Table 10.1 with Reaction 10.5 above gives

	log $K°$
$Fe_3O_4(\text{magnetite}) + 8H^+ + 2e^- \rightleftharpoons 3Fe^{2+} + 4H_2O$	35.69
$3Fe^{2+} + 9H_2O \rightleftharpoons 3Fe(OH)_3(\text{soil}) + 9H^+ + 3e^-$	$3(-15.74)$
$Fe_3O_4(\text{magnetite}) + 5H_2O \rightleftharpoons 3Fe(OH)_3(\text{soil}) + H^+ + e^-$	-11.53

$$\text{(10.8)}$$

from which

$$pe + pH = 11.53 \qquad (10.9)$$

Thus at $pe + pH = 11.53$, magnetite and soil-Fe can coexist. At higher redox soil-Fe is stable, whereas at lower redox magnetite is stable. The stability relationships of these minerals as well as other Fe(II) minerals are shown in Fig. 10.7.

In the $pe + pH$ range of 20.61 to 11.53 soil-Fe controls the solubility of Fe^{2+}. In this redox range magnetite and all other Fe(II) minerals shown in Fig. 10.7 are unstable and cannot form. At $pe + pH = 11.53$ magnetite and soil-Fe can coexist. Attempts to decrease $pe + pH$ farther cause soil-Fe to dissolve and magnetite to precipitate, but the $pe + pH$ during this transition tends to remain poised at 11.53. If the electrons are supplied more rapidly than the transformation can occur, the redox will temporarily drop below this value without the transformation being completed.

Below $pe + pH = 11.53$ magnetite is the stable phase until $FeCO_3$ (siderite) forms depending on the partial pressure of $CO_2(g)$. At $CO_2(g)$ partial pressures of $10^{-0.5}$, $10^{-1.5}$, $10^{-2.5}$, or $10^{-3.5}$ atm, siderite can form at $pe + pH$ of 5.22, 3.69, 2.19, or 0.69, respectively (Fig. 10.7). At these redox levels magnetite can transform to siderite. These transformations also tend to keep the redox of soils poised at the respective $pe + pH$ values at which the transformations occur. When $CO_2(g)$ is $10^{-3.52}$ atm and $H_4SiO_4°$ is

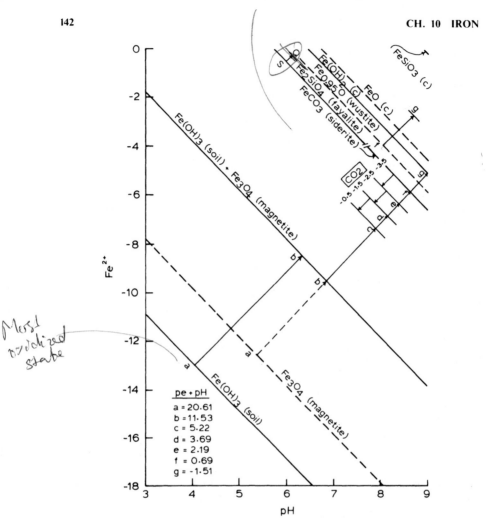

Fig. 10.7 The effect of redox, $CO_2(g)$, and silica on the solubility and stability of soil-Fe, magnetite, and various Fe(II) minerals.

controlled by soil-Si, siderite and Fe_2SiO_4 (fayalite) can coexist at $pe + pH = 0.69$. If $CO_2(g)$ drops lower, fayalite becomes the stable phase. The mineral $FeSiO_3(c)$ is much less stable than fayalite and is much too soluble to form in soils. The solubility line for fayalite shifts slightly with changes in $H_4SiO_4^\circ$ as indicated by equilibrium with Soil-Si (S) and quartz (Q), respectively.

Only if $CO_2(g)$ is excluded and $H_4SiO_4^\circ$ drops below 10^{-6} M can $Fe(OH)_2(c)$ form. Even then the $pe + pH$ must drop to -1.51 which is below the redox stability range of water. For these reasons it is virtually impossible for $Fe(OH)_2(c)$ to form in soils. The mineral $Fe_{0.95}O$(wustite) contains 0.85

moles of Fe(II) and 0.10 of Fe(III). Although wustite is more stable than $Fe(OH)_2(c)$ at $pe + pH$ of 20.61, it becomes less stable as redox drops. Wustite can never be stable in soils. The mineral FeO(c) is also metastable to $Fe(OH)_2(c)$ at 25 C.

Figure 10.7 indicates that soil-Fe, magnetite, or siderite can control the solubility of Fe^{2+} in soils depending upon redox and $CO_2(g)$. When soil-Fe is present, log Fe^{2+} increases one unit for each unit decrease in $pe + pH$. When magnetite is present log Fe^{2+} increases 2/3 of a log unit for each unit decrease in $pe + pH$. When siderite is present, Fe^{2+} is controlled by $CO_2(g)$ and decreases one log unit for each log unit increase in $CO_2(g)$.

A convenient plot depicting the solubility and stability relationships of iron minerals is given in Fig. 10.8. In this diagram, log Fe^{2+} is plotted against $pe + pH$. The lines are drawn for pH 7, but the short lines and arrows indicate that all lines shift up 2 log units for each unit decrease in pH and down 2 log units for each unit increase in pH.

An advantage of Fig. 10.8 is that it includes all Fe(III) oxides, not just soil-Fe. As the solubilities of these oxides decrease, lower redox levels are required before these oxides can be transformed to magnetite. The transformation of goethite to magnetite requires $pe + pH$ to be lower than 3.49, whereas soil-Fe can transform to magnetite at $pe + pH$ of 11.53. Freshly precipitated $Fe_3(OH)_8$(amorp) has been hypothesized to form in reduced soils. The solubility value reported for this precipitate (see the appendix) was used to plot the $Fe_3(OH)_8$ solubility line shown in Fig. 10.8. The reported solubility appears to be too high as it would require a $pe + pH < 6$ before even amorphous Fe oxide could dissolve to precipitate it.

It is hypothesized that mixed Fe(II) and Fe(III) oxides very likely precipitate in soils when they are highly reduced and that this amorphous product is less soluble than that reported for $Fe_3(OH)_8$(fresh precipitate) but more soluble than magnetite (Fig. 10.8). Evidence for such a product needs further investigation. In this text magnetite is used as the precipitating iron mineral in this redox range even though an amorphous product is more likely.

Of the Fe(III) oxides included in Fig. 10.8 only goethite is actually stable. Yet the other Fe(III) oxides are known to coexist metastably for extended periods of time. Because of the extremely low solubilities of iron oxides in the pH range of soils, the transformations among these species occur very slowly. Seldomly reduced, well-drained soils are expected to accumulate goethite and hematite with time whereas soils subject to frequent reduction will maintain more soluble iron trapped in the more soluble Fe(III) minerals or as soil-Fe.

In water-saturated soils the pore spaces are filled with water and the diffusion of gases is restricted. The depletion of $O_2(g)$ and the accumulation of $CO_2(g)$ favors the formation of $FeCO_3$(siderite) as the stable iron mineral.

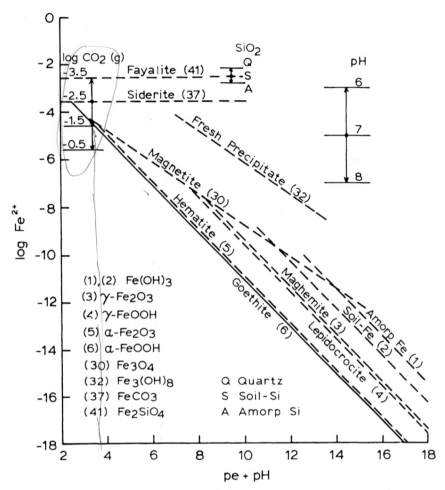

Fig. 10.8 The effect of $pe + pH$, $CO_2(g)$, and silica on the solubility of iron minerals at pH 7 showing two log units shift for each unit change in pH.

Levels of $CO_2(g)$ corresponding to 0.3 atm ($10^{-0.5}$) are not uncommon. Under such conditions even goethite can dissolve to precipitate siderite. As oxidizing conditions return, siderite will dissolve and the released iron will oxidize and precipitate amorphous iron oxide to repeat the slow recrystallization process toward stable Fe(III) oxides.

A third way of depicting the stability relationships of the various iron oxides with changes in redox is demonstrated in Fig. 10.9. Shown here are the sequence of mineral transformations that can occur in soils when they

EFFECT OF REDOX ON THE STABILITY OF IRON MINERALS

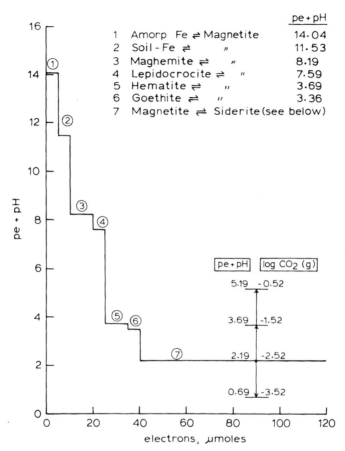

Fig. 10.9 The calculated electron titration curve of a hypothetical soil containing 15 μmoles of each of the Fe(III) oxides.

are reduced. For this example, a hypothetical soil containing 15 μmole of each of the six different Fe(III) iron oxides defined by Reactions 1 through 6 of Table 10.1 was considered. The $pe + pH$ at which each oxide becomes metastable with respect to Fe_3O_4(magnetite) is indicated by number. During the seventh transformation magnetite is converted to $FeCO_3$ (siderite). At 0.003 atm CO_2(g) this transformation occurs at $pe + pH$ 2.19. As CO_2(g) rises above $10^{-1.52} = 0.03$ atm, hematite and goethite can transform directly to siderite without first forming magnetite.

The relationships depicted in Fig. 10.9 are based on equilibrium conditions. The redox plateaus are difficult to detect experimentally because the rates of mineral transformations are generally much slower than the

rate at which the electrons are supplied. Other redox reactions that occur in soils are superimposed on those shown here. Because iron oxides comprise a major fraction of the reducible components in soils, they generally provide a major sink for electrons when soils are reduced. Recent studies in our laboratory (Sadiq, 1977; Lindsay and Sadiq, 1980) show that electron titration of soils can be used effectively to characterize the solubility relationships of iron, manganese, and other reducible soil components.
ponents.

The Fe(II) minerals: $FeCl_2$(lawrencite), $FeSO_4$(c), and $FeSO_4 \cdot 7H_2O$ (Reactions 36, 38, and 39 of Table 10.1) are much too soluble to form in soils as seen from their log $K°$ values. These minerals can be discounted as being important in soils.

10.7 HYDROLYSIS AND COMPLEXES OF Fe(II)

Ferrous iron hydrolyzes in aqueous solution to give the hydrolysis species defined by Reactions 42 through 46 of Table 10.1. Also Fe^{2+} reacts with various anions to form complexes defined by Reactions 47 through 52 of Table 10.1. These solution species are plotted together in Fig. 10.10 at $pe + pH$ of 11.53, corresponding to the redox at which soil-Fe and magnetite can coexist.

Below pH 6.75, Fe^{2+} is the major Fe(II) species in solution. Between pH 6.75 and 9.30, $FeOH^+$ is the predominant species while above pH 9.3, $Fe(OH)_2°$ is the major solution species. The hydrolysis species $Fe(OH)_3^-$ and $Fe(OH)_4^{2-}$ are quite insignificant in the pH range of soils. As the activity of Fe^{2+} rises and falls with changes in redox, so do the activities of its hydrolysis species. A log unit increase in Fe^{2+} gives a corresponding increase in all monomeric iron hydrolysis species included in Fig. 10.10. The $Fe_3(OH)_4^{2+}$ species (Reaction 46 of Table 10.1) is not sufficiently stable to be included in this plot.

A unit increase in log Fe^{2+} causes all lines in Fig. 10.10 to move upward by one log unit, while a unit decrease in log Fe^{2+} brings a corresponding decrease in activity of all species. From this figure it is easy to estimate the importance of each Fe(II) complex when the anionic activities are known.

Total soluble Fe(II) can be represented by the equation

$$[\text{Total Fe(II)}] = [Fe^{2+}] + [FeOH^+] + [Fe(OH)_2°] + [FeSO_4°]$$
$$+ [FeH_2PO_4^+] + [FeHPO_4°] \quad (10.10)$$

The $FeSO_4°$ ion pair is important even at 10^{-4} M SO_4^{2-}. The $FeH_2PO_4^+$ species becomes important below pH 7 when $H_2PO_4^-$ exceeds 10^{-5} M, and

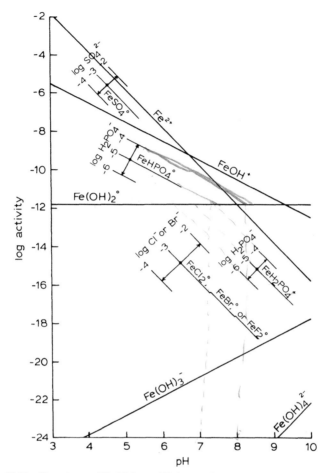

Fig. 10.10 Complexes of Fe(II) in equilibrium with Soil-Fe at $pe + pH$ of 11.53.

$FeHPO_4^°$ becomes slightly significant in the pH range of 7 to 8 when either $H_2PO_4^-$ or HPO_4^{2-} exceeds 10^{-5} M.

Although several species contribute significantly to total Fe(II) in solution as shown in Fig. 10.10, only $FeOH^+$ is of major importance above pH 6.75.

As indicated earlier Fig. 10.10 was constructed on the basis that Fe^{2+} is in equilibrium with soil-Fe and magnetite, which occurs at $pe + pH$ of 11.53. For each unit increase in $pe + pH$, the lines in this figure drop by one log unit as soil-Fe controls Fe^{2+} according to Eq. 10.7. As redox goes below $pe + pH$ of 11.53, magnetite controls Fe^{2+} according to the reaction

$$\log Fe^{2+} = 11.90 - 0.67(pe + pH) - 2pH \qquad (10.11)$$

indicating an upward shift of 0.67 log units for each unit decrease in $pe + pH$.

Figure 10.10 is useful in estimating the significance of the Fe(II) complexes in solution as a function of pH, $pe + pH$, and the activities of various anions. The complexes in Fig. 10.10 and 10.5 can be summed to give an equation for total iron in solution. The species that must be included vary with pH, redox, and the anionic activities of $H_2PO_4^-$ and SO_4^{2-}.

A comparison of Fig. 10.10 and 10.5 shows that in near neutral soils $FeOH^+$ is the major Fe(II) species and it is slightly higher than $Fe(OH)_2^+$ which is the major Fe(III) species. The exact redox at which these two species coincide can be obtained from the reactions:

		$\log K°$
$Fe^{3+} + 2H_2O \rightleftharpoons Fe(OH)_2^+ + 2H^+$		-5.69
$Fe^{2+} \rightleftharpoons Fe^{3+} + e^-$		-13.04
$FeOH^+ + H^+ \rightleftharpoons Fe^{2+} + H_2O$		6.74
$FeOH^+ + H_2O \rightleftharpoons Fe(OH)_2^+ + e^- + H^+$		-11.99	(10.12)

Thus the activities of $FeOH^+$ and $Fe(OH)_2^+$ are equal when

$$pe + pH = 11.99 \tag{10.13}$$

According to these reactions redox does not really affect total soluble iron in soils until $pe + pH$ drops below 12. This conclusion is independent of which solid phases govern iron solubility.

The solubility relationships of iron in soil are rather complex, but they can be understood by developments such as those presented in this chapter. Further investigations are needed to examine many of the theoretical developments proposed herein. Iron is also a component of many ferrite minerals in which various divalent metal cations can be included (see Chapters 13 and 14). Since iron is present in soils in much greater quantities than many of the divalent metals such as zinc, copper, etc. it is highly unlikely that the ferrite minerals will control iron solubility, but iron is expected to have an important influence on the solubility of many divalent metal cations.

REFERENCES

Breeman, N. van. 1976. Genesis and solution chemistry of acid sulfate soils in Thailand. Agric. Res. Rept. (Versl. landbouwk. Onderz.) 848. Wageningen, The Netherlands.

Lindsay, W. L. and M. Sadiq. 1980. The use of pe + pH as a redox parameter in soils. Unpublished.

Norvell, W. A. and W. L. Lindsay. 1981. Estimation of iron (III) solubility from EDTA chelate equilibria in soils. Soil Sci. Soc. Am. J. 45; (In Press).

Ponnamperuma, F. N., E. M. Tianco, and T. Loy. 1967. Redox equilibria in flooded soils: I. The iron hydroxide systems. Soil Sci. 103:374–382.

Sadiq, M. 1977. Use of electron titration to study Fe and Mn in soils. Ph.D. Thesis, Department of Agronomy, Colorado State University.

Schwertmann, U. and R. M. Taylor. 1977. Iron oxides. *In* J. B. Dixon and S. B. Weed, Eds., Minerals in Soil Environments. Soil Science Society of America, Madison, Wisconsin.

PROBLEMS

10.1 Develop the necessary equations and plot Fe^{3+} and $Fe(OH)_2^+$ in equilibrium with goethite as a function of pH. Indicate how these species shift if soil-Fe rather than goethite governs the solubility of iron.

10.2 Calculate the highest pH at which jarosite can form if SO_4^{2-} activity is $10^{-2.3}$ M, K^+ is $10^{-2.5}$ M, and $Fe(OH)_3$(amorp) is present. How does this pH shift if log Fe^{3+} + 3pH is decreased by 1 unit?

10.3 Using the Davies' estimate of activity coefficients at an ionic strength of 0.05 develop the necessary equations to calculate total Fe(III) in solution at pH 4 when the activity of SO_4^{2-} is 10^{-2} M, $H_2PO_4^-$ is $10^{-5.2}$ M, Cl^- is 10^{-2} M, and Fe^{3+} is governed by $Fe(OH)_3$(soil). Estimate the mole fraction of each species that contributes greater than 1% to total soluble Fe(III).

10.4 Develop the equations and plot log Fe^{2+} versus pH for
a. Equilibrium with $Fe(OH)_3$(soil) at pe + pH of 15.
b. Siderite in equilibrium with CO_2(g) at 0.01 atm.

10.5 Stipulate the equilibrium conditions that would be necessary for fayalite, quartz, magnetite, and siderite to coexist.

10.6 From Fig. 10.8 discuss conditions that very likely prevent goethite and hematite from becoming the predominant iron minerals in soils.

10.7 Modify the electron titration curve in Fig. 10.9 for a soil having a CO_2 partial pressure of 0.1 atm.

10.8 Prepare a plot of Fe^{2+}, $FeSO_4^\circ$, and $FeOH^+$ in equilibrium with siderite at 0.01 atm CO_2(g) when SO_4^{2-} activity is 10^{-3} M.

10.9 Prepare an activity plot of $Fe(OH)_2^+$ and $FeOH^+$ versus pH in equilibrium with soil-Fe at pe + pH = 15. Show how these lines shift sequentially as redox is lowered so that magnetite and eventually siderite at 0.03 atm of CO_2(g) controls iron solubility.

10.10 Prepare a plot of log Fe^{3+} versus pe + pH for a soil at pH 7 as soil-Fe, magnetite, and siderite at 0.3 and 0.003 atm CO_2(g) each governs iron solubility. Show how these relationships shift for a soil of pH 6.0.

ELEVEN

MANGANESE

The manganese content of the lithosphere is approximately 900 ppm, and soils generally contain from 20 to 3000 ppm with an average of 600 ppm (Table 1.1). This average would supply 10^{-1} M manganese in the soil solution at 10% moisture if it all dissolved. Manganese is less abundant than either aluminum or iron.

The chemistry of manganese in soils is complex because three oxidation states are involved: Mn(II), Mn(III), and Mn(IV). Manganese forms hydrated oxides with mixed valency states. Considerable work has been done to characterize the thermodynamic data on manganese (Bricker, 1965; Zordon and Hepler, 1968) and to study the reactions of manganese in flooded soils (Ponnamperuma et al., 1969; Patrick and Turner, 1968; Gotoh and Patrick, 1972) and in nonflooded soils (Bohn, 1970).

In this chapter attempts will be made to show how manganese solubility relationships are affected by pH, redox, and complexation. The solubility of manganese phosphates, chelates, sulfides, and molybdates are considered in Chapters 12, 15, 17, and 22, respectively.

11.1 EFFECT OF REDOX AND pH ON MANGANESE SOLUBILITY

Common minerals of manganese includes oxides, carbonates, silicates, and sulfates (Taylor et al., 1964; McKenzie, 1977). The solubilities of several manganese minerals are given by Reactions 1 through 15 of Table 11.1 and are plotted by Fig. 11.1.

TABLE 11.1 EQUILIBRIUM REACTIONS OF MANGANESE SPECIES AT 25°C

Reaction No.	Chemical Reaction	log $K°$
	Oxides and Hydroxides	
1	β-MnO$_2$(pyrolusite) + 4H$^+$ + 2e$^-$ ⇌ Mn^{2+} + 2H$_2$O	41.89
2	γ-MnO$_{1.9}$(nsutite) + 3.8H$^+$ + 1.8e$^-$ ⇌ Mn^{2+} + 1.9H$_2$O	38.89
3	δ-MnO$_{1.8}$(birnessite) + 3.6H$^+$ + 1.6e$^-$ ⇌ Mn^{2+} + 1.8H$_2$O	35.38
4	γ-MnOOH(manganite) + 3H$^+$ + e$^-$ ⇌ Mn^{2+} + 2H$_2$O	25.27
5	Mn$_2$O$_3$(bixbyite) + 6H$^+$ + 2e$^-$ ⇌ 2Mn^{2+} + 3H$_2$O	51.46
6	Mn$_3$O$_4$(hausmannite) + 8H$^+$ + 2e$^-$ ⇌ 3Mn^{2+} + 4H$_2$O	63.03
7	Mn(OH)$_2$(pyrochroite) + 2H$^+$ ⇌ Mn^{2+} + 2H$_2$O	15.19
8	MnO(manganosite) + 2H$^+$ ⇌ Mn^{2+} + H$_2$O	18.39

TABLE 11.1 (*Continued*)

Reaction No.	Chemical Reaction	log $K°$
	Other Mn(II) Minerals	
9	$MnCO_3(\text{rhodochrosite}) + 2H^+ \rightleftharpoons Mn^{2+} + CO_2(g) + H_2O$	8.08
10	$MnSiO_3(\text{rhodonite}) + 2H^+ + H_2O \rightleftharpoons Mn^{2+} + H_4SiO_4°$	10.25
11	$Mn_2SiO_4(\text{tephroite}) + 4H^+ \rightleftharpoons 2Mn^{2+} + H_4SiO_4°$	24.45
12	$MnSO_4(c) \rightleftharpoons Mn^{2+} + SO_4^{2-}$	3.43
13	$MnSO_4 \cdot H_2O(c) \rightleftharpoons Mn^{2+} + SO_4^{2-} + H_2O$	0.47
	Other Mn(III) Minerals	
14	$Mn_2(SO_4)_3(c) \rightleftharpoons 2Mn^{3+} + 3SO_4^{2-}$	−11.56
15	$Mn_2(SO_4)_3(c) + 2e^- \rightleftharpoons 2Mn^{2+} + 3SO_4^{2-}$	39.54
	Redox Reactions	
16	$Mn^{4+} + e^- \rightleftharpoons Mn^{3+}$	25.51
17	$Mn^{3+} + e^- \rightleftharpoons Mn^{2+}$	25.55
18	$Mn^{4+} + 2e^- \rightleftharpoons Mn^{2+}$	51.06
19	$Mn^{2+} + 2e^- \rightleftharpoons \alpha\text{-Mn}(c)$	−40.40
20	$MnO_4^{2-} + 8H^+ + 4e^- \rightleftharpoons Mn^{2+} + 4H_2O$	118.31
21	$MnO_4^- + 8H^+ + 5e^- \rightleftharpoons Mn^{2+} + 4H_2O$	127.71
	Hydrolysis Reactions	
22	$Mn^{2+} + H_2O \rightleftharpoons MnOH^+ + H^+$	−10.95
23	$Mn^{2+} + 2H_2O \rightleftharpoons Mn(OH)_2° + 2H^+$	
24	$Mn^{2+} + 3H_2O \rightleftharpoons Mn(OH)_3^- + 3H^+$	−34.00
25	$Mn^{2+} + 4H_2O \rightleftharpoons Mn(OH)_4^{2-} + 4H^+$	−48.29
26	$2Mn^{2+} + H_2O \rightleftharpoons Mn_2OH^{3+} + H^+$	−10.60
27	$2Mn^{2+} + 3H_2O \rightleftharpoons Mn_2(OH)_3^+ + 3H^+$	−23.89
28	$Mn^{3+} + 3H_2O \rightleftharpoons MnOH^{2+} + 3H^+$	0.40
	Mn^{2+} Complexes	
29	$Mn^{2+} + Cl^- \rightleftharpoons MnCl^+$	0.61
30	$Mn^{2+} + 2Cl^- \rightleftharpoons MnCl_2°$	0.04
31	$Mn^{2+} + CO_2(g) + H_2O \rightleftharpoons MnHCO_3^+ + H^+$	−6.02
32	$Mn^{2+} + CO_2(g) + H_2O \rightleftharpoons MnCO_3° + 2H^+$	−18.87
33	$Mn^{2+} + SO_4^{2-} \rightleftharpoons MnSO_4°$	2.26

EFFECT OF REDOX AND pH ON MANGANESE SOLUBILITY

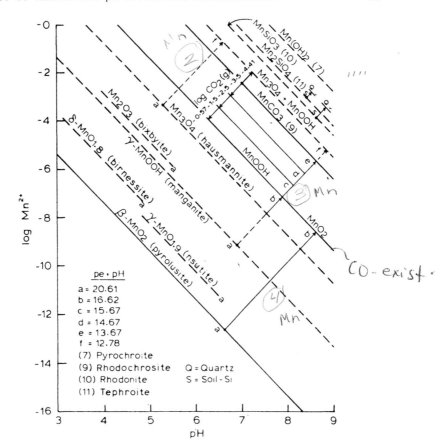

Fig. 11.1 The effect of redox, $CO_2(g)$, and silica on the solubility and stability of manganese minerals in soils.

The most stable manganese mineral under well-oxidized conditions is pyrolusite. The solubilities of $\beta\text{-}MnO_2$(pyrolusite), $\gamma\text{-}MnOOH$(manganite), Mn_3O_4(hausmannite), and other minerals are represented at $pe + pH$ of 20.61 by lines with a small a in Fig. 11.1. As $pe + pH$ decreases, all of these solubility lines move upward, but they move at different rates. For each unit decrease in $pe + pH$, $\beta\text{-}MnO_2$ moves up two log units, $\gamma\text{-}MnOOH$ one unit, and Mn_3O_4 two-thirds unit, reflecting the number of electrons required to reduce manganese to Mn(II). The solubility lines of the more oxidized minerals eventually surpass those of the less oxidized ones. When $pe + pH$ reaches 16.62, the solubility lines of $\beta\text{-}MnO_2$ and $\gamma\text{-}MnOOH$ coincide at line b. At this redox both minerals can coexist at equilibrium. This redox value is obtained by combining Reactions 1 and 4 of Table 11.1:

$$\log K°$$

$$\beta\text{-MnO}_2(\text{pyrolusite}) + 4\text{H}^+ + 2e^- \rightleftharpoons \text{Mn}^{2+} + 2\text{H}_2\text{O} \qquad 41.89$$

$$\text{Mn}^{2+} + 2\text{H}_2\text{O} \rightleftharpoons \gamma\text{-MnOOH(manganite)} + 3\text{H}^+ + e^- \qquad -25.27$$

$$\beta\text{-MnO}_2(\text{pyrolusite}) + \text{H}^+ + e^- \rightleftharpoons \gamma\text{-MnOOH(manganite)} \qquad 16.62 \quad (11.1)$$

$$pe + pH = 16.62 \qquad (11.2)$$

Equations 11.1 and 11.2 show that upon reduction a soil tends to be poised at $pe + pH$ of 16.62. At this redox pyrolusite dissolves, Mn^{4+} is reduced to Mn^{3+}, and manganite is precipitated. Since Eq. 11.2 is based on equilibrium conditions, the electrons must be supplied sufficiently slow that near equilibrium conditions are maintained, otherwise the redox will drop below 16.62 before the reactions are completed. With further reduction the solubility line of manganite moves up as indicated by the dashed lines and arrows and eventually coincides with that of MnCO_3(rhodochrosite). This occurs at lines b, c, d, e, or f when log $\text{CO}_2(g)$ is -0.57, -1.5, -2.5, -3.5, and -4.41 atm, respectively. The corresponding $pe + pH$ values for each of these lines are also shown in Fig. 11.1. For example, when $\text{CO}_2(g)$ is $10^{-3.5}$ atm and $pe + pH$ reaches 13.67, manganite and rhodochrosite can coexist. If $\text{CO}_2(g)$ is fixed and $pe + pH$ continues to drop, Mn^{2+} solubility will be governed by rhodochrosite while manganite will dissolve. If $\text{CO}_2(g)$ falls below $10^{-4.41}$ atm, manganite will control Mn^{2+} solubility until Mn_3O_4(hausmannite) forms at line f, corresponding to a $pe + pH$ of 12.78. Lower CO_2 and redox levels are necessary for MnSiO_3(rhodonite) and/or Mn_2SiO_4(tephroite) to form. When $\text{H}_4\text{SiO}_4°$ is controlled by quartz at 10^{-4} M, the two silicates have approximately equal solubilities.

For Mn(OH)_2(pyrochroite) to form, it is necessary for $\text{CO}_2(g)$ to be $<10^{-7.11}$ atm and for $\text{H}_4\text{SiO}_4°$ to be $<10^{-5.94}$ M. The probability of these conditions existing in soils is extremely rare. Soils that attain low redox levels are usually waterlogged. Under these conditions, the $\text{CO}_2(g)$ generally exceeds that found in air ($10^{-3.52}$ atm).

The solubility of MnO(manganosite) is much too high for this mineral to form in soils (Reaction 8 of Table 11.1) as >1 M concentration of Mn^{2+} would be required. The solubility of Mn^{2+} in soils appears to be governed by pyrolusite above $pe + pH$ of 16.62, by manganite at slightly lower redox, and by rhodochrosite at still lower redox depending upon CO_2.

The mineral Mn_2O_3(bixbyite) contains manganese in the Mn(III) oxidation state and is slightly more soluble than γ-MnOOH (Fig. 11.1). For these reasons bixbyite is always metastable to manganite and is expected to slowly revert to it. Bixbyite may form temporarily when highly reduced soils containing rhodochrosite are oxidized providing bixbyite can precipitate faster than manganite or other higher valency oxides.

The minerals birnessite ($MnO_{1.8}$) and nsutite ($MnO_{1.9}$) contain mainly Mn(IV) with only small amounts of Mn(III). These minerals are slightly more soluble than pyrolusite. As reduction occurs, the solubility lines of these minerals move upward, but they move slower than pyrolusite. The pyrolusite line coincides with nsutite at pe + pH of 15.00 and with birnessite at 16.28. Since pyrolusite transforms to manganite at pe + pH of 16.62, nsutite and birnessite can never be stable phases.

It is possible for mixed valency manganese oxides to form in reduced soils as they become oxidized. In the pe + pH range of 15 to 16 the Mn(IV) minerals, manganite and rhodochrosite may all coexist. This probably accounts for the fact that manganese minerals in soils are so difficult to characterize (McKenzie, 1977).

A second way of showing the effect of redox on the stability of manganese minerals is demonstrated by Fig. 11.2, which is drawn for pH 7.0. In the pe + pH range near 16.6 β-MnO_2(pyrolusite), γ-MnOOH(manganite), γ-$MnO_{1.9}$(nsutite), and δ-$MnO_{1.8}$(birnessite) have similar solubilities. At redox levels above pe + pH of 16.62 pyrolusite is the stable phase, whereas below this redox manganite is stable. Nsutite and birnessite are unstable at both higher and lower redox.

When manganite is present, Mn^{2+} increases one log unit for each unit decrease in pe + pH. The relationship is readily obtained from Reaction 4 of Table 11.1 which gives

$$\log Mn^{2+} = 25.27 - (pe + pH) - 2pH \tag{11.3}$$

In the presence of $10^{-3.52}$ atm of $CO_2(g)$, manganite is stable from pe + pH of 16.62 to 13.67 below which $MnCO_3$(rhodocrosite) precipitates. The formation of rhodochrosite at pH 7 fixes Mn^{2+} at $10^{-2.4}$ M. Increasing log $CO_2(g)$ by one unit decreases log Mn^{2+} in equilibrium with $MnCO_3$ by one unit and increases the pe + pH at which manganite and rhodochrosite can coexist by one unit. If $CO_2(g)$ is $> 10^{-0.52}$ atm (0.3 atm), γ-MnOOH cannot form regardless of redox.

The minerals Mn_2O_3(bixbyite), Mn_3O_4(hausmannite), and $Mn(OH)_2$ (pyrochrosite) are not stable in normal soils because $CO_2(g)$ seldomly drops below that of normal air ($10^{-3.52}$ atm).

The relationships depicted in Fig. 11.2 are helpful in understanding the effects of pH, redox, and $CO_2(g)$ on the solubility and stability of manganese

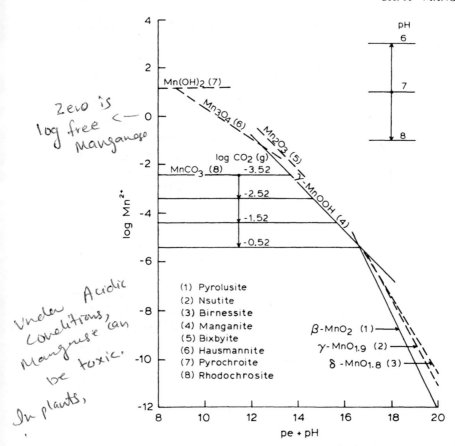

Fig. 11.2 The effect of redox and $CO_2(g)$ on the solubility and stability of manganese minerals at pH 7 showing shifts of two log units for each unit change in pH.

minerals. All lines in this diagram shift upward by 2 log units for each unit decrease in pH as shown by the short lines and arrows in the upper right of the diagram. Obviously manganese becomes very soluble under acid, low redox conditions. Fortunately low redox is only attained in submerged or waterlogged soils where the pH shifts toward neutral and $CO_2(g)$ rises. These shifts favor the precipitation of rhodochrosite which limits Mn^{2+} solubility. If it were not for these shifts, manganese toxicities would be prevalent in highly reduced soils.

As mentioned earlier the average manganese content of soils (600 ppm) provides $10^{-1.0}$ M manganese if it were to completely dissolve in the soil solution at 10% moisture (Table 1.1). Similarly, if 10 g of soil with this manganese content were extracted with 100 ml of solution, the maximum concentration of manganese would be $0.1 \times 1/100 = 10^{-3}$ M. Keeping in mind the total amount of an element present in soils helps to limit the discussion of solubilities to realistic ranges. For example, Fig. 11.2 can be used to predict that rhodochrosite will not form in a soil of pH 6 in equilibrium with atmospheric $CO_2(g)$. At this pH and $CO_2(g)$ rhodochrosite would require $10^{-0.4}$ M Mn^{2+} which is greater than the $10^{-1.0}$ M total manganese in this soil. Increasing $CO_2(g)$, however, permits rhodochrosite to form.

The sulfate minerals $MnSO_4(c)$, $MnSO_4 \cdot H_2O(c)$, and $Mn_2(SO_4)_2(c)$, whose solubilities are given by Reactions 12 through 15 of Table 11.1, are too soluble to precipitate in soils at the sulfate and redox levels normally encountered in soils. Thus manganese sulfates are not expected to be found in soils.

11.2 SOLUTION SPECIES OF MANGANESE

Manganese forms many solution species. Those expected to be of greatest importance in soils are included in Reactions 16 through 33 of Table 11.1 and are plotted in Figs. 11.3 and 11.4. These plots were developed at pe + pH of 16.62, the redox at which manganite and pyrolusite coexist. At this redox it is evident that Mn^{2+} is the predominant solution species, and the hydrolysis species of manganese are of only minor importance in soils. In alkaline soils at high $CO_2(g)$, $MnHCO_3^+$ may become significant. Even at 10^{-2} M Cl^-, $MnCl^+$ contributes only slightly to total soluble manganese. Both $MnCl_2^{\circ}$ and $MnCO_3^{\circ}$ are present at very low levels.

Selection of pe + pH of 16.62 for plotting the solution species in Figs. 11.3 and 11.4 has several advantages: (1) it is representative of a large number of moderate-to-well-drained soils, (2) shifts to lower redox can be projected along the manganite equilibrium lines, and (3) shifts to higher redox can be projected along the pyrolusite equilibrium line (see Fig. 11.2). As redox changes many of the solution species in these figures shift depending upon the difference in oxidation state of manganese in the solid phase controlling manganese solubility and the solution species. This shift can be calculated from the formula

$$\Delta \log a_i = \Delta(pe + \text{pH})n(Z_i - Z_s) \quad (11.4)$$

where $\Delta \log a_i$ is the change in log activity of solution species (i) with change in redox, $\Delta(pe + \text{pH})$; Z_i is the valency of manganese in the solution species (i); n is the number of manganese atoms in that species; and Z_s is the valency of manganese in the solid phase, that is $+3$ for manganite or $+4$ for pyrolusite. To demonstrate the use of Eq. 11.4 let us consider a soil having a

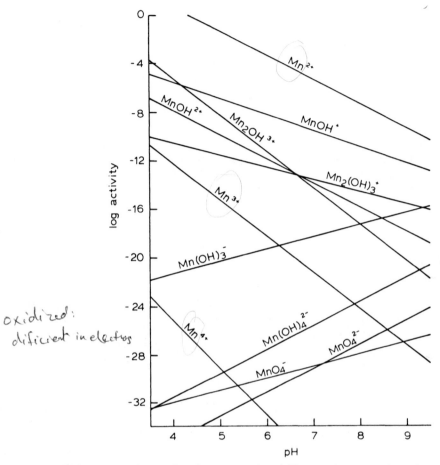

Fig. 11.3 Solution species of manganese in equilibrium with manganite and pyrolusite at $pe + pH$ of 16.62.

$pe + pH$ of 15.62 and look at the species $MnOH^+$. According to Eq. 11.4:

$$\Delta \log MnOH^+ = (15.62 - 16.62)(1)(2 - 3) = 1$$

Thus the $MnOH^+$ line in Fig. 11.3 moves upward by one log unit. If $pe + pH$ increases to 18.62, pyrolysite becomes the stable phase and Eq. 11.4 gives

$$\Delta \log MnOH^+ = (18.62 - 16.62)(1)(2 - 4) = -4$$

indicating that the $MnOH^+$ line in Fig. 11.3 shifts downward by four log units. In this manner Figs. 11.3 and 11.4 along with Eq. 11.4 can be used to estimate the activity of manganese solution species throughout the redox range in which either manganite or pyrolusite are stable.

SOLUTION SPECIES OF MANGANESE

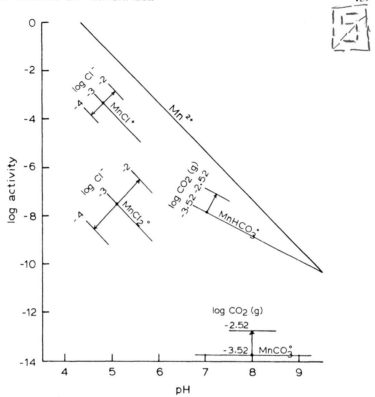

Fig. 11.4 Chloride, carbonate, and bicarbonate complexes of Mn^{2+} in equilibrium with manganite and pyrolusite at $pe + pH$ of 16.62.

Should the redox drop sufficiently that $MnCO_3$(rhodochrosite) forms (Fig. 11.2), then Eq. 11.4 must be modified to include two shifts: one while manganite is the stable phase and the other while rhodochrosite is the stable phase. The expression becomes rather complex and adds little useful information. As soon as rhodochrosite forms, Mn^{2+} and other divalent species are fixed by pH and $CO_2(g)$ while all higher valency manganese species decrease.

The equilibrium $pe + pH$ at which manganite and rhodochrosite can coexist depends upon $CO_2(g)$ and is obtained by combining Reactions 4 and 9 of Table 11.1 to give

$$\gamma\text{-MnOOH(manganite)} + H^+ + e^- + CO_2(g) \rightleftharpoons$$
$$MnCO_3(\text{rhodochrosite}) + H_2O \quad (11.5)$$
$$\log K° = 17.19$$

from which

$$pe + \text{pH} = 17.19 + \log CO_2(g) \tag{11.6}$$

Thus the redox at which manganite and rhodochrosite can coexist is $CO_2(g)$ dependent.

There are other known manganese minerals in addition to those considered here. When reliable solubility values are available, they can be included in these diagrams. The present development, however, clearly shows the importance of pH and redox on manganese solubility.

REFERENCES

Bohn, H. L. 1970. Comparisons of measured and theoretical Mn^{2+} concentrations in soil suspensions. Soil Sci. Soc. Am. Proc. 34:195-197.

Bricker, O. 1965. Some stability relations in the system $Mn-O_2-H_2O$ at 25°C and one atmosphere total pressure. Am. Miner. 50:1296-1354.

Gotoh, S. and W. H. Patrick, Jr. 1972. Transformation of manganese in a waterlogged soil as affected by redox potential and pH. Soil Sci. Soc. Am. Proc. 36:738-742.

McKenzie, R. M. 1977. The manganese oxides and hydroxides. *In* J. B. Dixon and S. B. Weed, Eds., *Minerals in Soil Environments*, pp. 181-193. Soil Science Society of America, Madison, WI.

Patrick, W. H. Jr. and F. T. Turner. 1968. Effect of redox potential on manganese transformation in waterlogged soil. Nature 220:476-478.

Ponnamperuma, F. N., T. A. Loy, and E. M. Tiaco. 1969. Redox equilibria in flooded soils. II. The MnO_2 system. Soil Sci. 108:48-57.

Taylor, R. M., R. M. McKenzie, and K. Norrish. 1964. The mineralogy and chemistry of manganese in some Australia soils. Aust. J. Soil Res. 2:235-248.

Zordan. T. A., and L. G. Hepler. 1968. Thermochemistry and oxidation potentials of manganese and its compounds. Chem. Rev. 68:737-745.

PROBLEMS

11.1 Develop the equilibrium relationship to prove that nsutite can never be a stable phase in soils.

11.2 Under what equilibrium conditions can tephroite, rhodonite, and rhodochrosite coexist?

11.3 Derive the necessary equations and plot at pH 7 $\log Mn^{2+}$ vs $(pe + \text{pH})$ for the following
 a. MnO_2(pyrolusite).
 b. MnOOH(manganite).
 c. $MnCO_3$ (rhodochrosite) with CO_2 at 0.3 and 0.003 atm.

PROBLEMS

11.4 Calculate a theoretical electron titration curve for manganese when 10 g of soil containing 1162 ppm of pyrolusite is suspended in 100 ml of solution buffered at pH 7 and slowly reduced with a platinum cathode. Assume that $CO_2(g)$ is fixed at $10^{-3.52}$ atm and that activity coefficients are unity.

11.5 How does the Mn^{4+} line in Fig. 11.3 change
   a. For a redox of 20.61.
   b. For a redox of 14 and a $CO_2(g)$ of $10^{-3.52}$ atm.
   c. For a redox of 14 and a $CO_2(g)$ of $10^{-1.52}$ atm.

11.6 What equilibrium $pe + pH$ would be imposed on a soil of pH 6 by extraction with 10^{-2} M $KMnO_4$?

Soil Poise.

TWELVE

PHOSPHATES

The phosphorus content of the lithosphere is approximately 1200 ppm, and that of soils generally ranges from 200 to 5000 ppm with an average of 600 ppm (Table 1.1).

The phosphorus content of soils is intermediate between the major and the trace elements constituents. For this reason it is convenient to consider phosphorus reactions in two categories: (1) reactions involving cations that are more abundant than phosphorus and may control phosphorus solubility and (2) reactions involving cations that are less abundant than phosphorus and normally cannot control phosphorus solubility.

Only reactions of the first category, which include aluminum, iron, calcium, magnesium, potassium, and manganese, are considered in this chapter. Other trace element reactions with phosphorus are given in subsequent chapters. Solution complexes of phosphorus are considered with their respective cations regardless of the abundance of the cations. Complexes that may contribute significantly to total phosphorus in solution are briefly summarized in this chapter. Most readers will appreciate this treatment as they come to realize the many phosphorus minerals and complexes that form in soils.

12.1 ORTHOPHOSPHORIC ACID

It will be shown that orthophosphate is the stable form of phosphorus in soils. For this reason orthophosphoric acid provides a good starting point for considering phosphorus reactions in soils. The dissociation reactions of orthophosphoric acid are given by Reactions 1 through 4 of Table 12.1 and are plotted in the mole fraction diagram in Fig. 12.1. An example of how this diagram was developed is given for $H_2PO_4^-$:

$$MF_{H_2PO_4^-} = \frac{H_2PO_4^-}{H_3PO_4^\circ + H_2PO_4^- + HPO_4^{2-} + PO_4^{3-}} \quad (12.1)$$

where $MF_{H_2PO_4^-}$ is the mole fraction of total orthophosphate present as $H_2PO_4^-$. Each term on the right side of this equation can be expressed in terms of $H_2PO_4^-$, H^+, and an appropriate equilibrium constant. When this is done, $H_2PO_4^-$ can be factored from both numerator and denominator and cancelled. When activity coefficients are considered as unity, the resulting simplified expression becomes:

$$MF_{H_2PO_4^-} = \frac{1}{10^{2.15}(H^+) + 1 + \dfrac{10^{-7.20}}{(H^+)} + \dfrac{10^{-19.55}}{(H^+)^2}} \quad (12.2)$$

TABLE 12.1 EQUILIBRIUM CONSTANTS FOR VARIOUS PHOSPHATE REACTIONS AT 25°C

Reaction No.	Equilibrium Reaction	$\log K°$
	Orthophosphoric Acid	
1	$H_3PO_4^° \rightleftharpoons H^+ + H_2PO_4^-$	−2.15
2	$H_2PO_4^- \rightleftharpoons H^+ + HPO_4^{2-}$	−7.20
3	$HPO_4^{2-} \rightleftharpoons H^+ + PO_4^{3-}$	−12.35
4	$H_2PO_4^- \rightleftharpoons 2H^+ + PO_4^{3-}$	−19.55
5	$2H_2PO_4^- \rightleftharpoons (H_2PO_4)_2^{2-}$	−0.35
	Aluminum Phosphates	
6	$AlPO_4(\text{berlinite}) + 2H^+ \rightleftharpoons Al^{3+} + H_2PO_4^-$	0.50
7	$AlPO_4 \cdot 2H_2O(\text{variscite}) + 2H^+ \rightleftharpoons Al^{3+} + H_2PO_4^- + 2H_2O$	−2.50
8	$H_6K_3Al_5(PO_4)_8 \cdot 18H_2O(\text{K-taranakite}) + 10H^+ \rightleftharpoons 3K^+ + 5Al^{3+} + 8H_2PO_4^- + 18H_2O$	−22.30
9	$H_6(NH_4)_3Al_5(PO_4)_8 \cdot 18H_2O(\text{NH}_4\text{-taranakite}) + 10H^+ \rightleftharpoons 3NH_4^+ + 5Al^{3+} + 8H_2PO_4^- + 18H_2O$	−19.10
	Iron Phosphates	
10	$FePO_4(c) + 2H^+ \rightleftharpoons Fe^{3+} + H_2PO_4^-$	−5.37
11	$FePO_4 \cdot 2H_2O(\text{strengite}) + 2H^+ \rightleftharpoons Fe^{3+} + H_2PO_4^- + 2H_2O$	−6.85
12	$Fe_3(PO_4)_2 \cdot 8H_2O(\text{vivianite}) + 4H^+ \rightleftharpoons 3Fe^{2+} + 2H_2PO_4^- + 8H_2O$	3.11

Calcium Phosphates

13	$Ca(H_2PO_4)_2 \cdot H_2O(MCP) \rightleftharpoons Ca^{2+} + 2H_2PO_4^- + H_2O$	−1.15
14	$CaHPO_4 \cdot 2H_2O(brushite) + H^+ \rightleftharpoons Ca^{2+} + H_2PO_4^- + 2H_2O$	0.63
15	$CaHPO_4(monetite) + H^+ \rightleftharpoons Ca^{2+} + H_2PO_4^-$	0.30
16	$Ca_4H(PO_4)_3 \cdot 2.5H_2O(octacalcium\ phosphate) + 5H^+ \rightleftharpoons 4Ca^{2+} + 3H_2PO_4^- + 2.5H_2O$	11.76
17	$\alpha\text{-}Ca_3(PO_4)_2(c) + 4H^+ \rightleftharpoons 3Ca^{2+} + 2H_2PO_4^-$	13.61
18	$\beta\text{-}Ca_3(PO_4)_2(c) + 4H^+ \rightleftharpoons 3Ca^{2+} + 2H_2PO_4^-$	10.18
19	$Ca_5(PO_4)_3OH(hydroxapatite) + 7H^+ \rightleftharpoons 5Ca^{2+} + 3H_2PO_4^- + H_2O$	14.46
20	$Ca_5(PO_4)_3F(fluorapatite) + 6H^+ \rightleftharpoons 5Ca^{2+} + 3H_2PO_4^- + F^-$	−0.21

Magnesium Phosphates

21	$MgHPO_4 \cdot 3H_2O(newberryite) + H^+ \rightleftharpoons Mg^{2+} + H_2PO_4^- + 3H_2O$	1.38
22	$MgKPO_4 \cdot 6H_2O(c) + 2H^+ \rightleftharpoons Mg^{2+} + K^+ + H_2PO_4^- + 6H_2O$	8.93
23	$MgNH_4PO_4 \cdot 6H_2O(struvite) + 2H^+ \rightleftharpoons Mg^{2+} + NH_4^+ + H_2PO_4^- + 6H_2O$	6.40
24	$Mg_3(PO_4)_2(c) + 4H^+ \rightleftharpoons 3Mg^{2+} + 2H_2PO_4^-$	24.51
25	$Mg_3(PO_4)_2 \cdot 8H_2O(bobierrite) + 4H^+ \rightleftharpoons 3Mg^{2+} + 2H_2PO_4^- + 8H_2O$	14.10
26	$Mg_3(PO_4)_2 \cdot 22H_2O(c) + 4H^+ \rightleftharpoons 3Mg^{2+} + 2H_2PO_4^- + 22H_2O$	16.01

Manganese Phosphates

27	$MnHPO_4(c) + H^+ \rightleftharpoons Mn^{2+} + H_2PO_4^-$	−5.74
28	$Mn_3(PO_4)_2(c) + 4H^+ \rightleftharpoons 3Mn^{2+} + 2H_2PO_4^-$	11.78

(Continued)

TABLE 12.1 (Continued)

Reaction No.	Equilibrium Reaction	log $K°$
	Potassium Phosphates	
29	$KH_2PO_4(c) \rightleftharpoons K^+ + H_2PO_4^-$	−0.21
30	$K_2HPO_4(c) + H^+ \rightleftharpoons 2K^+ + H_2PO_4^-$	10.80
	Phosphorous Acid	
31	$H_3PO_3^° \rightleftharpoons H^+ + H_2PO_3^-$	−1.50
32	$H_2PO_3^- \rightleftharpoons H^+ + HPO_3^{2-}$	−6.79
33	$H_2PO_4^- + 2H^+ + 2e^- \rightleftharpoons H_2PO_3^- + H_2O$	−10.87
	Elemental Phosphorus and Phosphine	
34	$H_3PO_4^° + 5H^+ + 5e^- \rightleftharpoons P(c, white) + 4H_2O$	−35.21
35	$H_2PO_4^- + 9H^+ + 8e^- \rightleftharpoons PH_3(g) + 4H_2O$	−36.73
36	$PH_3(g) \rightleftharpoons PH_3^°$	2.09
	Pyrophosphoric Acid	
37	$H_4P_2O_7^° \rightleftharpoons H^+ + H_3P_2O_7^-$	−0.80
38	$H_3P_2O_7^- \rightleftharpoons H^+ + H_2P_2O_7^{2-}$	−2.28
39	$H_2P_2O_7^{2-} \rightleftharpoons H^+ + HP_2O_7^{3-}$	−6.70
40	$HP_2O_7^{3-} \rightleftharpoons H^+ + P_2O_7^{4-}$	−9.41
41	$H_2P_2O_7^{2-} + H_2O \rightleftharpoons 2H_2PO_4^-$	5.71

Pyrophosphate Minerals

42	$\beta\text{-}Ca_2P_2O_7(c) \rightleftharpoons 2Ca^{2+} + P_2O_7^{4-}$	−14.70
43	$Fe_2P_2O_7(c) \rightleftharpoons 2Fe^{2+} + P_2O_7^{4-}$	−17.60

Pyrophosphate Complexes

44	$Ca^{2+} + P_2O_7^{4-} \rightleftharpoons CaP_2O_7^{2-}$	6.80
45	$Ca^{2+} + H^+ + P_2O_7^{4-} \rightleftharpoons CaHP_2O_7^-$	13.01
46	$Ca^{2+} + H_2O + P_2O_7^{4-} \rightleftharpoons CaOHP_2O_7^{3-} + H^+$	−5.10

Orthophosphate Complexes

47	$Fe^{2+} + H_2PO_4^- \rightleftharpoons FeH_2PO_4^+$	2.70
48	$Fe^{3+} + H_2PO_4^- \rightleftharpoons FeH_2PO_4^{2+}$	5.43
49	$Fe^{2+} + H_2PO_4^- \rightleftharpoons FeHPO_4^\circ + H^+$	−3.60
50	$Fe^{3+} + H_2PO_4^- \rightleftharpoons FeHPO_4^+ + H^+$	3.71
51	$Ca^{2+} + H_2PO_4^- \rightleftharpoons CaH_2PO_4^+$	1.40
52	$Ca^{2+} + H_2PO_4^- \rightleftharpoons CaHPO_4^\circ + H^+$	−4.46
53	$Ca^{2+} + H_2PO_4^- \rightleftharpoons CaPO_4^- + 2H^+$	−13.09
54	$Mg^{2+} + H_2PO_4^- \rightleftharpoons MgHPO_4^c + H^+$	−4.29

Other Reactions

55	$CaF_2(\text{fluorite}) \rightleftharpoons Ca^{2+} + 2F^-$	−10.41

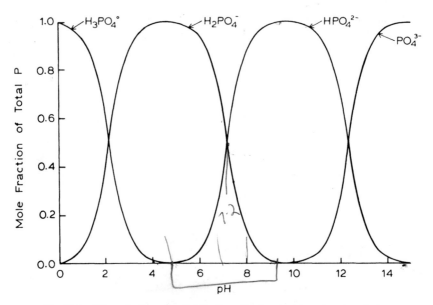

Fig. 12.1 Effect of pH on the distribution of orthophosphate ions in solution.

Similar mole fraction expressions can be developed for each of the orthophosphoric acid species plotted together in Fig. 12.1. This plot shows the relative abundance of the different orthophosphoric acid species as a function of pH. The formation constant (log $K°$) relating two species is numerically equal to the pH at which the reacting species have equal activities. For example,

$$HPO_4^{2-} + H^+ \rightleftharpoons H_2PO_4^- \qquad \log K° = 7.20 \qquad (12.3)$$

$$\frac{(H_2PO_4^-)}{(HPO_4^{2-})(H^+)} = 10^{7.20}$$

$$\log \frac{(H_2PO_4^-)}{(HPO_4^{2-})} = \log K° - pH = 7.20 - pH \qquad (12.4)$$

Thus when pH = log $K°$, the activity ratio of the reacting species is unity. A decrease in pH of one unit increases the ratio $H_2PO_4^-/HPO_4^{2-}$ by 10-fold, while increasing pH by one unit decreases this ratio 10-fold. Equation 12.4 and similar expressions for the other orthophosphate ions permit easy calculation of the relative abundance of the orthophosphate species in solution.

In concentrated phosphate solutions, two $H_2PO_4^-$ ions combine to form the dimer $(H_2PO_4)_2^{2-}$ according to Reaction 5 of Table 12.1. The mole fraction of this species is concentration dependent. For example, at 0.01 M $H_2PO_4^-$, the ratio

$$\frac{(H_2PO_4)_2^{2-}}{(H_2PO_4^-)} = 10^{-0.35}(H_2PO_4^-) = 10^{-2.35} = 0.0045 \quad (12.5)$$

For solutions containing <0.01 M $H_2PO_4^-$, the dimer constitutes <1% of the total phosphorus. Since total phosphorus in soils is generally less than 0.01 M, the dimer species can be safely discounted as being significant in soils.

Total orthophosphoric acid in soil solution can be represented by the equation *ortho-phosphorus*

$$[Total\ P] = [H_3PO_4^0] + [H_2PO_4^-] + [HPO_4^{2-}] + [PO_4^{3-}] \quad (12.6)$$

$$= \frac{(H_3PO_4^0)}{\gamma_{H_3PO_4^0}} + \frac{(H_2PO_4^-)}{\gamma_{H_2PO_4^-}} + \frac{(HPO_4^{2-})}{\gamma_{HPO_4^{2-}}} + \frac{(PO_4^{3-})}{\gamma_{PO_4^{3-}}} \quad (12.7)$$

Expressing each term on the right side of Eq. 12.7 in terms of $H_2PO_4^-$, H^+, and an appropriate equilibrium constant upon rearranging gives:

$$(H_2PO_4^-) = \frac{[Total\ P]}{10^{2.15}(H^+) + \dfrac{1}{\gamma_{H_2PO_4^-}} + \dfrac{10^{-7.20}}{\gamma_{HPO_4^{2-}}(H^+)} + \dfrac{10^{-19.55}}{\gamma_{PO_4^{3-}}(H^+)^2}} \quad (12.8)$$

Equation 12.8 can be used to calculate the activity of $H_2PO_4^-$ from total soluble phosphate, pH, and ionic strength, which permits calculation of activity coefficients (Eq. 2.11 or 2.13). The assumption made in using Eq. 12.8 is that only orthophosphoric acid species contribute significantly to total phosphorus in solution. If other phosphorus species are present, they must also be included.

12.2 ALUMINUM PHOSPHATES

Aluminum is a major constituent in soils and forms a number of insoluble phosphate minerals. For these reasons it must be considered as a potential controller of phosphate solubility. Reactions 6 through 9 of Table 12.1 summarize the solubilites of a few aluminum phosphate minerals whose solubilities are known. These minerals are plotted together in Fig. 12.2.

For this development kaolinite in equilibrium with quartz was considered to control Al^{3+} activity (See Fig. 5.7). An example of how this figure was developed is shown for $AlPO_4 \cdot 2H_2O$(variscite) by combining Reaction 7

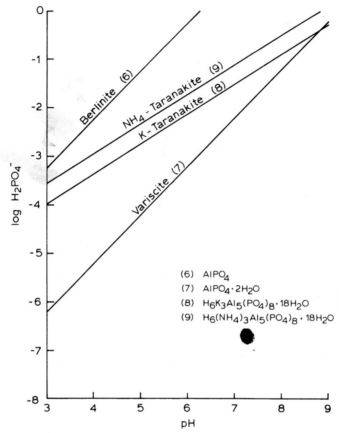

Fig. 12.2 The solubility of several aluminum phosphates in equilibrium with kaolinite and quartz when K^+ and NH_4^+ are $10^{-3}\,M$.

H_2SiO_4

of Table 12.1 with Reactions 6 and 42 of Table 5.1

		log $K°$
$AlPO_4 \cdot 2H_2O(var) + 2H^+ \rightleftharpoons Al^{3+} + H_2PO_4^- + 2H_2O$		-2.50
$Al^{3+} + H_4SiO_4° + 0.5H_2O \rightleftharpoons 0.5Al_2Si_2O_5(OH)_4(kaol) + 3H^+$		$0.5(-5.45)$
$SiO_2(q) + 2H_2O \rightleftharpoons H_4SiO_4°$		-4.00
$AlPO_4 \cdot 2H_2O(var) + SiO_2(q) + 0.5H_2O \rightleftharpoons H_2PO_4^- + H^+ + 0.5Al_2Si_2O_5(OH)_4(kaol)$		-9.23

(12.9)

ALUMINUM PHOSPHATES

$$(H^+)(H_2PO_4^-) = 10^{-9.23}$$

$$\log H_2PO_4^- = -9.23 + pH \qquad (12.10)$$

Equation 12.10 is plotted as the variscite line in Fig. 12.2. Equations for the other lines in this figure were determined similarly. In plotting the taranakites a reference level of 10^{-3} M for K^+ and NH_4^+ was used.

The solubilities of the aluminum phosphate minerals decrease in the order

$AlPO_4$(berlinite) > $H_6(NH_4)_3Al_5(PO_4)_8 \cdot 18 H_2O$($NH_4$-taranakite)
> $H_6K_3Al_5(PO_4)_8 \cdot 18 H_2O$(K-taranakite) > $AlPO_4 \cdot 2 H_2O$(variscite).

Berlinite is an anhydrous mineral that is stable only at high temperatures. At 25°C it is more than 1000 times more soluble than variscite, and for this reason is not expected to form in soils.

Both potassium and ammonium taranakites have been identified in soils as reaction products of phosphate fertilizer (Lindsay et al., 1962). They are especially prevalent from fertilizers containing NH_4^+ and K^+. The high activity of these cations in such fertilizers temporarily depresses the solubility lines below that of variscite (Fig. 12.2), making them more stable than variscite. Later as the concentrations of NH_4^+ and K^+ in the soil-fertilizer reaction zone decrease, these minerals again become unstable relative to variscite.

Another way of representing the solubilities of aluminum phosphate minerals is to use double function parameters involving H^+, $H_2PO_4^-$, and Al^{3+}. For variscite these relationships can be developed as follows:

		log $K°$
$AlPO_4 \cdot 2 H_2O$(variscite) + $2H^+$ ⇌ Al^{3+} + $H_2PO_4^-$ + $2H_2O$		-2.50
H^+ ⇌ H^+		0.00
$AlPO_4 \cdot 2 H_2O$(variscite) + $3H^+$ ⇌ Al^{3+} + H^+ + $H_2PO_4^-$ + $2H_2O$		-2.50

$$\qquad (12.11)$$

$$\frac{(Al^{3+})(H^+)(H_2PO_4^-)}{(H^+)^3} = 10^{-2.50}$$

$$\log H_2PO_4^- - pH = -2.50 - (\log Al^{3+} + 3pH) \qquad (12.12)$$

A plot of Eq. 12.12 is shown in Fig. 12.3 as the variscite line. The solubility of gibbsite developed earlier (Eq. 3.2) appears as a vertical line at $\log Al^{3+} + 3pH = 8.04$, whereas lines for kaolinite in equilibrium with various silica phases are represented along the horizontal axis.

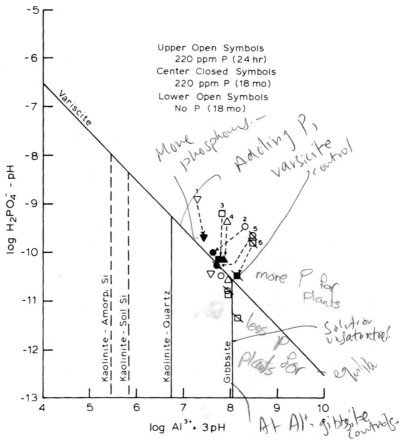

Fig. 12.3 Changes in the solubilities of aluminum and phosphate in six acid New York soils during an 18-month period following the addition of 220 ppm of phosphorus as MCP relative to known minerals (adapted from Lindsay et al., 1959).

Points in this diagram that lie above the variscite line represent supersaturation with respect to variscite, whereas points lying below the line represent undersaturation. Points to the right of the gibbsite or kaolinite lines represent supersaturation with respect to these phases, whereas points to the left represent undersaturation. Points of intersection represent equilibrium with respect to the intersecting phases.

Lindsay et al. (1959) fertilized six New York acid soils with MCP (monocalcium phosphate monohydrate) and allowed the soils to wet and dry in the greenhouse for a period of 18 months. No plants were grown during this period. At various times 100-g samples of the soil (air-dry basis) were removed and shaken with 200-ml of 0.01 M $CaCl_2$ for 24 hr. The filtrates were analyzed for pH, total phosphorus, and total aluminum. The activity of Al^{3+} was calculated using Eq. 3.10 and the activity of $H_2PO_4^-$ was calculated from Eq. 12.8 using appropriate terms for the pH range of acid soils. Equation 2.13 was used to estimate activity coefficients at 0.03 ionic strength. The results are plotted in Fig. 12.3.

The pH of these six soils in 0.01 M $CaCl_2$ extracts ranged from 3.7 to 5.0. The unfertilized soils fell very near or slightly below the variscite line indicating possible equilibrium with this phosphate mineral. Adding 220 ppm P as MCP raised the phosphate status of these soils by approximately 1.0 to 1.5 log units. During the ensuing 18 months, the solubility points moved down again toward the variscite line. These results suggest that the initial reaction products of phosphate fertilization were 10 to 30 times more soluble than variscite and may have been an amorphous aluminum phosphate. During the ensuing 18 months, the phosphate solubility decreased and approached that of variscite.

Solubility measurements such as these are no direct proof that variscite was formed, but the results support the hypothesis that it *could* have formed. The values of log Al^{3+} + 3pH for these soils changed slightly during the 18-month period and appeared to approach a value of approximately 7.8, which is only slightly below that of gibbsite (8.04).

The possibility that iron phosphates may have formed in these acid soils is examined in the next section.

12.3 IRON PHOSPHATES

The solubilities of $FePO_4(c)$ and $FePO_4 \cdot 2H_2O$(strengite) are given by Reactions 10 and 11 of Table 12.1 and are plotted in Fig. 12.4.

The activity of Fe^{3+} used to prepare this plot is that in equilibrium with soil-Fe (Chapter 10). An example of how this figure was developed is shown for $FePO_4 \cdot 2H_2O$(strengite) by combining Reaction 11 of Table 12.1 with Reaction 2 of Table 10.1:

$$
\begin{array}{lr}
 & \log K° \\
FePO_4 \cdot 2H_2O\text{(strengite)} + 2H^+ \rightleftharpoons Fe^{3+} + H_2PO_4^- + 2H_2O & -6.85 \\
Fe^{3+} + 3H_2O \rightleftharpoons 3H^+ + Fe(OH)_3\text{(soil)} & -2.70 \\
\hline
FePO_4 \cdot 2H_2O\text{(strengite)} + H_2O \rightleftharpoons H^+ + H_2PO_4^- + Fe(OH)_3\text{(soil)} & -9.55 \\
\end{array}
$$

(12.13)

$$(H^+)(H_2PO_4^-) = 10^{-9.55}$$

$$\log H_2PO_4^- = -9.55 - \text{pH} \quad (12.14)$$

Equation 12.14 is plotted as the strengite solubility line in Fig. 12.4. The solubility line for $FePO_4(c)$ was developed similarly and lies 1.47 log units above strengite indicating a 30-fold greater solubility. The anhydrous mineral is stable at high temperatures but is not stable in soils.

Iron phosphate minerals such as strengite can also be plotted in terms of the double function parameters involving the activities of H^+, $H_2PO_4^-$, and Fe^{3+}. This can be done as follows:

$$\begin{array}{lr} & \log K° \\ FePO_4 \cdot 2H_2O(\text{strengite}) + 2H^+ \rightleftharpoons Fe^{3+} + H_2PO_4^- + 2H_2O & -6.85 \\ H^+ \rightleftharpoons H^+ & 0.00 \\ \hline FePO_4 \cdot 2H_2O(\text{strengite}) + 3H^+ \rightleftharpoons Fe^{3+} + H^+ + H_2PO_4^- + 2H_2O & -6.85 \end{array}$$

(12.15)

$$\frac{(Fe^{3+})(H^+)(H_2PO_4^-)}{(H^+)^3} = 10^{-6.85}$$

$$\log H_2PO_4^- - pH = -6.85 - (\log Fe^{3+} + 3pH) \qquad (12.16)$$

A plot of Eq. 12.16 is shown as the strengite solubility line in Fig. 12.5. The solubilities of various Fe(III) oxides discussed in Chapter 10 are included along the horizontal axis at their respective $\log Fe^{3+} + 3pH$ values. Unfortunately most solubility measurements of phosphate in soils have not been accompanied by appropriate measurements of Fe^{3+} which are necessary for use with this diagram.

Let us again examine the phosphate solubility measurements of Lindsay et al. (1959) for the six New York acid soils discussed in the previous section. These solubility measurements are plotted in Fig. 12.5 on the basis that Fe^{3+} is controlled by soil-Fe since actual Fe^{3+} activities were not measured. The data points were arbitrarily displaced to each side of the soil-Fe line so that the six soils could be distinguished.

The data points in Fig. 12.5 show that the initial reaction products of 220 ppm P as MCP were approximately in equilibrium with strengite and soil-Fe. After 18 months, however, the points moved below the strengite line indicating that the final reaction products were less soluble than strengite. The soils to which no phosphate was added fell even lower suggesting that strengite was not the ultimate reaction product of phosphate in these soils.

Again it must be emphasized that iron solubility measurements were not included in this study. Nevertheless Fig. 12.5 illustrates how iron solubility in

IRON PHOSPHATES

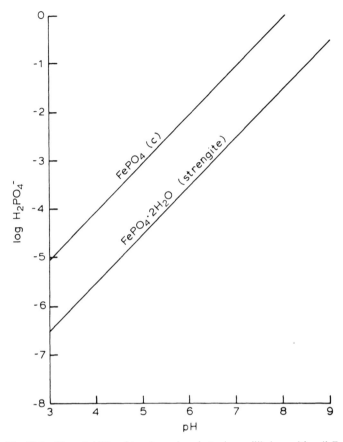

Fig. 12.4 The solubility of two iron phosphates in equilibrium with soil-Fe.

soils can affect the level of phosphate maintained by strengite. Since minerals such as goethite and hematite lower Fe^{3+} activity far below that of amorphous or soil-Fe (Fig. 10.1), the more insoluble iron oxides permit strengite to maintain higher levels of phosphate.

The question is frequently asked: "which is more stable in soils, strengite or variscite?" The answer demands some elaboration, and Fig. 12.6 is designed to provide the answer. If soil-Fe controls Fe^{3+} and kaolinite-quartz controls Al^{3+}, strengite is more stable by only 0.3 log unit of $H_2PO_4^-$. If kaolinite-soil-Si controls Al^{3+}, then variscite is 1.2 log units more soluble. As soils weather and gibbsite controls Al^{3+}, variscite becomes more stable than strengite. Weathering tends to increase the stability of aluminum

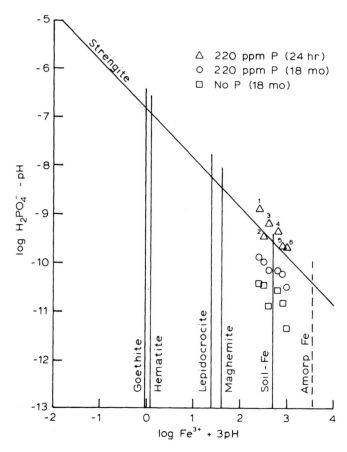

Fig. 12.5 The solubility of strengite in equilibrium with various Fe(III) oxides. The data points were taken from Lindsay et al. (1959) and reflect changes in phosphate solubility after fertilization. The data points were arbitrarily displaced from the soil-Fe line.

phosphates and decrease the stability of iron phosphates. Figure 12.6 depicts these solubility relationships and shows that a given phosphate mineral has a range of solubilities depending upon the activities of the other ions in that mineral. Added to these complexities is the fact that most initial precipitates of phosphate are amorphous and may show solubilities that are 10 to 100 times greater than the crystalline minerals. For these reasons it is understandable why many researchers in the past have become disenchanted with the use of solubility product principles to explain phosphate reactions in soils.

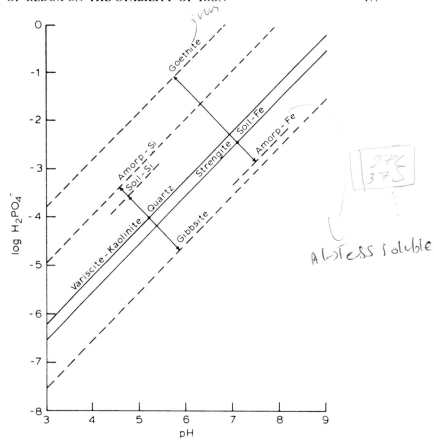

Fig. 12.6 The effect of various parameters on the solubilities of variscite and strengite in soils.

12.4 EFFECT OF REDOX ON THE STABILITY OF IRON AND ALUMINUM PHOSPHATES

When soils are reduced, the solubility of iron changes, and this in turn affects the solubility of iron phosphates (Williams and Patrick, 1973). Figure 12.7 was constructed to show the effect of redox on the solubility of iron and aluminum phosphates in soils.

Let us consider an acid soil in which Al^{3+} is controlled by kaolinite-quartz, and Fe^{3+} is controlled by soil-Fe. Of the minerals considered so far, strengite would be the stable phosphate phase and $\log H_2PO_4^- - pH$ for this soil would remain at -9.55 in moderate to well-oxidized soils. As the soil $pe + pH$ value drops below 11.53, magnetite becomes more stable

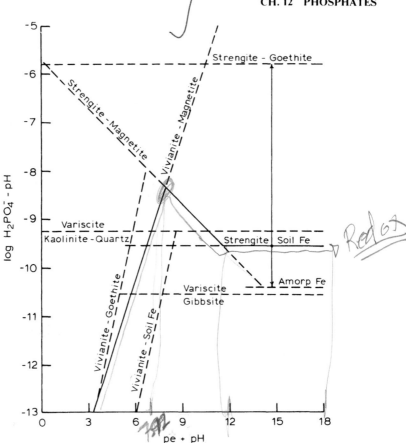

Fig. 12.7 The redox at which strengite and variscite can transform into vivianite.

than soil-Fe (Chapter 10). This depresses Fe^{3+} activity and enables strengite to support a higher level of phosphate as shown by the reaction:

$$3\,FePO_4 \cdot 2\,H_2O(\text{strengite}) + e^- \rightleftharpoons$$
$$Fe_3O_4(\text{magnetite}) + 3\,H_2PO_4^- + 2\,H^+ + 2\,H_2O \quad (12.17)$$

for which $\log K° = -17.10$. At equilibrium

$$\log H_2PO_4^- - pH = -5.71 - 0.33(pe + pH) \quad (12.18)$$

Equation 12.18 is plotted as the strengite-magnetite line in Fig. 12.7 and shows that phosphate solubility, expressed as $\log H_2PO_4^- - pH$, increases 0.33 log unit for each unit decrease in $pe + pH$.

As reduction continues, a redox level is reached at which the ferrous phosphate $Fe_3(PO_4)_2 \cdot 8\,H_2O$(vivianite) becomes more stable than strengite.

The vivianite-magnetite line in Fig. 12.7 is defined by the reaction:

$$Fe_3(PO_4)_2 \cdot 8H_2O(\text{vivianite}) \rightleftharpoons$$
$$Fe_3O_4(\text{magnetite}) + 2H_2PO_4^- + 4H^+ + 2e^- \quad (12.19)$$
$$\log K^\circ = -32.58$$

which reduces to

$$\log H_2PO_4^- - pH = -16.29 + (pe + pH) \quad (12.20)$$

Combining Eq. 12.20 and 12.18 shows that both strengite and vivianite can coexist with magnetite at $pe + pH = 7.92$. As redox goes below this point, vivianite becomes the stable phosphate phase, and strengite can convert to vivianite.

Let us consider a second soil which is more highly weathered in which Fe^{3+} is controlled by goethite and Al^{3+} is controlled by gibbsite. In this soil variscite would be the stable phosphate mineral, and the $\log H_2PO_4^- - pH$ would be fixed at 10.54 until the vivianite-goethite equilibrium line is reached. This would not occur until $pe + pH$ drops to 4.96. Below this redox variscite can be converted to vivianite as goethite controls iron solubility.

Let us consider a third soil which contains strengite and soil-Fe, but assume that magnetite does not form. The $\log H_2PO_4 - pH$ of this soil will be fixed at -9.55 until a $pe + pH$ of 8.34 is reached. At lower redox levels vivianite can precipitate as strengite dissolves. The redox below which variscite-gibbsite converts to vivianite-soil Fe is $pe + pH$ 7.68.

As pointed out in Chapter 10 iron oxides formed under fairly reduced conditions may contain both ferric and ferrous iron. These oxides have not been adequately characterized. Nevertheless, strengite and variscite can be transformed to vivianite in the $pe + pH$ range of 4 to 8.3 regardless of which oxides determine iron solubility. This redox range is bounded by the vivianite-goethite and the vivianite-soil-Fe solubility lines in Fig. 12.7.

When soils are flooded, two principle factors operate to increase the availability of phosphorus in acid soils. First, the pH of reduced soils generally rises toward neutral which increases the solubility of iron and aluminum phosphates (Fig. 12.2 and 12.4). Secondly, as $pe + pH$ drops below 8.34, depending upon which iron oxides control iron and which minerals control Al^{3+}, strengite and variscite convert to vivianite. A rice plant is still able to obtain sufficient phosphorus under the highly reduced conditions in the presence of vivianite because in the immediate vicinity of a root, the redox is higher than that of the bulk soil because O_2 is supplied through the stem of a rice plant to its roots. In this way plants are able to absorb phosphorus from a medium even though vivianite suppresses the solubility of phosphate in the bulk of the soil to very low levels (Fig. 12.7).

Figure 12.7 is useful in explaining the beneficial effects of flooding acid soils for increasing nutrient availability. After a flooded soil is allowed to drain and O_2 reenters, oxidation again takes place and the reverse processes occur. That is, vivianite dissolves, Fe^{2+} is oxidized to Fe(III), and phosphate precipitates. Since amorphous Fe(III) oxides form initially, the strengite line is depressed below that of variscite favoring the precipitation of iron rather than aluminum phosphates.

Most redox changes in soils occur fast enough that initial reaction products are likely to be amorphous rather than crystalline. Magnetite may not be the actual form of iron precipitated in reduced soil. Instead an amorphous product slightly more soluble than magnetite may form. Much has been written about $Fe_3(OH)_8$(ferrosic oxide) as the possible form of iron oxide in reduced soils (Arden, 1950; Ponnamperuma et al., 1967), but the reported solubilities are too high (Chapter 10). Further research is needed to characterize the solubility relationships and mixed valency states of iron in highly reduced soils.

12.5 SOLUBILITY OF CALCIUM PHOSPHATES

The solubilities of several calcium phosphates found in soils are given by Reactions 13 through 20 of Table 12.1. These minerals are plotted in Fig. 12.8, which is similar to diagrams published earlier (Lindsay and Moreno, 1960; Lindsay and Vlek, 1977).

In this diagram only the most abundant orthophosphate species is plotted. Below pH 7.20, $H_2PO_4^-$ is most abundant, whereas above this pH, HPO_4^{2-} is most abundant. The activity of Ca^{2+} used to develop this diagram was $10^{-2.5}$ M or equilibrium with $CaCO_3$(calcite) at 0.0003 atm of $CO_2(g)$ at high pH. The activity of F^- used to fix the fluorapatite line was 10^{-4} M, which is the equilibrium level of CaF_2(fluorite).

From Fig. 12.8 it is evident that strengite and variscite are more stable in highly acid soils than any of the calcium phosphates. In general all of the calcium phosphates decrease in solubility with increase in pH. In calcareous soils when Ca^{2+} is depressed, phosphorus solubility again increases. For the $CO_2(g)$ level represented in Fig. 12.8, this begins at pH 7.88 (Eq. 7.14).

The solubilities of calcium phosphates generally decrease in the order

$CaHPO_4 \cdot 2H_2O$(brushite, DCPD) > $CaHPO_4$(monetite, DCP)

> $Ca_4H(PO_4)_3 \cdot 2.5H_2O$(octacalcium phosphate, OCP)

> β-$Ca_3(PO_4)_2$(β-tricalcium phosphate, β-TCP)

> $Ca_5(PO_4)_3OH$(hydroxyapatite, HA)

> $Ca_5(PO_4)_3F$(fluorapatite, FA).

SOLUBILITY OF CALCIUM PHOSPHATES

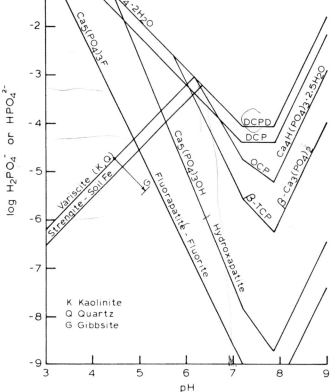

Fig. 12.8 The solubility of calcium phosphates compared to strengite and varsicite when Ca^{2+} is $10^{-2.5}$ M or is fixed by calcite and $CO_2(g)$ at 0.0003 atm.

Although brushite and monetite are readily formed under acid conditions, they transform to more basic calcium phosphates at higher pH.

In the pH range of 6 to 6.5 several of the phosphate minerals shown in Fig. 12.8 can coexist. These minerals include variscite, strengite, DCPD, DCP, OCP, and β-TCP, which all tend to maintain approximately $10^{-3.25}$ M $H_2PO_4^-$ in this pH range. This happens to be the pH range in which phosphates in soils are generally most available to plants.

Figure 12.8 helps to explain why liming acid soils to pH 6.5 increases phosphorus availability by enabling iron and aluminum phosphates to convert to calcium phosphates. It also helps to explain why finely divided hydroxy-, fluoro-, and carbonate-apatite are available in acid soils below pH 6 but are unavailable to plants growing on alkaline soils.

Figure 12.8 can be used to estimate phosphate solubility relationships that differ from those used to construct this diagram. For non-calcareous soils, the effect of different levels of Ca^{2+} can be obtained by the relationship

$$\Delta \log(H_2PO_4^- \text{ or } HPO_4^{2-}) = -\Delta \log Ca^{2+} \left(\frac{Ca}{P}\right)_i \quad (12.21)$$

where $(Ca/P)_i$ is the molar ratio of calcium to phosphorus in mineral (i). For example, if Ca^{2+} activity were 10^{-3} M instead of the $10^{-2.5}$ used to construct Fig. 12.8, the solubility line for octacalcium phosphate (OCP) would be shifted by $0.5(\frac{4}{3}) = 0.67$ log unit. For calcareous soils, Ca^{2+} activity is controlled by $CO_2(g)$ according to Eq. 7.14, so shifts in the calcium phosphates lines in Fig. 12.8 with changes in $CO_2(g)$ follow the relationship

$$\Delta \log HPO_4^{2-} = \Delta \log CO_2(g) \left(\frac{Ca}{P}\right)_i \quad (12.22)$$

For example, if 0.003 instead of 0.0003 atm of $CO_2(g)$ were used to construct Fig. 12.8, the solubility line for octacalcium phosphate would shift upward by $1.0(\frac{4}{3}) = 1.33$ log units of HPO_4^{2-}. Shifts of the variscite and strengite lines were discussed in the previous section. Thus Fig. 12.8 can be used to estimate solubility relationships of the minerals included therein at any Ca^{2+} or $CO_2(g)$ level.

Calcium phosphates can also be plotted in terms of double function parameters consisting of H^+, $H_2PO_4^-$, and Ca^{2+} (Aslyng, 1954). An example calculation for $CaHPO_4 \cdot 2H_2O$(brushite) is

			$\log K°$
$CaHPO_4 \cdot 2H_2O$(brushite) $+ H^+$	$\rightleftharpoons$	$Ca^{2+} + H_2PO_4^- + 2H_2O$	0.63
H^+	$\rightleftharpoons$	H^+	0.00
$CaHPO_4 \cdot 2H_2O$(brushite) $+ 2H^+$	$\rightleftharpoons$	$Ca^{2+} + H^+ + H_2PO_4^- + 2H_2O$	0.63

$$\frac{(Ca^{2+})(H^+)(H_2PO_4^-)}{(H^+)^2} = 10^{0.63} \quad (12.23)$$

$$\log H_2PO_4^- - pH = +0.63 - (\log Ca^{2+} + 2pH) \quad (12.24)$$

SOLUBILITY OF CALCIUM PHOSPHATES

Equation 12.24 is shown in Fig. 12.9 as the brushite solubility line. The other lines in this figure were developed similarly. Points above the lines represent supersaturation while points below represent undersaturation. Intersecting lines represent points of coexistence of the intersecting minerals. The solubility of $CaCO_3$ (calcite) is shown by the vertical lines along the abscissa defined by the equation:

$$\log Ca^{2+} + 2pH = 9.74 - \log CO_2(g) \qquad (12.25)$$

The $CO_2(g)$ levels of 0.0003 and 0.01 atm are shown in Fig. 12.9. Most calcareous soils lie within this range of $\log Ca^{2+} + 2pH$ values. Higher

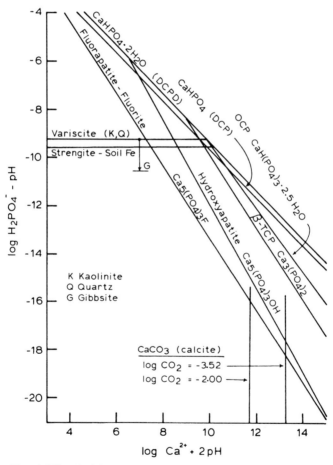

Fig. 12.9 The solubility of calcium, iron, and aluminum phosphates showing limits imposed by calcite and $CO_2(g)$.

values can only be achieved if $CO_2(g)$ is reduced below that of the atmosphere which is not likely to occur in natural soils.

The double function parameters used to construct Fig. 12.9 may seem awkward, but they have the advantage that no arbitrary activity of Ca^{2+} is needed. This diagram is most useful for comparing solubility measurements having variable Ca^{2+} activities. The double parameter axes can easily be converted to $\log H_2PO_4^-$ and $\log Ca^{2+}$ for any given pH.

The fluorapatite lines in Fig. 12.8 and 12.9 were drawn on the basis that CaF_2(fluorite) fixes F^- activity. If other minerals control F^- activity, the phosphate in equilibrium with fluorapatite will change. For example, if $KMg_3AlSi_3O_{10}F_2$(fluorphlogopite) is present, the solubility of fluorapatite can be modified as shown by the fluorapatite-fluorphlogopite line in Fig. 12.10. The equation for this line was obtained from Reaction 20 of Table 12.1 and Reactions 31 and 6 of Table 5.1 giving

$$\log H_2PO_4^- = -0.92 + 0.17 \log K^+ + 0.50 \log Mg^{2+}$$
$$- 1.67 \log Ca^{2+} 0.33 \log H_4SiO_4^\circ - 1.17 \, pH \quad (12.26)$$

Equation 12.26 is plotted in Fig. 12.10 for $\log K^+ = \log Mg^{2+} = 10^{-3}\,M$, $Ca^{2+} = 10^{-2.5}\,M$ or equilibrium with $CaCO_3$(calcite) at 0.0003 atm CO_2 and in equilibrium with quartz. Fluorite, fluorphlogopite, and fluorapatite can all coexist at pH 6.6. At lower pH fluorphlogopite is unstable, while at high pH, fluorite is unstable. This means that above pH 6.6 fluorphlogopite depresses F^- activity and allows fluorapatite to maintain a higher level of phosphate which approaches that of hydroxyapatite. As seen from Eq. 12.26 changes in the activities of K^+, Mg^{2+}, Ca^{2+}, and $H_4SiO_4^\circ$ can cause the fluorapatite-fluorphlogopite line to shift slightly.

Based on the average phosphorus and fluoride contents of soils (Table 1.1) there is more than enough fluoride to precipitate soil phosphorus as fluorapatite. The extra fluoride would be available to form a F^- controlling mineral such as fluorite or fluorphlogopite. Since the fluorine content of soils is highly variable, there may be many soils in which there is not enough fluoride to match the phosphorus present, especially in highly fertilized or low fluorine soils. In such cases fluorapatite may control the solubility of F^- and depress its activity sufficiently that fluorapatite and hydroxyapatite can coexist. Isomorphous substitutions of OH^-, F^-, PO_4^{3-}, and CO_3^{2-} are known to occur in apatite structures (Chien and Black, 1976). Added to this complexity is the fact that phosphate forms poorly crystalline minerals at temperatures commonly found in soils. For these reasons, phosphate solubility in slightly acid to alkaline soils may approach that of hydroxyapatite, but generally soils remain supersaturated to this mineral indefinitely.

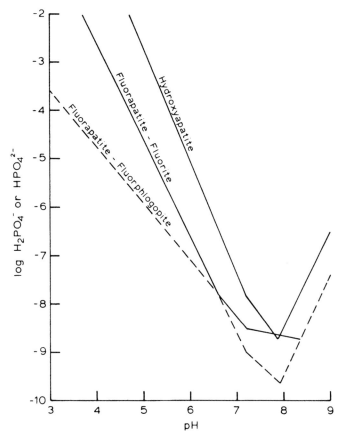

Fig. 12.10 The solubility of fluorapatite in equilibrium with fluorite or fluorphlogopite compared to that of hydroxyapatite (see text).

There may be other insoluble phosphate minerals that are important fixation products of phosphate in soils. As these minerals are recognized and the solubilities are established, they, too, can be included.

12.6 EFFECT OF REDOX ON THE STABILITY OF CALCIUM PHOSPHATES

When near-neutral and calcareous soils are subjected to reduction, phosphorus availability to plants may or may not increase (Patrick and Mahapatra, 1968; Ponnamperuma, 1972). Figure 12.7 in conjunction with

Fig. 12.8 can be used to estimate the redox below which a given calcium phosphate mineral can be transformed to vivianite. For example, let us consider a soil of pH 7.0 showing equilibrium with $Ca_4H(PO_4)_3 \cdot 2.5H_2O$(OCP). From Fig. 12.8 the log $H_2PO_4^-$ of this soil is -4.4, which corresponds to a log $H_2PO_4^-$ – pH of -11.4. From Fig. 12.7, a $pe + pH < 5.0$ would be required for vivianite to precipitate when magnetite fixes iron solubility. If soil-Fe fixes iron solubility, a $pe + pH < 7.0$ is required.

Using the same procedure for hydroxyapatite in a soil at pH 7.0, the log $H_2PO_4^-$ is -7.7 which corresponds to log $H_2PO_4^-$ – pH of -14.7. This would require a $pe + pH$ of <1.5 in the case of magnetite and <4.0 in the case of soil-Fe in order for vivianite to precipitate.

Only under extremely reducing conditions can insoluble basic calcium phosphates like hydroxyapatite be solubilized in soils by transformation to reduced phosphate minerals like vivianite. The possibility that reduction may affect other reactions that either directly or indirectly involve phosphorus must be kept in mind. The reported inconsistent benefits of reducing calcareous soils on increasing available phosphorus is explainable from these theoretical considerations. As pointed out in Section 12.4 reducing acid soils is nearly always beneficial in making phosphorus more available.

12.7 SOLUBILITY OF MAGNESIUM PHOSPHATES

The solubilities of several magnesium phosphates are given by Reactions 21 through 26 of Table 12.1 and are plotted in Fig. 12.11. To develop this diagram Mg^{2+} was fixed at 10^{-3} M and Ca^{2+} at $10^{-2.5}$ M.

All of the magnesium phosphates represented here are more soluble than $CaHPO_4 \cdot 2H_2O$(brushite) and $Ca_4H(PO_4)_3 \cdot 2.5H_2O$(OCP). Included in this comparison are $MgHPO_4 \cdot 3H_2O$(newberryite), $MgKPO_4 \cdot 6H_2O$(c), $MgNH_4PO_4 \cdot 6H_2O$(struvite), $Mg_3(PO_4)_2$(c), $Mg_3(PO_4)_2 \cdot 8H_2O$(bobierrite), and $Mg_3(PO_4)_2 \cdot 22H_2O$(c). The mineral, $Mg_3(PO_4)_2$(c), is too soluble to appear in this diagram.

Solubilities have been reported for the following additional magnesium minerals: $Mg_5(PO_4)_3OH$(Mg-hydroxyapatite), $Mg_5(PO_4)_3F$(Mg-fluorapatite), $Mg_4O(PO_4)_2$(Mg-oxyapatite), and $Mg_2(PO_4)F$(wagnerite). These minerals were considered in a previous review (Lindsay and Vlek, 1977), but they are omitted here because of uncertainties in the solubility measurements.

Magnesium phosphates have been found as initial reaction products of phosphate fertilizers in soils (Lindsay et al., 1962; Taylor et al., 1963a, b), but they later disappear as more stable minerals are formed. Magnesium phosphates can be discounted as permanent fixation products of phosphorus in soils. Instead these phosphates should constitute useful fertilizers for

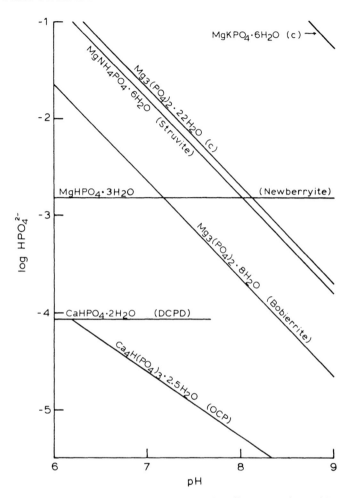

Fig. 12.11 The instability of magnesium phosphates in soils compared to DCPD and OCP when Ca^{2+} is $10^{-2.5}$ M and Mg^{2+} is 10^{-3} M.

supplying readily available phosphorus to plants. More work is needed to reexamine the reported solubilities of some of the more basic magnesium phosphates that may form in soils even though present information indicates that they are probably too soluble to be important.

12.8 MANGANESE PHOSPHATES

Solubilities have been reported for $MnHPO_4(c)$ and $Mn_3(PO_4)_2(c)$ (Reactions 27 and 28 of Table 12.1). Since Mn^{2+} solubility in soils is redox dependent, a special solubility diagram is constructed in Fig. 12.12 to consider the stabilities of manganese phosphates relative to other phosphate minerals. In this figure the solubility of strengite in equilibrium with soil-Fe is shown as the upper horizontal line at log $H_2PO_4^-$ − pH of 9.55. The activity of Mn^{2+} is controlled by pyrolusite (P) above pe + pH of 16.62 and by manganite (M) or rhodocrosite (R) at lower redox depending upon CO_2 (Chapter 11). These controls cause $Mn_3(PO_4)_2(c)$ to be more soluble than strengite at pe + pH values > 15 but less soluble below this redox.

Based on the solubility reported for $MnHPO_4(c)$, this mineral is very stable in soils. As shown in Fig. 12.12, it is more stable than strengite at all pe + pH values < 19. In the pH range of 5.55 to 7.0 either $MnHPO_4(c)$ or hydroxyapatite may be stable depending upon pH and redox. In the pH

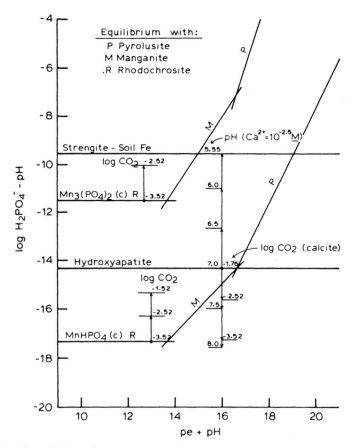

Fig. 12.12 The solubility of manganese phosphates relative to strengite and hydroxyapatite.

range above 7 the equilibrium relationships are even more complex. Increasing pH in the absence of calcite lowers hydroxyapatite solubility, whereas in its presence increasing CO_2 depresses pH and increases hydroxyapatite solubility. On the other hand, lowering redox depresses the solubility of $MnHPO_4(c)$ until $MnCO_3$(rhodochrosite) forms, and then its solubility is controlled by CO_2.

In calcareous soils at $10^{-2.52}$ atm of $CO_2(g)$, $MnHPO_4(c)$ can be more stable than hydroxyapatite depending on redox. If the molecular ratio of Mn/P in such a soil is much greater than one, $MnHPO_4(c)$ could control phosphate solubility. On the other hand, if this ratio is less than one, both $MnHPO_4(c)$ and hydroxyapatite could coexist, and the solubilities of both phosphate and Mn^{2+} would be controlled by hydroxyapatite. Since the phosphorus in most soils is more soluble than hydroxyapatite, it is possible that phosphate may depress the activity of Mn^{2+} below the equilibria levels depicted in Fig. 11.2.

Many other important deductions could be made regarding manganese and phosphorus solubilities in soils based on the solubility data presented in Figure 12.12. Since there is very little information on the possibility that manganese phosphates may be important in soils, further speculation seems unwarranted until the solubilities of manganese phosphates, and especially $MnHPO_4(c)$, are reexamined and their solubility relationships in soils are investigated.

12.9 OTHER ORTHOPHOSPHATES

The solubilities of $KH_2PO_4(c)$ and $K_2HPO_4(c)$ given by Reactions 29 and 30 of Table 12.1 indicate that these minerals are highly soluble in soils. Their formation would require greater than $1 M$ phosphate. Similar phosphates of sodium and ammonium also are too soluble to form in soils.

The trace element phosphates other than manganese are discussed separately in the chapters dealing with those elements. Phosphate also reacts with other cations such as strontium, barium, yttrium, and the rare-earth elements. The solubility relationships of these phosphates minerals in soils can not be examined until the solubility relationship of the cations are known.

12.10 REDUCED FORMS OF PHOSPHORUS

Phosphorus is generally considered to be present in soils in oxidation state $+5$. Reactions 31 through 36 of Table 12.1 demonstrate the insignificance of reduced phosphorus species in soil environments.

Phosphorus in $H_3PO_3^\circ$(phosphorous acid) is present in +3 oxidation state. This acid dissociates similarly to that of phosphoric acid with equal amounts of $H_2PO_3^-$ and HPO_3^{2-} being present at pH 6.79. The redox dependency of phosphorous acid can be seen from the reaction

$$\text{log } K^\circ$$

$$H_2PO_4^- + 2H^+ + 2e^- \rightleftharpoons H_2PO_3^- + H_2O \quad -10.87 \quad (12.27)$$

$$\log \frac{H_2PO_3^-}{H_2PO_4^-} = -10.87 - 2(pe + pH) \quad (12.28)$$

According to Eq. 12.28 the ratio of $H_2PO_3^-/H_2PO_4^-$ is $10^{-10.87}$ when $pe + pH = 0$. For all practical purposes phosphorous acid species in soils can be safely discounted since the $pe + pH$ of soils is nearly always greater than zero.

There are rumors that flashes of light have been seen from old cemetaries at night. Some have speculated that gaseous PH_3(phosphine) might be formed from decomposing bones, which spontaneously ignites upon exposure to the air. The following equations are useful for examining this possibility

$$H_2PO_4^- + 9H^+ + 8e^- \rightleftharpoons PH_3(g) + 4H_2O \quad -36.73 \quad (12.29)$$

$$\log PH_3(g) = -36.73 + \log H_2PO_4^- - pH - 8(pe + pH) \quad (12.30)$$

Thus at pH 7 with $pe + pH$ at zero and $10^{-4} M$ $H_2PO_4^-$ there could only be $10^{-47.73}$ atm of $PH_3(g)$. Only if $pe + pH$ drops below -5 could appreciable $PH_3(g)$ be generated. For this reason it is virtually impossible for $PH_3(g)$ to form in soils or in cemetaries for that matter.

According to Reaction 34 of Table 12.1, elemental phosphorus cannot form in soils. Thus phosphorus in soils is only expected to occur in the +5 oxidation state. If compounds of lower oxidation state are added to soils, the phosphorus will readily oxidized to orthophosphate, even in highly reduced soils in the $pe + pH$ range of 0 to 8.

12.11 STABILITY OF POLYPHOSPHATES IN SOILS

In recent years polyphosphates have been developed for use as fertilizers. Since they contain less water than orthophosphates, they are more economical to transport. Removing water from orthophosphoric acid gives

$$2H_3PO_4 \rightleftharpoons H_4P_2O_7(\text{pyro P}) + H_2O \quad (12.31)$$

$$3H_3PO_4 \rightleftharpoons H_5P_3O_{10}(\text{tri-poly P}) + 2H_2O \quad (12.32)$$

$$4H_3PO_4 \rightleftharpoons H_{x+3}P_xO_{3x+1}(\text{poly P}) + (x-1)H_2O \quad (12.33)$$

where x is the moles of phosphorus in the polyphosphate species. The first member of this species is $H_4P_2O_7$(pyrophosphoric acid) named from pyro meaning fire, since heating orthophosphates drives off water and converts them to pyrophosphates. As more water is removed, polyphosphates having higher chain lengths are formed (Table 12.2). Removal of sufficient water to give 87% P_2O_5 gives 99.2% polyphosphates, of which 76% has >9 phosphorus atoms per species.

In aqueous systems protons dissociate from polyphosphates. The dissociation constants for $H_4P_2O_7^\circ$ are shown by Reactions 37 through 40 of Table 12.1. These species can be plotted in mole fraction diagrams similar to those for orthophosphates given in Fig. 12.1. In the pH range of soils $H_2P_2O_7^{2-}$ and $HP_2O_7^{3-}$ are of greatest abundance and have equal activities at pH 6.70. Longer chained polyphosphates also dissociate protons according to specific dissociation constants.

Frequently NH_4^+, K^+, or various micronutrients are added to polyphosphate fertilizers. Some polyphosphates form complexes that permit rather high concentrations of nutrients to be held by liquid polyphosphate fertilizer. In soils various cations participate in these reactions and precipitate as complex polyphosphates. Reaction products containing Ca^{2+} and Mg^{2+} from the soil and NH_4^+ and K^+ from added fertilizers are common reactants.

One of the end members in the reaction sequence of polyphosphates in soils is β-$Ca_2P_2O_7$. The stability relationships of pyrophosphates in soils relative to orthophosphates can be obtained from the reactions

		log $K°$
β-$Ca_2P_2O_7(c) \rightleftharpoons 2Ca^{2+} + P_2O_7^{4-}$		-14.70
$P_2O_7^{4-} + 2H^+ \rightleftharpoons H_2P_2O_7^{2-}$		16.11
$2Ca^{2+} \rightleftharpoons 2\,\text{soil-Ca}$		$2(2.50)$
$H_2P_2O_7^{2-} + H_2O \rightleftharpoons 2H_2PO_4^-$		5.71
β-$Ca_2P_2O_7(c) + 2H^+ + H_2O \rightleftharpoons 2\,\text{soil-Ca} + 2H_2PO_4^-$		12.12

(12.34)

$$\log H_2PO_4^- = 6.06 - pH \qquad (12.35)$$

Equation 12.35 is plotted as the uppermost line in Fig. 12.13 and indicates that the level of $H_2PO_4^-$ required for equilibrium with β-$Ca_2P_2O_7(c)$ and soil-Ca is $10^{3.26}$ time higher than that permitted by $CaHPO_4 \cdot 2H_2O(c)$. This means that Reaction 12.34 will proceed to the right as β-$Ca_2P_2O_7(c)$ dissolves, the products hydrolyze to $H_2PO_4^-$, and $CaHPO_4 \cdot 2H_2O$ or any

TABLE 12.2 THE CALCULATED DISTRIBUTION OF PHOSPHATE SPECIES IN SUPERPHOSPHORIC ACID

Acid Composition ($\% P_2O_5$)	Distribution of Phosphorus According to the Number of P Atoms Present (%)								
	1	2	3	4	5	6	7	8	9	>9
73	81.4	18.3	0.3	0.0					
75	60.2	35.8	3.6	0.3	0.0				
77	38.5	46.9	11.5	2.5	0.51	0.1	0.0	0.0	
80	15.5	40.2	22.9	11.6	5.5	2.5	1.1	0.5	0.2	0.1
84	3.9	11.5	12.5	12.1	10.9	9.5	8.0	6.6	5.4	19.7
87	0.8	1.6	2.2	2.6	3.0	3.2	3.4	3.6	3.6	76.0

Source: Fleming (1969).

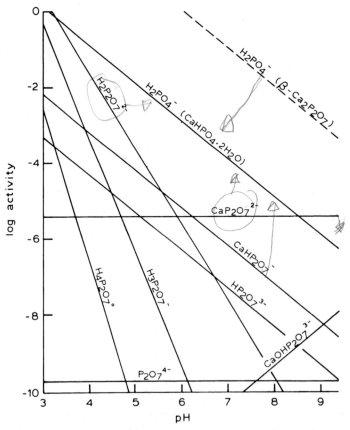

Fig. 12.13 Solution species of pyrophosphate in equilibrium with $\beta\text{-}Ca_2P_2O_7(c)$ and $10^{-2.5} M$ Ca^{2+} showing the instability of pyrophosphates relative to $CaHPO_4 \cdot 2H_2O(DCPD)$.

other more insoluble phosphates precipitate (Fig. 12.8). Thus pyrophosphate is unstable in soils and eventually converts to orthophosphates. For pyrophosphate to be stable it would be necessary to form pyrophosphate minerals of sufficient stability that the equilibrium level of $H_2PO_4^-$ maintained by that mineral would be below that maintained by the least soluble orthophosphate mineral. Longer chained polyphosphates are expected to be even less stable than pyrophosphate.

Also included in Fig. 12.13 are many of the solution species and complexes of pyrophosphate in equilibrium with β-$Ca_2P_2O_7(c)$. The solution species $H_2P_2O_7^{2-}$, $CaHP_2O_7^-$, and $CaP_2O_7^{2-}$ are the major ions in solution. Since the rate-limiting step in the conversion of pyrophosphate to orthophosphate is the hydrolysis step, the pyrophosphate species in solution are expected to approach the equilibrium levels depicted in Fig. 12.13. The $H_2PO_4^-$ line depicting equilibrium with β-$Ca_2P_2O_7(c)$ is drawn with a dashed line at a metastable level that cannot be maintained in soils.

Pyrophosphate reacts with Fe^{2+} to form $Fe_2P_2O_7(c)$; its solubility is given by Reaction 43 of Table 12.1. The conditions under which this mineral is more stable than β-$Ca_2P_2O_7(c)$ can be obtained by combining the following equilibria:

		log $K°$
$Fe_2P_2O_7(c) \rightleftharpoons 2Fe^{2+} + P_2O_7^{4-}$		-17.60
$2Fe^{2+} + \frac{8}{3}H_2O \rightleftharpoons \frac{2}{3}Fe_3O_4(\text{magnetite}) + \frac{4}{3}e^- + \frac{16}{3}H^+$		$\frac{2}{3}(-35.69)$
$2Ca^{2+} + P_2O_7^{4-} \rightleftharpoons \beta\text{-}Ca_2P_2O_7(c)$		14.70
$2\,\text{soil-Ca} \rightleftharpoons 2Ca^{2+}$		$2(-2.5)$
$Fe_2P_2O_7(c) + \frac{8}{3}H_2O + 2\,\text{soil-Ca} \rightleftharpoons \frac{2}{3}Fe_3O_4(\text{magnetite}) + \frac{4}{3}e^- + \frac{16}{3}H^+ + \beta\text{-}Ca_2P_2O_7(c)$		-31.69

(12.36)

from which

$$pe + pH = 23.77 - 3\,pH \qquad (12.37)$$

A plot of Eq. 12.37 is shown as line a in Fig. 12.14. This line designates the redox and pH conditions at which $Fe_2P_2O_7(c)$ is more stable than β-$Ca_2P_2O_7(c)$. For this development Fe^{2+} was fixed by Fe_3O_4(magnetite) below $pe + pH$ of 11.53 and Ca^{2+} was fixed by soil-Ca at $10^{-2.5}$ M Ca^{2+}. Line b in this figure was developed similarly except that soil-Fe was used to control Fe^{2+} above $pe + pH$ of 11.53, since soil-Fe is more stable than magnetite in this redox range. Thus $Fe_2P_2O_7(c)$ can form in preference to β-$Ca_2P_2O_7(c)$ only at low redox and pH.

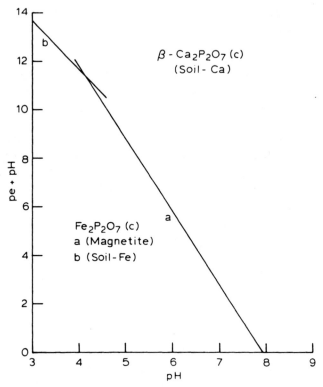

Fig. 12.14 Redox and pH conditions under which $Fe_2P_2O_7(c)$ and $\beta\text{-}Ca_2P_2O_7(c)$ are stable when Ca^{2+} is fixed by soil-Ca and Fe^{2+} is fixed by either soil-Fe or magnetite.

Since most soils tend to approach pH 7 as they become reduced, a $pe + pH$ < 3 would be required to convert $\beta\text{-}Ca_2P_2O_7(c)$ to $Fe_2P_2O_7(c)$. Since very few soils reach this low redox, the formation of $Fe_2P_2O_7(c)$ in soils is highly unlikely.

The availability of pyrophosphates to plants is not expected to differ much from that of orthophosphates since they are eventually converted to orthophosphates. Pyrophosphate (Fig. 12.13) maintains quite adequate levels of soluble phosphorus under acid soil conditions. Polyphosphates may have some advantage over orthophosphates in acid soils if particle size and placement effects are optimized. It is generally known that polyphosphates hydrolyze most rapidly under acid conditions. This is expected since protons are utilized (Reaction 12.13). Also the polyphosphate ions, which are the reactants in the hydrolysis reaction, are maintained at much greater activities under acid conditions.

12.12 ORTHOPHOSPHATE COMPLEXES IN SOLUTION

Since polyphosphates are unstable in soils and eventually convert to orthophosphates, it is useful to examine the solution complexes of orthophosphate in soil solution. These complexes are given by Reactions 47 through 54, and each species is plotted in Fig. 12.15 as a fraction of the total phosphorus in solution. This plot was developed as follows:

$$[\text{Total P}] = [\text{FeH}_2\text{PO}_4^+] + [\text{FeH}_2\text{PO}_4^{2+}] + [\text{FeHPO}_4^\circ] + [\text{FeHPO}_4^+]$$
$$+ [\text{CaH}_2\text{PO}_4^+] + [\text{CaHPO}_4^\circ] + [\text{CaPO}_4^-] + [\text{MgHPO}_4^\circ]$$
$$+ [\text{H}_3\text{PO}_4^\circ] + [\text{H}_2\text{PO}_4^-] + [\text{HPO}_4^{2-}]$$

(12.38)

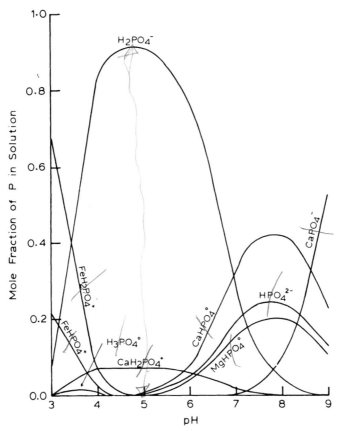

Fig. 12.15 The effect of pH on the phosphate species in soil solution when $Ca^{2+} = 10^{-2.5}$ M, $Mg^{2+} = 10^{-3}$ M. Fe^{3+} is controlled by soil-Fe and redox is set at $pe + pH$ of 11.53.

The mole fraction of any species, as for example, $[FeH_2PO_4^+]$ to the total phosphorus, can be expressed

$$MF_{FeH_2PO_4^+} = \frac{[FeH_2PO_4^+]}{[Total\ P]} \quad (12.39)$$

In order to evaluate the importance of each of the complexes on the right side of Eq. 12.38 under conditions expected in soils, the following conditions were selected: $Ca^{2+} = 10^{-2.5}\ M$, $Mg^{2+} = 10^{-3}\ M$, Fe^{3+} is controlled by soil-Fe, and Fe^{2+} is controlled by soil-Fe at $pe + pH$ of 11.53. This redox was arbitrarily selected as a reference point below which magnetite is stable and above which soil-Fe is stable. To simplify calculations for this development, activity coefficients were taken as unity. Each term on the right-hand side of Eq. 12.38 was expressed in terms of a formation constant, pH, and $H_2PO_4^-$ activity. For example, the equation for $FeH_2PO_4^+$ was obtained as follows.

	$\log K°$
$Fe^{2+} + H_2PO_4^- \rightleftharpoons FeH_2PO_4^+$	2.70
$Fe^{3+} + e^- \rightleftharpoons Fe^{2+}$	13.04
$Fe(OH)_3(soil) + 3H^+ \rightleftharpoons Fe^{3+} + 3H_2O$	2.70
$Fe(OH)_3(soil) + H_2PO_4^- + e^- + 3H^+ \rightleftharpoons FeH_2PO_4^+ + 3H_2O$	18.44

(12.40)

At $pe + pH = 11.53$ the reaction gives:

$$FeH_2PO_4^+ = 10^{6.91 - 2\,pH}(H_2PO_4^-) \quad (12.41)$$

Similar expressions were developed for each of the terms in Eq. 12.38 and substituted into Eq. 12.39. The $(H_2PO_4^-)$ term is common to both numerator and denominator and cancels leaving only formation constants and pH terms. Thus Eq. 12.39 becomes

$$MF_{FeH_2PO_4^+} = \frac{10^{6.91-2\,pH}}{\begin{array}{c}10^{6.91-2\,pH} + 10^{8.13-3\,pH} + 10^{0.61-pH} + 10^{6.41-2\,pH} \\ + 10^{-1.10} + 10^{-6.96+pH} + 10^{-15.59+2\,pH} + 10^{-7.29+pH} \\ + 10^{2.15-pH} + 1 + 10^{-7.20+pH}\end{array}} \quad (12.42)$$

Equation 12.42 is plotted as the $FeH_2PO_4^+$ line in Fig. 12.15. Similar equations were developed for each of the other species.

The predominant phosphate species in solution in the pH range of 3.5 to 7.0 is $H_2PO_4^-$. Below pH 4 the ferrous phosphate complex, $FeH_2PO_4^+$, increases rapidly. This complex becomes even more significant at lower redox, but diminishes at higher redox. The ferrous complex $FeHPO_4^o$ is not significant at $pe + pH$ of 11.53 so it does not appear in Fig. 12.15. The ferric complex $FeHPO_4^+$ increases below pH 4, whereas the line for $FeH_2PO_4^{2+}$ barely touches the lower left-hand corner of this diagram.

At higher pH calcium and magnesium complexes become important. In fact $CaHPO_4^o$ comprises 0.42 of the total phosphorus in solution at pH 7.8 compared to 0.24 for HPO_4^{2-} and 0.20 for $MgHPO_4^o$. At pH values above 8, $CaPO_4^-$ rapidly becomes important, but the presence of $CaCO_3$(calcite) can be expected to reduce this ion.

It is obvious from Fig. 12.15 that solution complexes must be considered in estimating phosphate activities from total phosphorus in solution. Computer programs can be developed to estimate solution complexes from elemental analyses of the soil solution. Ionic strength estimates also can be made to convert from concentrations to activities. To do this, it is necessary to recognize the major solution complexes and have reliable formation constants for them. Specific ion electrodes can be used to measure directly the activities of certain ions, but further investigations are needed to obtain reliable estimates of ion activities in soils in order to test important solubility relationships such as those developed in this chapter.

12.13 REACTIONS OF PHOSPHATE FERTILIZERS WITH SOILS

Reactions of phosphate fertilizers with soils have received considerable attention over the years. Particular interest has centered around concentrated superphosphate, which for many years was considered the standard fertilizer to which all other phosphate fertilizers were compared. The major phosphate mineral in this fertilizer is monocalcium phosphate monohydrate (MCP). The reactions of MCP with soils have been widely studied (Brown and Lehr, 1959; Lehr et al., 1959; Lindsay and Stephenson, 1959a,b) and provide an interesting insight to the dissolution and precipitation reactions by which fertilizer phosphorus is converted to more stable phosphorus minerals in soils.

Let us consider a fertilizer granule of MCP in the soil (Fig. 12.16). Initially liquid water moves into the granule from the surrounding soil, as water vapor condenses inside the granule. A solution is formed from the dissolving MCP that moves by capillarity out of the granule into the surrounding soil. In the case of MCP the solution that forms within the dissolving granule is supersaturated with respect to $CaHPO_4 \cdot 2H_2O$(DCPD) and $CaHPO_4$(DCP). As a result these minerals precipitate at the granule site during the dissolution of MCP. Precipitation of DCPD gives a solution of pH 1.48, while the

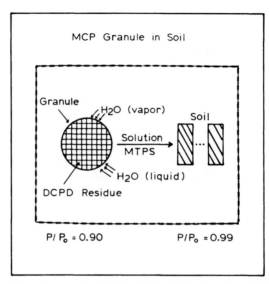

Fig. 12.16 Dissolution of a granule of MCP in soils with the formation of MTPS and a residue of DCPD.

precipitation of DCP gives a pH of 1.01. These pH values are much lower than 4.68 expected for the congruent dissolution of $Ca(H_2PO_4)_2 \cdot H_2O$ into Ca^{2+} and $H_2PO_4^-$ ions (Fig. 12.1).

The actual solution compositions that accompany the dissolution of MCP can be seen from the solubility diagram of the $CaO-P_2O_5-H_2O$ system shown in Fig. 12.17 taken from Lindsay and Stephenson (1959a). Shown here are solubility isotherms for DCP, DCPD, and MCP. Composition of the solution phase resulting from the dissolution of MCP would normally follow the dotted line labeled "congruent dissolution of MCP" in which 2 moles of phosphorus enters solution for each mole of calcium. Experimental findings obtained by shaking MCP in water for periods varying from 15 sec to 17 days are shown in Fig. 12.17 and show departure from the congruent dissolution line. After an hour the solution composition was still at MTPS (metastable triple point solution), which is the solution approaching equilibrium with dissolving MCP and precipitating DCPD. The solution composition remained at MTPS for more than 24 hr, then slowly shifted up the MCP isotherm to TPS (stable triple point solution) which represents final equilibrium with MCP and DCP. The solution MTPS is termed "metastable" because it is supersaturated with respect to DCP. At MTPS, DCP precipitates and DCPD dissolves. Between MTPS and TPS, DCP precipitates and MCP dissolves. Thus TPS represents the stable triple point solution in which both MCP and DCP are stable.

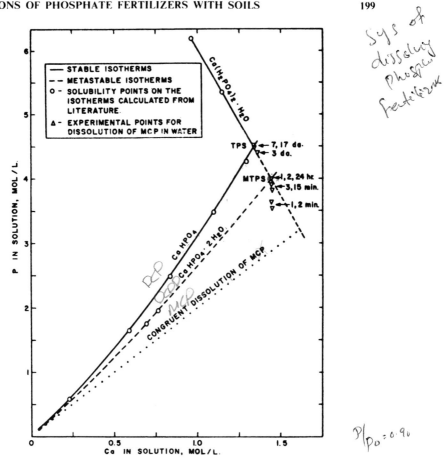

Fig. 12.17 Solubility isotherms in the concentrated region of the CaO—P_2O_5—H_2O system at 25°C. The experimental points show the dissolution of MCP with time (Lindsay and Stephenson, 1959a).

The compositions of metastable triple point solution (MTPS) and stable triple point solution (TPS) are represented here:

Triple Point	pH	P (moles/liter)	Ca (moles/liter)
MTPS	1.48	3.98	1.44
TPS	1.01	4.49	1.34

Source: Lindsay and Stephenson (1959a).

The solution resulting from the dissolution of MCP is highly concentrated and acidic. The water vapor pressure of these solutions represents a relative humidity near 90% compared to 98 to 100% for most soils. This vapor pressure gradient accounts for the movement of water vapor into the granule simultaneously with movement of MTPS out of the granule.

During dissolution of MCP, the precipitation of DCPD releases $H_3PO_4^\circ$ according to the reaction

$$Ca(H_2PO_4)_2 \cdot H_2O(c) + H_2O \rightleftharpoons CaHPO_4 \cdot 2H_2O + H_3PO_4^\circ$$

(12.43)

The released $H_3PO_4^\circ$ accounts for the low pH of MTPS. Similarly precipitation of DCP to give TPS reflects even greater release of $H_3PO_4^\circ$ as the pH drops to 1.01.

After observing MCP fertilizer granules in soils, Brown and Lehr (1959) and Lehr et al (1959) showed that both DCPD and DCP precipitate at the fertilizer site. Apparently DCPD forms more readily than DCP. Some initially formed DCPD later dissolves as DCP eventually forms. Because of the rapid nature of the initial reactions, Lindsay et al. (1959) concluded that MTPS rather than TPS is most likely the solution that moves out of a dissolving granule of MCP.

An example of how the solubility diagram of the concentrated region of the $CaO-P_2O_5-H_2O$ system can be used to understand solubility relationships involving MTPS, TPS, DCPD, DCP, and MCP is shown in Fig. 12.18. Points on this diagram represent solution compositions in terms of moles per liter of Ca and P. Regions to the right of the DCP, MCP, and DCPD isotherms represent regions of supersaturation in which solids will precipitate. If MCP precipitates, the solution composition moves to lower concentrations along a line of 2:1 slope showing removal of 2 moles of phosphorus for each mole of calcium to form solid phase $Ca(H_2PO_4)_2 \cdot H_2O$. If either DCP or DCPD precipitate, the solution composition moves along a line of 1:1 slope reflecting the removal of 1 mole of phosphorus for each mole of calcium.

Systems whose *total* composition would fall in Region A will precipitate MCP as the only stable phase. In Region B MCP may precipitate initially, but it will partially convert to DCP. The final solution will be TPS with both MCP and DCP present. In Region C MCP precipitates initially, then partially converts to DCPD forming MTPS. As DCP slowly precipitates, DCPD dissolves and the solution composition again rests at TPS with both MCP and DCP present. In Region D MCP precipitates initially, then converts in part to DCPD forming MTPS. DCP slowly precipitates as the solution composition moves up the MCP line until all MCP is dissolved,

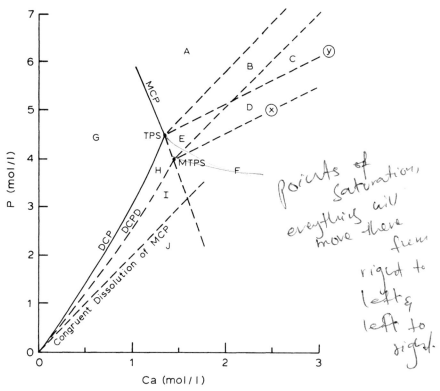

Fig. 12.18 Stability fields and solubility isotherms in the concentrated region of the $CaO-P_2O_5-H_2O$ system at 25°C (adapted from Fig. 12.17, see text for explanation).

then the solution composition moves to the DCP line leaving this mineral as the only stable solid phase. In Region E MCP precipitates initially, then converts to DCP, and the solution composition moves to the DCP line again with DCP as the only solid phase. In Region F MCP precipitates initially, then converts to DCPD. Slowly with time DCPD converts to DCP, and the final solution composition will lie on the DCP line with DCP as the only solid phase present. In Region G no solid phases will precipitate and the solutions are stable. In Region H only DCP will precipitate and the solution will end up on the DCP line. In Regions I and J, DCPD precipitates initially moving the solution composition to the DCPD line. Slowly DCPD converts to DCP, and the solution ends up on the DCP line with DCP as the only solid phase.

From Fig. 12.18 it is possible to estimate the fraction of MCP in a fertilizer granule that precipitates as DCPD as MTPS forms and moves into the surrounding soil. Extrapolation of the congruent dissolution line for MCP

(2:1 slope) to point x where precipitation of DCPD (1:1 slope) produces MTPS gives the necessary information. Thus, 21.6% of the P in dissolving MCP is converted to DCPD by forming MTPS. If TPS were the solution moving out of the granule, sufficient MCP would dissolve to reach point y and precipitation of DCP to TPS represents 28.8% of phosphorus dissolving from MCP. The equation that permits the calculation of these fractions is

$$f = \frac{P - 2Ca}{2P - 2Ca} \quad (12.44)$$

where f is the fraction of P from MCP that precipitates as DCP or DCPD at the granule site and P and Ca are the molar concentrations of phosphorus and calcium in the solution that moves out into the soil.

The reactions of TPS and MTPS with soils have been studied in detail (Lindsay and Stephenson, 1959b; Lindsay et al., 1959). In these studies 2:1 suspensions of TPS or MTPS with soil were shaken for various periods. The suspensions were then filtered, and the resulting filtrate was reacted with new soil to simulate the successive reactions of fertilizer solution with soil as it slowly moves out of a dissolving fertilizer granule.

Typical results of reacting TPS with soils are shown in Fig. 12.19. The very low pH of TPS (pH 1.01) causes iron and aluminum, as well as other cations, to come into solution. Slowly these neutralization reactions with soil bases cause pH to increase. Evidence that precipitation reactions are also occurring is reflected by the curves in Fig. 12.19, which show loss of aluminum and iron

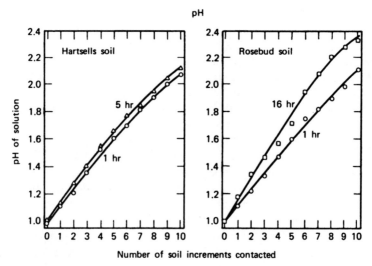

(a)

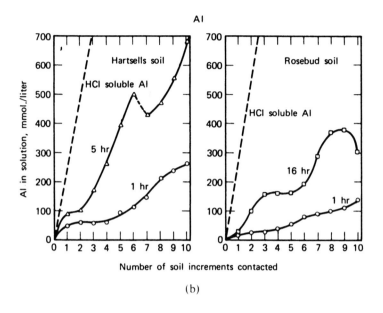

(b)

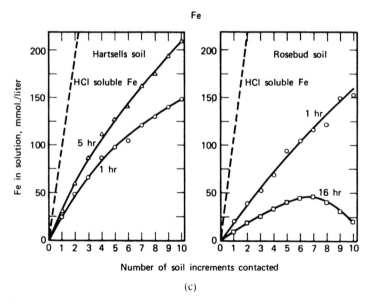

(c)

Fig. 12.19 Increase in pH and dissolution of Al and Fe from two soils during the reaction of TPS with successive increments of soil (from Lindsay and Stephenson, 1959b).

from solution as the reactions proceed. Aluminum attained 0.7 M and iron 0.2 M during reaction with the acid Hartsell soil (Fig. 12.19).

By sampling the filtrates during the reaction process just described, it has been possible to identify many of the initial reaction products of phosphate fertilizers in soils. Thirty-two phosphate minerals were identified in some of the initial studies in which soils were reacted with phosphate solutions (Lindsay et al., 1962). Since then further investigations have been made of the reaction products of fertilizers and many of these products have been characterized (Lehr et al., 1968).

Inclusion of cations such as NH_4^+, K^+, Ca^{2+}, Mg^{2+} in fertilizers enables these cations to be included among the initial reaction products. For example, MCP contains sufficient Ca^{2+} to precipitate half of the phosphorus as DCPD or DCP. In acid soils iron and aluminum generally precipitate the additional phosphorus. In calcareous soils Ca^{2+} is readily dissolved in the acid fertilizer reaction zone enabling most of the phosphorus to precipitate initially as DCPD and DCP.

The stability relationships of the reaction products of phosphate fertilizer in soils and the subsequent minerals to which they go can best be seen from the various stability diagrams developed and discussed earlier in this chapter.

REFERENCES

Arden, T. V. 1950. The solubility products of ferrous and ferrosic hydroxides. J. Chem. Soc. 882–885.

Aslyng, H. C. 1954. The Lime and Phosphate Potentials of Soils; The Solubility and Availability of Phosphates. Royal Veterinary and Agricultural College, Copenhagen, Denmark, Yearbook.

Bray, R. R. and L. T. Kurtz. 1945. Determination of total organic and available forms of phosphorus rocks. Soil Sci. 59:39–45.

Brown, W. E. and J. R. Lehr. 1959. Application of phase rule to the chemical behavior of monocalcium phosphate monohydrate in soils. Soil Sci. Soc. Am. Proc. 23:7–12.

Chien, S. H. and C. A. Black. 1976. Free energy of formation of carbonate apatites in some phosphate rocks. Soil Sci. Soc. Am. J. 40:234–239.

Fleming, J. D. 1969. Polyphosphates are revolutionizing fertilizers. Part I. What polyphosphates are: Farm Chem. 132:30–36.

Lehr, J. R., W. E. Brown, and E. H. Brown. 1959. Chemical behaviour of monocalcium phosphate monohydrate in soils. Soil Sci. Soc. Am. Proc. 23:3–12.

Lehr, J. R., E. H. Brown, A. W. Frazier, J. P. Smith, and R. D. Thrasher. 1967. Crystallographic properties of fertilizer compounds. Chem. Eng. Bull. 6, Tennessee Valley Authority, Muscle Shoals, Alabama.

Lindsay, W. L., A. W. Frazier, and H. F. Stephenson. 1962. Identification of reaction products from phosphate fertilizers in soils. Soil Sci. Soc. Am. Proc. 26:446–452.

Lindsay, W. L., J. R. Lehr, and H. F. Stephenson. 1959. Nature of the reactions of monocalcium phosphate monohydrate in soils: III. Studies with metastable triple-point solution. Soil Sci. Soc. Am. Proc. 23:342–345.

Lindsay, W. L. and E. C. Moreno. 1960. Phosphate phase equilibria in soils. Soil Sci. Soc. Am. Proc. 24:177–182.

Lindsay, W. L., M. Peech, and J. S. Clark. 1959. Solubility criteria for the existence of variscite in soils. Soil Sci. Soc. Am. Proc. 23:357–360.

Lindsay, W. L. and H. F. Stephenson. 1959a. The nature of the reactions of monocalcium phosphate monohydrate in soils: I. The solution that reacts with the soils. Soil Sci. Soc. Am. Proc. 23:12–18.

Lindsay, W. L. and H. F. Stephenson. 1959b. Nature of the reactions of monocalcium phosphate monohydrate in soils: II. Disolution and precipitation reactions involving iron, aluminum, manganese, and calcium. Soil Sci. Soc. Am. Proc. 23:18–22.

Lindsay, W. L. and P. L. G. Vlek. 1977. Phosphate minerals. *In* J. B. Dixon and S. B. Weed, Eds., Minerals in the Soil Environment, pp. 639–672. Soil Science Society of America, Madison, Wi.

Moreno, E. C., W. L. Lindsay, and G. Osborn. 1960. Reactions of dicalcium phosphate dihydrate in soils. Soil Sci. 90:58–68.

Olsen, S. R., C. V. Cole, F. S. Watanabe, and L. A. Dean. 1954. Estimation of available phosphorus in soils by extraction with sodium bicarbonate. U.S. Dept. Agric. Circular 939.

Patrick, W. H., Jr., S. Gotoh, and B. G. Williams. 1973. Strengite dissolution in flooded soils and sediments. Science 179:564–565.

Patrick, W. H., Jr. and I. C. Mahapatra. 1968. Transformations and availability to rice of nitrogen and phosphorus in waterlogged soils. Adv. Agron. 20:323–359.

Ponnamperuma, F. N., E. M. Tianco, and T. Loy. 1967. Redox equilibria in flooded soils: I. The iron hydroxide systems. Soil Sci. 103:374–382.

Ponnamperuma, F. N. 1972. The chemistry of submerged soils. Adv. Agron. 24:29–96.

Taylor, A. W., W. A. Frazier, E. L. Gurney, and J. P. Smith. 1963a. Solubility products of di- and trimanganesium phosphates and the dissociation of magnesium phosphate solutions. Trans. Faraday Soc. 59:1585–1589.

Taylor, A. W., A. W. Frazier, and E. L. Gurney. 1963b. Solubility products of magnesium ammonium and magnesium potassium phosphates. Trans. Faraday Soc. 59:1580–1584.

Williams, B. G. and W. H. Patrick, Jr. 1973. The dissolution of complex ferric phosphates under controlled Eh and pH conditions. Soil Sci. Soc. Am. Proc. 37:33–96.

PROBLEMS

12.1 Develop the equation for plotting the mole fraction of HPO_4^{2-} in Fig. 12.1.

12.2 Based on the dissociation constants of $H_3PO_4^\circ$ calculate and sketch an approximate pH titration curve for the neutralization of 10 mmol of $H_3PO_4^\circ$ with 1 N NaOH. Consider activity coefficients as unity and ignore dilution effects. Point out the pH regions of greatest buffering capacity.

12.3 Prepare a plot of the various orthophosphoric acid species in solution in equilibrium with variscite when Al^{3+} is governed by kaolinite and quartz. How would this plot change if Al^{3+} were controlled by gibbsite?

12.4 The following measurements were made in 0.01 M $CaCl_2$ extracts of an acid soil at various periods after adding MCP.

P added (ppm)	pH			Total Al (μM)			Total P (μM)		
	24 hr	6 mo	18 mo	24 hr	6 mo	18 mo	24 hr	6 mo	18 mo
0	4.11	4.08	4.10	248	195	146	0.52	0.28	0.52
110	4.09	4.10	4.09	157	143	131	2.89	0.80	1.04
220	4.07	4.16	4.11	115	100	90	8.84	1.52	1.61

Calculate the activities of Al^{3+} and $H_2PO_4^-$ and plot this information on a diagram similar to that shown in Fig. 12.3. Interpret and discuss the results in terms of equilibrium with variscite, gibbsite, and kaolinite-quartz.

12.5 Develop the necessary equations to plot a solubility line for strengite in equilibrium with maghemite in Fig. 12.6. From this figure discuss how the phosphate status of acid soils may be expected to change as these soils go through natural weathering processes from young to old soils.

12.6 On the basis of Fig. 12.7 suggest a way in which the phosphorus in a acid soil can be made more available without the addition of more phosphate fertilizer.

12.7 Consider rice growing in a neutral soil containing magnetite and vivianite in which $pe + pH = 4.5$. Next to the root the redox rises to $pe + pH = 9$. From Fig. 12.7 estimate the levels of $H_2PO_4^-$ available for the following:
   a. Leaching.
   b. Uptake by the rice root.

12.8 From Fig. 12.8 explain why soils in the pH range of 6 to 6.5 generally show the greatest availability of both soil and fertilizer phosphorus.

12.9 Account for the different slopes of the phosphate lines in Fig. 12.8. How would these lines shift if Ca^{2+} decreased to $10^{-3.5}$ M or attained equilibrium with calcite at 0.01 atm of $CO_2(g)$, which ever was smaller?

12.10 If equilibrium were attained with hydroxyapatite in a soil extracted for available phosphorus with the Olsen et al. (1954) bicarbonate soil test, how many ppm of extractable phosphorus (soil weight basis) would be extracted? In their procedure 5 g of soil is shaken for 30 minutes with 100 ml of 0.5 M $NaHCO_3$ solution adjusted to pH 8.5 with NaOH. The suspension is filtered and phosphorus is determined in the extract. Use the Davies' equation to estimate activity coefficients in solving this problem.

12.11 Develop the equation used to plot the β-TCP line in Fig. 12.9.

12.12 Under what conditions can fluorapatite and hydroxyapatite support the same level of phosphorus in soils? If this occurred in a soil of pH 6 with Ca^{2+} activity at $10^{-2.5}$ M, what can you say about the following:
   a. The phosphate level in this soil?
   b. The activity of F^- in this soil?

12.13 Hydroxyapatite attains equilibrium at log $H_2PO_4^-$ $-$ pH $= -12$. From Fig. 12.9 estimate the following parameters for a soil of pH 6.2:
   a. The activity of Ca^{2+}.
   b. The activity of HPO_4^{2-}.

12.14 From Fig. 12.7 and 12.8 estimate the approximate redox levels that are necessary in a submerged soil at pH 7.0 containing magnetite in order that a) $CaHPO_4$(monetite) and b) hydroxyapatite can be transformed into vivianite. On the basis of these estimates discuss the possibility of increasing phosphorus availability in calcareous soils through flooding.

12.15 Using Fig. 12.10 discuss the conditions under which fluorphlogopite, fluorite, fluorapatite, and/or hydroxyapatite are
   a. Stable.
   b. Unstable.

12.16 In Fig. 12.12 develop the necessary equations and calculate the $pe + pH$ at which $MnHPO_4(c)$, rhodochrosite, and manganite can all coexist at $CO_2(g)$ at 0.0003 atm. If this soil were also calcareous, could hydroxyapatite also attain equilibrium under these conditions? Explain.

12.17 If the solubility relationships of $MnHPO_4(c)$ indicated in Fig. 12.12 are correct, discuss briefly how phosphate can affect the solubility relationships of Mn^{2+} in Fig. 11.2.

12.18 Develop the equation used to plot the $CaP_2O_7^{2-}$ species in Fig. 12.13. Also develop the equation for this species in the presence of calcite when $CO_2(g)$ is 0.003 atm.

12.19 Develop the equation for line *b* in Fig. 12.14 and explain its significance.

12.20 Develop the equation used to plot the $CaHPO_4^0$ line in Fig. 12.15. Of what significance are the relationships shown in this figure?

12.21 Calculate the mole fraction of the soluble orthophosphate species in a soil at pH 7 having an *Eh* of -178 mv, 10^{-3} M Mg^{2+}, 0.1 atm of $CO_2(g)$, if the soil contains both calcite and vivianite.

12.22 On a diagram similar to that given in Fig. 12.18 sketch the pathway of solution compositions that would result from the indicated operations, and discuss the sequence of reactions that accompany the changes in solution composition.
  a. 0.5 moles of MCP are added to 1 liter of 2 M H_3PO_4.
  b. 1.5 moles of MCP are added to 1 liter of 3 M H_3PO_4.
  c. 1.0 moles of MCP are added to 1 liter of H_2O.
  d. 2.0 moles of MCP are added to 1 liter of H_2O.
  e. 2.8 moles of MCP are added to 1 liter of H_2O.
  f. Excess MCP is added to water.
  g. MTPS is allowed to stand in a closed beaker.
  h. MTPS reacts with acid soils.
  i. MTPS reacts with a calcareous soil.

12.23 From Fig. 12.18 estimate graphically the fraction of phosphorus in MCP that will be converted to DCPD if MTPS moves out into the soil. Also estimate the fraction of phosphorus in MCP that will be converted to DCP when TPS moves into the soil.

12.24 Develop a general equation that gives the fraction of MCP converted to either DCPD or DCP in arriving at any point along either the DCPD isotherm or the DCP isotherm.

12.25 Explain why MTPS is expected to be the solution that forms in a dissolving MCP fertilizer granule and moves into the surrounding soil.

12.26 Explain how H_2O can move into a dissolving MCP fertilizer granule at the same time that MTPS moves out into the surrounding soil.

12.27 Read the paper "Reactions of dicalcium phosphate dihydrate in soils" (Moreno et al., 1960) and be prepared to discuss the chemical reactions and mineral transformations that occur in fertilizer reaction zones in soils.

12.28 How many stable invarient point solutions are there in the

PROBLEMS 209

CaO—P_2O_5—H_2O system? Describe these by appropriate equilibrium reactions and discuss their significance.

12.29 The Bray and Kurtz (1945) soil test extractants for phosphorus consist of (1) 0.03 N NH_4F in 0.025 N HCl or (2) 0.03 N NH_4F in 0.1 N HCl. Discuss the chemical basis of these extractants and the types of soils in which they may be expected to be most successful.

THIRTEEN
ZINC

The zinc content of the lithosphere is approximately 80 ppm, and the common range for soils is 10–300 ppm with an average content of 50 ppm. Zinc is considered a trace element in soils. If all of the zinc present in an average soil were in solution at 10% moisture, it would comprise only $10^{-2.12}$ M zinc (Table 1.1).

The specific minerals controlling the activity of Zn^{2+} in soils are not known. Instead a reference solubility is given in this text which permits Zn^{2+} solubility relationships to be compared to those of known zinc minerals. The stability relationships of zinc chelates are considered in Chapter 15 and those of zinc sulfides in Chapter 17. A review showing how zinc solubility in soils affects the availability of zinc to plants was published earlier (Lindsay, 1972).

13.1 OXIDATION STATE OF ZINC

The oxidation state of zinc in natural environments like soils is exclusively Zn^{2+}. Metallic $Zn(c)$ forms only in highly reducing environments

$$Zn^{2+} + 2e^- \rightleftharpoons Zn(c) \qquad \log K^\circ = -25.80 \qquad (13.1)$$

from which

$$pe = -12.90 + \tfrac{1}{2} \log Zn^{2+} \qquad (13.2)$$

Thus if Zn^{2+} were one molar, a pe of -12.90 would be required to form $Zn(c)$. Since this is below the stability of water (Fig. 2.1), $Zn(c)$ can not form in soils.

13.2 SOLUBILITY OF ZINC MINERALS IN SOILS

The solubilities of various Zn minerals that may occur in soils are given by Reactions 2 through 17 of Table 13.1. Reaction 9 in this table gives the solubility of soil-Zn based on the experimental measurements of Norvell and Lindsay (1969, 1972). In their studies ZnEDTA and ZnDTPA were reacted with soils of different pH, and the total zinc remaining in solution was measured after 30 days. It will be shown in Chapter 15 that these measurements provide an estimation of the equilibrium constant for the reaction

$$\text{soil-Zn} + 2H^+ \rightleftharpoons Zn^{2+} \qquad \log K^\circ = 5.8 \qquad (13.3)$$

which gives

$$\log Zn^{2+} = 5.8 - 2pH \qquad (13.4)$$

Equation 13.4 provides a very useful reference solubility for Zn^{2+} in soils.

TABLE 13.1 EQUILIBRIUM CONSTANTS FOR VARIOUS REACTIONS INVOLVING ZINC

Reaction No.	Equilibrium Reaction	log $K°$
	Metal Zinc	
1	$Zn^{2+} + 2e^- \rightleftharpoons Zn(c)$	−25.80
	Oxides and Hydroxides	
2	$Zn(OH)_2(amorp) + 2H^+ \rightleftharpoons Zn^{2+} + 2H_2O$	12.48
3	$\alpha\text{-}Zn(OH)_2(c) + 2H^+ \rightleftharpoons Zn^{2+} + 2H_2O$	12.19
4	$\beta\text{-}Zn(OH)_2(c) + 2H^+ \rightleftharpoons Zn^{2+} + 2H_2O$	11.78
5	$\gamma\text{-}Zn(OH)_2(c) + 2H^+ \rightleftharpoons Zn^{2+} + 2H_2O$	11.74
6	$\varepsilon\text{-}Zn(OH)_2(c) + 2H^+ \rightleftharpoons Zn^{2+} + 2H_2O$	11.53
7	$ZnO(zincite) + 2H^+ \rightleftharpoons Zn^{2+} + H_2O$	11.16
	Carbonates	
8	$ZnCO_3(smithsonite) + 2H^+ \rightleftharpoons Zn^{2+} + CO_2(g) + H_2O$	7.91
	Soil-Zn and Zn-Fe Oxide	
9	$soil\text{-}Zn + 2H^+ \rightleftharpoons Zn^{2+}$	5.80
10	$ZnFe_2O_4(franklinite) + 8H^+ \rightleftharpoons Zn^{2+} + 2Fe^{3+} + 4H_2O$	9.85
	Silicates	
11	$ZnSiO_3(c) + 2H^+ + H_2O \rightleftharpoons Zn^{2+} + H_4SiO_4°$	*
12	$Zn_2SiO_4(willemite) + 4H^+ \rightleftharpoons 2Zn^{2+} + H_4SiO_4°$	13.15
	Chlorides	
13	$ZnCl_2(c) \rightleftharpoons Zn^{2+} + 2Cl^-$	7.07
	Sulfates	
14	$ZnSO_4(zinkosite) \rightleftharpoons Zn^{2+} + SO_4^{2-}$	3.41
15	$ZnO \cdot 2ZnSO_4(c) + 2H^+ \rightleftharpoons 3Zn^{2+} + 2SO_4^{2-} + H_2O$	19.12
16	$Zn(OH)_2 \cdot ZnSO_4(c) + 2H^+ \rightleftharpoons 2Zn^{2+} + SO_4^{2-} + 2H_2O$	7.50
	Phosphates	
17	$Zn_3(PO_4)_2 \cdot 4H_2O(hopeite) + 4H^+ \rightleftharpoons 3Zn^{2+} + 2H_2PO_4^- + 4H_2O$	3.80

TABLE 13.1 *(Continued)*

Reaction No.	Equilibrium Reaction	log $K°$
	Hydrolysis Species	
18	$Zn^{2+} + H_2O \rightleftharpoons ZnOH^+ + H^+$	−7.69
19	$Zn^{2+} + 2H_2O \rightleftharpoons Zn(OH)_2° + 2H^+$	−16.80
20	$Zn^{2+} + 3H_2O \rightleftharpoons Zn(OH)_3^- + 3H^+$	−27.68
21	$Zn^{2+} + 4H_2O \rightleftharpoons Zn(OH)_4^{2-} + 4H^+$	−38.29
	Chloride Complexes	
22	$Zn^{2+} + Cl^- \rightleftharpoons ZnCl^+$	0.43
23	$Zn^{2+} + 2Cl^- \rightleftharpoons ZnCl_2°$	0.00
24	$Zn^{2+} + 3Cl^- \rightleftharpoons ZnCl_3^-$	0.50
25	$Zn^{2+} + 4Cl^- \rightleftharpoons ZnCl_4^{2-}$	0.20
	Other Complexes	
26	$Zn^{2+} + H_2PO_4^- \rightleftharpoons ZnH_2PO_4^+$	1.60
27	$Zn^{2+} + H_2PO_4^- \rightleftharpoons ZnHPO_4° + H^+$	−3.90
28	$Zn^{2+} + NO_3^- \rightleftharpoons ZnNO_3^+$	0.40
29	$Zn^{2+} + 2NO_3^- \rightleftharpoons Zn(NO_3)_2°$	−0.30
30	$Zn^{2+} + SO_4^{2-} \rightleftharpoons ZnSO_4°$	2.33

* Reported value not reliable (see Norvell and Lindsay, 1970).

The solubilities of several zinc minerals are plotted in Fig. 13.1. In order of decreasing solubility these minerals include

$Zn(OH)_2(amorp) > \alpha\text{-}Zn(OH)_2 > \beta\text{-}Zn(OH)_2 > \gamma\text{-}Zn(OH)_2$
$> \varepsilon\text{-}Zn(OH)_2 > ZnCO_3(smithsonite) > ZnO(zincite)$
$> Zn_2SiO_4(willemite) > soil\text{-}Zn > ZnFe_2O_4(franklinite).$

These minerals represent a range in Zn^{2+} solubility of 10^8. Franklinite can be the most insoluble of these minerals, depending upon the Fe(III) oxides controlling Fe^{3+} activity. The crystalline zinc hydroxides (Reactions 3 through 6 of Table 13.1) have similar solubilities and lie between $Zn(OH)_2(amorp)$ and $ZnO(zincite)$. These minerals, along with $ZnCO_3$ (smithsonite), are much more soluble than $Zn_2SiO_4(willemite)$. The solubilities of all zinc minerals in Fig. 13.1 decrease 100-fold for each unit

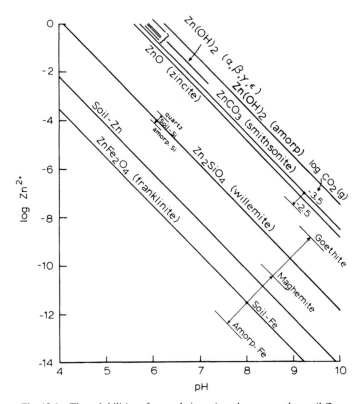

Fig. 13.1 The solubilities of several zinc minerals compared to soil-Zn.

increase in pH. The effect of different levels of $H_4SiO_4^0$ (controlled by quartz, soil-Si, or amorphous silica) and of $CO_2(g)$ on the solubilities of these zinc minerals is also indicated.

Earlier Lindsay and Norvell (1969) hypothesized that $ZnSiO_3(c)$ in equilibrium with SiO_2(amorp) may account for the Zn^{2+} solubility in soils depicted by Eq. 13.4. Further investigations into the solubility of $ZnSiO_3(c)$ reported in the literature showed that value to be in error (Norvell and Lindsay, 1970). From Fig. 13.1 it appears that $ZnFe_2O_4$(franklinite) may account for the solubility of Zn^{2+} depicted by soil-Zn. The soil-Zn line can be expected to shift somewhat for different soils. Likewise the solubility of franklinite shifts depending on the activity of Fe^{3+}. For example, $Fe(OH)_3$(amorp) depresses the solubility of franklinite, whereas crystalline Fe(III) oxides such as maghemite or goethite lower Fe^{3+} and permit higher equilibrium levels of Zn^{2+} in Fig. 13.1. Added to this complexity is the fact that franklinite may be amorphous and other cations of similar size may partially substitute for

SOLUBILITY OF ZINC MINERALS IN SOILS

Zn^{2+}. Thus it is easy to see why Zn^{2+} solubility in soils can result in a range of solubilities. Iron oxides form analogous minerals with other trace elements like Cu^{2+}, Mn^{2+}, Co^{2+}, etc. Further investigations are needed to examine the more general applicability of the soil-Zn level of soluble Zn^{2+} in soils, and the possibility that franklinite in equilibrium with Fe(III) oxides possibly control Zn^{2+} solubility in soils.

All of the $Zn(OH)_2$ minerals, ZnO(zincite), and $ZnCO_3$(smithsonite) shown in Fig. 13.1 are too soluble to persist in soils. They are about 10^5 times more soluble than soil-Zn. These minerals make good zinc fertilizers in soils because they dissolve sufficiently to maintain levels of Zn^{2+} that are adequate for plants. The mineral Zn_2SiO_4(willemite) is of intermediate solubility, but it is too soluble to account for the soil-Zn found in most soils.

The solubilities of various zinc chlorides, sulfates, oxysulfates, and phosphates are given by Reactions 13 through 17 of Table 13.1. The log $K°$ values of these minerals indicate that greater than 1 M levels of Zn^{2+} would be necessary for $ZnCl_2(g)$, $ZnSO_4$(zinkosite), and $ZnO \cdot 2ZnSO_4$(c) to form in soils. Hence these minerals are not expected in soils.

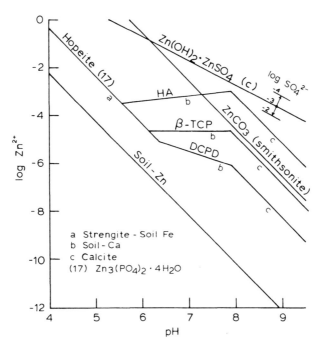

Fig. 13.2 The solubilities of hopeite and $ZnO \cdot ZnSO_4$(c) compared to other zinc minerals when phosphate is fixed by various iron and calcium phosphates (drawn for $CO_2(g) = 0.0003$ atm).

The minerals $Zn(OH)_2 \cdot ZnSO_4$(c) and $Zn_3(PO_4)_2 \cdot 4H_2O$(hopeite) are plotted in Fig. 13.2 along with soil-Zn and $ZnCO_3$(smithsonite) for comparison. Obviously $Zn(OH)_2$. $ZnSO_4$ is too soluble to be of much significance in soils even at high levels of SO_4^{2-}. Hopeite is plotted in this figure on the basis that $H_2PO_4^-$ is fixed by strengite-soil-Fe at low pH and by $CaHPO_4 \cdot 2H_2O$(DCPD), β-$Ca_3(PO_4)_2$(β-TCP) or $Ca_5(PO_4)_3OH$(HA) at higher pH when Ca^{2+} is fixed either by soil-Ca at $10^{-2.5}$ M Ca^{2+} or by $CaCO_3$(calcite) and CO_2(g). The lines for hopeite shift upward as more insoluble phosphates control phosphate solubility as shown here for DCPD, β-TCP, and hydroxyapatite.

The conclusion that can be drawn from Fig. 13.2 is that hopeite is generally more soluble than soil-Zn, but it is less soluble than zinc oxides, hydroxides, or carbonates. In soils it can be expected to furnish available zinc to plants. Since soils normally contain more phosphorus than zinc, it is more realistic to consider how phosphate may control zinc solubility than how zinc could control phosphorus solubility. Figure 13.2 was drawn with this objective in mind.

13.3 ZINC SPECIES IN SOLUTION

The zinc hydrolysis species in aqueous solution are given by Reactions 18 through 21 of Table 13.1 and are plotted in Fig. 13.3 in equilibrium with soil-Zn.

The predominant zinc species in solution below pH 7.7 is Zn^{2+}, although $ZnOH^+$ is more prevalent above this pH. The neutral species $Zn(OH)_2^\circ$ is the major species above pH 9.11, whereas the species $Zn(OH)_3^-$ and $Zn(OH)_4^{2-}$ are never major solution species in the pH range of soils.

Zinc also forms complexes with chloride, phosphate, nitrate, and sulfate. Equilibrium reactions for these species are given by Reactions 22 through 30 of Table 13.1 and are plotted in Fig. 13.4.

This plot gives a summary view of the zinc complexes in solution and the conditions that affect their stabilities. The complex $ZnSO_4^\circ$ is very important in soils and can contribute significantly to total zinc in solution. According to Reaction 30 of Table 13.1 the activity of this species equals that of Zn^{2+} when SO_4^{2-} is $10^{-2.33}$ M. Inclusion of sulfate in zinc fertilizer is often beneficial because this complex increases the solubility and mobility of Zn^{2+} in soils.

The complexes $ZnNO_3^+$, $Zn(NO_3)_2^\circ$, $ZnCl^+$, $ZnCl_2^\circ$, $ZnCl_3^-$, $ZnCl_4^{2-}$, and $ZnH_2PO_4^+$ do not contribute significantly to total zinc in solution. In neutral and calcareous soils the species $ZnHPO_4^\circ$ may be significant depending upon pH and the activity of phosphate. For example, when

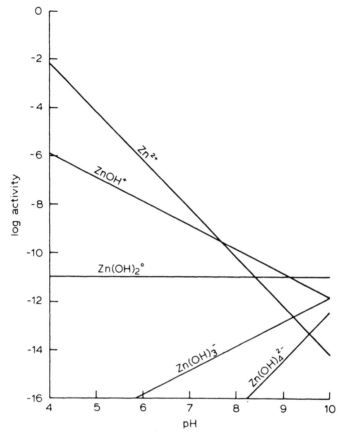

Fig. 13.3 The hydrolysis species of Zn^{2+} in equilibrium with soil-Zn.

$H_2PO_4^- = 10^{-4}$ M, the activity of $ZnHPO_4^0$ is approximately equal to that of $ZnOH^+$. To the extent that lines in Fig. 13.4 representing the various complexes fall below that of Zn^{2+}, their contribution to total zinc in solution is less than that of free Zn^{2+}.

The zinc species contributing significantly to total inorganic zinc in solution $[Zn_{inorg}]$ can be represented as

$$[Zn_{inorg}] = [Zn^{2+}] + [ZnSO_4^0] + [ZnOH^+] + [Zn(OH)_2^0] \\ + [ZnHPO_4^0] \quad (13.5)$$

Substituting activities gives

$$[Zn_{inorg}] = \frac{(Zn^{2+})}{\gamma_{Zn^{2+}}} + (ZnSO_4^0) + \frac{(ZnOH^+)}{\gamma_{ZnOH^+}} + (Zn(OH)_2^0) + (ZnHPO_4^0) \quad (13.6)$$

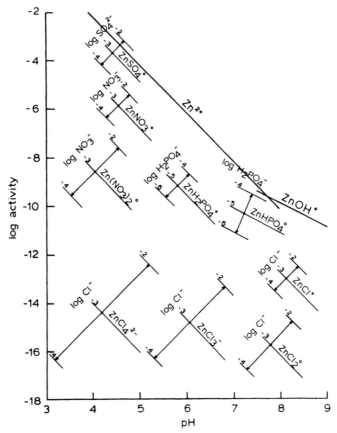

Fig. 13.4 Chloride, nitrate, phosphate, and sulfate complexes of Zn^{2+} in equilibrium with soil-Zn.

When the terms on the right side of this equation are expressed in terms of their equilibrium constants taken from Table 13.1, Eq. 13.6 can be rearranged to give:

$$(Zn^{2+}) = \frac{[Zn_{inorg}]}{\frac{1}{\gamma_{Zn^{2+}}} + 10^{2.33}(SO_4^{2-}) + \frac{10^{-7.69}}{\gamma_{ZnOH} \cdot (H^+)} + \frac{10^{-16.80}}{(H^+)^2} + \frac{10^{-3.90}(H_2PO_4^-)}{(H^+)}}$$

(13.7)

Thus the activity of Zn^{2+} can be estimated from Eq. 13.7 when the following are known: (1) total inorganic zinc, (2) pH, (3) ionic strength, (4) SO_4^{2-}

activity, and (5) $H_2PO_4^-$ activity. If there are other inorganic complexes that contribute significantly to soluble zinc, they must also be included in Eqs. 13.5 through 13.7. Hodgson et al. (1966) found that zinc was present in soil solution as organic as well as inorganic species. In this case

$$[\text{Total Soluble Zn}] = [\text{Zn}_{\text{inorganic}}] + [\text{Zn}_{\text{organic}}] \quad (13.8)$$

and some independent estimate must be made for $[\text{Zn}_{\text{organic}}]$ in order to estimate $[\text{Zn}_{\text{inorg}}]$ from [Total Soluble Zn]. One way of doing this is to extract soils in the presence and absence of charcoal black which adsorbs the organic complexes. This method was used successfully to separate organic and inorganic forms of calcium in soils (Moreno et al., 1960). Methods of estimating or fixing $H_2PO_4^-$ and SO_4^{2-} activities are given in Chapters 12 and 17, respectively.

Estimates of Zn^{2+} activities in soils can be used to test solubility relationships depicted in Fig. 13.1. The method of Norvell and Lindsay (1981) to estimate Fe^{3+} activities in soils can be modified to measure Zn^{2+} activities.

REFERENCES

Hodgson, J. F., W. L. Lindsay, and J. F. Trierweiler. 1966. Micronutrient cation complexing in soil solution: II. Complexing of zinc and copper in displaced solution from calcareous soils. Soil Sci. Soc. Am. Proc. 30: 723–725.

Lindsay, W. L. 1972. Zinc in soils and plant nutrition. Adv. Agron. 24: 147–186.

Lindsay, W. L. and W. A. Norvell. 1969. Equilibrium relationships of Zn^{2+}, Fe^{3+}, Ca^{2+}, and H^+ with EDTA and DTPA in soils. Soil Sci. Soc. Am. Proc. 33: 62–68.

Moreno, E. C., W. L. Lindsay, and G. Osborn. 1960. Reactions of dicalcium phosphate dihydrate in soils. Soil Sci. 90: 58–68.

Norvell, W. A. and W. L. Lindsay. 1969. Reactions of EDTA complexes of Fe, Zn, Mn, and Cu with soils. Soil Sci. Soc. Am. Proc. 33: 86–91.

Norvell, W. A. and W. L. Lindsay. 1970. Lack of evidence for $ZnSiO_3$ in soils. Soil Sci. Soc. Am. Proc. 34: 360–361.

Norvell, W. A. and W. L. Lindsay. 1972. Reactions of DTPA chelates of Fe, Zn, Cu, and Mn with soils. Soil Sci. Soc. Am. Proc. 36: 778–783.

Norvell, W. A. and W. L. Lindsay. 1981. Estimation of iron(III) solubility from EDTA chelate equilibria in soils. Soil Sci. Soc. Am. J. 45: (in press).

PROBLEMS

13.1 At what partial pressure of $CO_2(g)$ can $ZnCO_3$(smithsonite) and ε-$Zn(OH)_2$(c) coexist?

13.2 Develop the equations and plot the solubility of $ZnFe_2O_4$(franklinite) in equilibrium with lepidocrocite. Is franklinite in equilibrium with this mineral more or less soluble than willemite in equilibrium with soil-Si?

13.3 Under what conditions can smithsonite, hopeite, and β-tricalcium phosphate coexist in a calcareous soil?

13.4 Develop the equations and show how the hopeite lines in Fig. 13.2 change if phosphate were controlled by octacalcium phosphate and $CO_2(g)$ varies from 0.0003 to 0.01 atm.

13.5 Estimate the activity of Zn^{2+} in a calcareous, gypsiferous soil of pH 7.8 which is in equilibrium with β-TCP and atmospheric $CO_2(g)$. Total soluble zinc in this soil dropped from $10^{-7.2}$ to $10^{-7.6}$ M when carbon black was included in the extract maintained at 0.025 ionic strength.

13.6 Outline an experiment that would test the hypothesis that franklinite may be the mineral responsible for the soil-Zn line in Fig. 13.1.

FOURTEEN

COPPER

The average copper content of the lithosphere is reported to be 70 ppm, whereas soils generally range from 2 to 100 ppm. The average content for soils is estimated at 30 ppm, which represents a maximum concentration of $10^{-2.33}$ M Cu in the soil solution at 10% moisture (Table 1.1). Copper is slightly less abundant in soils than zinc.

The minerals governing the solubility of Cu^{2+} in soils are not known. In this chapter a reference solubility based on experimentally determined soil-Cu is used to show the instability of most copper minerals in soil environments. Competition between Cu^{2+} and other metal cations for chelating agents is given in Chapter 15, and the precipitation of copper sulfides is given in Chapter 17.

14.1 SOLUBILITY OF Cu(II) MINERALS IN SOILS

The solubilities of various Cu(II) minerals are given by Reactions 1 through 15 of Table 14.1, and the solubilities of Cu(II) oxides, hydroxides, carbonates, cupric ferrite, and soil-Cu are plotted in Fig. 14.1. Reaction 7 of this table gives the solubility expression for soil-Cu, which is taken from the experimental measurements of Norvell and Lindsay (1969, 1972). In their studies CuEDTA and CuDTPA were reacted with soils of different pH. The chelated Cu remaining in solution after 30 days was measured and used to estimate an equilibrium constant for the reaction

$$\text{soil-Cu} + 2H^+ \rightleftharpoons Cu^{2+} \qquad \log K° = 2.8 \qquad (14.1)$$

which gives

$$\log Cu^{2+} = 2.8 - 2pH \qquad (14.2)$$

Equation 14.2 is plotted as the soil-Cu line in Fig. 14.1 and is used as the reference solubility for Cu^{2+} in soils. The activity of Cu^{2+} maintained by soil-Cu is about 10^{-3} that of Zn^{2+} in equilibrium with soil-Zn (Eq. 13.4).

All minerals included in Fig. 14.1 are more soluble than soil-Cu. The order of decreasing solubility is

$$CuCO_3(c) > Cu_3(OH)_2(CO_3)_2(\text{azurite}) > Cu(OH)_2(c)$$
$$> Cu_2(OH)_2CO_3(\text{malachite}) > CuO(\text{tenorite})$$
$$> CuFe_2O_4(\text{cupric ferrite}) > \text{soil-Cu}.$$

These minerals have a solubility range of 10^9: Increasing CO_2 decreases the solubility of the carbonate minerals. The solubility of cupric ferrite is influenced by Fe^{3+} and lies close to that of soil-Cu. It is possible that soil-Cu may indeed be cupric ferrite. The solubility relationships upon which the

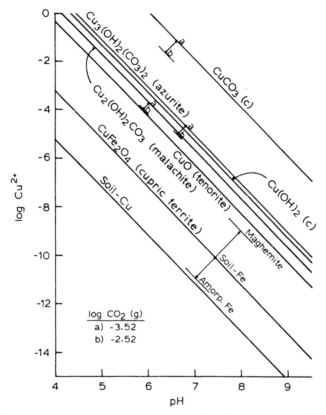

Fig. 14.1 The solubilities of various copper minerals compared to soil-Cu.

two are based are not sufficiently well known to permit a final conclusion at this time. The possibility that Cu^{2+} and Fe^{3+} are combined in a double oxide such as cupric ferrite is of great significance in soils and merits further investigation.

Copper forms several sulfate and oxysulfate minerals shown by Reactions 8 through 12 of Table 14.1. The minerals $CuSO_4$(chalcocyanite) and $CuSO_4 \cdot 5H_2O$(c) are too soluble to form in soils because they would require $> 1\ M\ Cu^{2+}$. The oxysulfate minerals are less soluble and are plotted in Fig. 14.2. These minerals decrease in solubility in the order $CuO \cdot CuSO_4$(c) > $Cu_4(OH)_6SO_4 \cdot 1.3H_2O$(c) > $Cu_4(OH)_6SO_4$(bronchantite) > soil-Cu. The oxysulfate minerals are also too soluble to remain in soils and will dissolve to form soil-Cu.

The copper orthophosphates shown by Reactions 13 and 14 of Table 14.1 are plotted in Fig. 14.3. Since copper in soils is generally less abundant

TABLE 14.1 EQUILIBRIUM CONSTANTS FOR VARIOUS REACTIONS INVOLVING COPPER

Reaction No.	Equilibrium Reaction	$\log K°$
	Cu(II) Oxides, Hydroxides, and Carbonates	
1	$CuO(\text{tenorite}) + 2H^+ \rightleftharpoons Cu^{2+} + H_2O$	7.66
2	$Cu(OH)_2(c) + 2H^+ \rightleftharpoons Cu^{2+} + 2H_2O$	8.68
3	$CuCO_3(c) + 2H^+ \rightleftharpoons Cu^{2+} + CO_2(g) + H_2O$	8.52
4	$Cu_2(OH)_2CO_3(\text{malachite}) + 4H^+ \rightleftharpoons 2Cu^{2+} + CO_2(g) + 3H_2O$	12.99
5	$Cu_3(OH)_2(CO_3)_2(\text{azurite}) + 6H^+ \rightleftharpoons 3Cu^{2+} + 2CO_2(g) + 4H_2O$	19.57
	Cu(II) Ferrite and Soil-Cu	
6	$\alpha\text{-}CuFe_2O_4(\text{cupric ferrite}) + 8H^+ \rightleftharpoons Cu^{2+} + 2Fe^{3+} + 4H_2O$	10.13
7	$\text{Soil-Cu} + 2H^+ \rightleftharpoons Cu^{2+}$	2.80
	Cu(II) Sulfates	
8	$CuSO_4(\text{chalcocyanite}) \rightleftharpoons Cu^{2+} + SO_4^{2-}$	3.72
9	$CuSO_4 \cdot 5H_2O(c) \rightleftharpoons Cu^{2+} + SO_4^{2-} + 5H_2O$	-2.61
10	$CuO \cdot CuSO_4(c) + 2H^+ \rightleftharpoons 3Cu^{2+} + SO_4^{2-} + H_2O$	11.50
11	$Cu_4(OH)_6SO_4(\text{bronchantite}) + 6H^+ \rightleftharpoons 4Cu^{2+} + SO_4^{2-} + 6H_2O$	15.35
12	$Cu_4(OH)_6SO_4 \cdot 1.3H_2O(c) + 6H^+ \rightleftharpoons 4Cu^{2+} + SO_4^{2-} + 7.3H_2O$	17.27
	Cu(II) Phosphates	
13	$Cu_3(PO_4)_2(c) + 4H^+ \rightleftharpoons 3Cu^{2+} + 2H_2PO_4^-$	2.24
14	$Cu_3(PO_4)_2 \cdot 2H_2O(c) + 4H^+ \rightleftharpoons 3Cu^{2+} + 2H_2PO_4^- + 2H_2O$	0.34
15	$Cu_2P_2O_7(c) \rightleftharpoons 2Cu^{2+} + P_2O_7^{4-}$	-15.22
	Cu(II) Hydrolysis	
16	$Cu^{2+} + H_2O \rightleftharpoons CuOH^+ + H^+$	-7.70
17	$Cu^{2+} + 2H_2O \rightleftharpoons Cu(OH)_2^0 + 2H^+$	-13.78
18	$Cu^{2+} + 3H_2O \rightleftharpoons Cu(OH)_3^- + 3H^+$	-26.75
19	$Cu^{2+} + 4H_2O \rightleftharpoons Cu(OH)_4^{2-} + 4H^+$	-39.59
20	$2Cu^{2+} + 2H_2O \rightleftharpoons Cu_2(OH)_2^{2+} + 2H^+$	-10.68
	Cu(II) Complexes	
21	$Cu^{2+} + Cl^- \rightleftharpoons CuCl^+$	0.40
22	$Cu^{2+} + 2Cl^- \rightleftharpoons CuCl_2^0$	-0.12
23	$Cu^{2+} + 3Cl^- \rightleftharpoons CuCl_3^-$	-1.57

TABLE 14.1 (*Continued*)

Reaction No.	Equilibrium Reaction	log $K°$
24	$Cu^{2+} + CO_2(g) + H_2O \rightleftharpoons CuHCO_3^+ + H^+$	−5.73
25	$Cu^{2+} + CO_2(g) + H_2O \rightleftharpoons CuCO_3^° + 2H^+$	−11.43
26	$Cu^{2+} + 2CO_2(g) + 2H_2O \rightleftharpoons Cu(CO_3)_2^{2-} + 4H^+$	−26.48
27	$Cu^{2+} + NO_3^- \rightleftharpoons CuNO_3^+$	0.50
28	$Cu^{2+} + 2NO_3^- \rightleftharpoons Cu(NO_3)_2^°$	−0.40
29	$Cu^{2+} + H_2PO_4^- \rightleftharpoons CuH_2PO_4^+$	1.59
30	$Cu^{2+} + H_2PO_4^- \rightleftharpoons CuHPO_4^° + H^+$	−4.00
31	$Cu^{2+} + 2H^+ + P_2O_7^{4-} \rightleftharpoons CuH_2P_2O_7^°$	18.67
32	$Cu^{2+} + H^+ + P_2O_7^{4-} \rightleftharpoons CuHP_2O_7^-$	14.78
33	$Cu^{2+} + P_2O_7^{4-} \rightleftharpoons CuP_2O_7^{2-}$	6.64
34	$2Cu^{2+} + P_2O_7^{4-} \rightleftharpoons Cu_2P_2O_7^°$	−0.03
35	$Cu^{2+} + SO_4^{2-} \rightleftharpoons CuSO_4^°$	2.36
	Redox Reactions	
36	$Cu^{2+} + e^- \rightleftharpoons Cu^+$	2.62
37	$Cu^+ + e^- \rightleftharpoons Cu(c)$	8.87
38	$Cu^{2+} + 2e^- \rightleftharpoons Cu(c)$	11.49
	Cu(I) Minerals	
39	$Cu_2O(\text{cuperite}) + 2H^+ \rightleftharpoons 2Cu^+ + H_2O$	−2.17
40	$CuOH(c) + H^+ \rightleftharpoons Cu^+ + H_2O$	−0.70
41	$\alpha\text{-}Cu_2Fe_2O_4(\text{cuprous ferrite}) + 8H^+ \rightleftharpoons 2Cu^+ + 2Fe^{3+} + 4H_2O$	−13.53
42	$Cu_2SO_4(c) \rightleftharpoons 2Cu^+ + SO_4^{2-}$	−1.95
	Cu(I) Complexes	
43	$Cu^+ + Cl^- \rightleftharpoons CuCl^°$	2.70
44	$Cu^+ + 2Cl^- \rightleftharpoons CuCl_2^-$	5.51
45	$Cu^+ + 3Cl^- \rightleftharpoons CuCl_3^{2-}$	5.70
46	$2Cu^+ + 4Cl^- \rightleftharpoons Cu_2Cl_4^{2-}$	13.10

than phosphorus (Table 1.1), this plot was developed to show how phosphorus may affect the solubility of Cu^{2+}. The mineral $Cu_3(PO_4)_2 \cdot 2H_2O(c)$ is always more stable than $Cu_3(PO_4)_2(c)$ as it lowers Cu^{2+} by 0.63 log unit. Details of only the dihydrate mineral are shown in Fig. 14.3. In this plot phosphate is fixed by strengite and soil-Fe at low pH, by hydroxyapatite, β-TCP, or DCPD with soil-Ca at intermediate pH, and by the same minerals with calcite and $CO_2(g)$ at high pH.

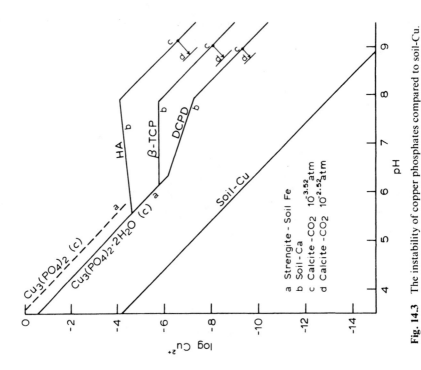

Fig. 14.3 The instability of copper phosphates compared to soil-Cu.

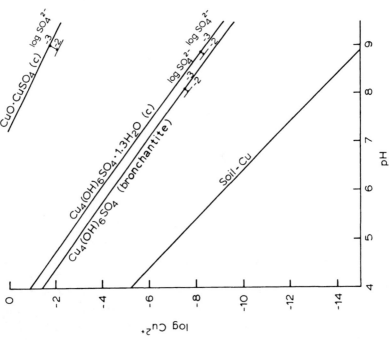

Fig. 14.2 The instability of several copper oxysulfates compared to soil-Cu.

SOLUBILITY OF Cu(II) MINERALS IN SOILS

Both copper phosphates are more soluble than soil-Cu especially as phosphate solubility is depressed by hydroxyapatite. These copper phosphates can dissolve sufficiently in soils to provide both available copper and phosphorus for plants.

The solubility of $Cu_2P_2O_7(c)$ is given by Reaction 15, and the stabilities of several copper pyrophosphate complexes are given by Reactions 31 through 34 of Table 14.1. These stability relationships are plotted in Fig. 14.4 under the conditions that pyrophosphate is fixed by $\beta\text{-}Ca_2P_2O_7(c)$ and soil-Ca (Section 12.10) and Cu^{2+} is fixed by soil-Cu. The complexes $CuHP_2O_7^-$, $CuH_2P_2O_7^\circ$, $CuP_2O_7^{2-}$, and $Cu_2P_2O_7^\circ$ are all less abundant than the $H_2P_2O_7^{2-}$, $CaHP_2O_7^-$, and $CaP_2O_7^{2-}$ shown in the upper part of

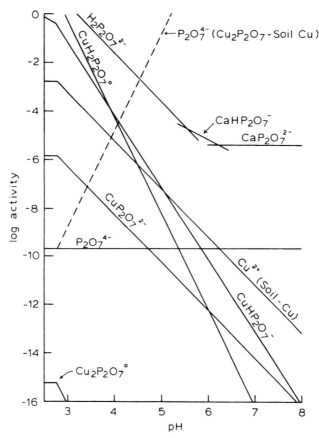

Fig. 14.4 The insignificance of copper pyrophosphate complexes in soils when $\beta\text{-}Ca_2P_2O_7(c)$ and soil-Ca fix $P_2O_7^{4-}$. The instability of $Cu_2P_2O_7(c)$ relative to $\beta\text{-}Ca_2P_2O_7(c)$ is also shown.

the diagram. Thus copper pyrophosphate complexes are not expected to contribute significantly to total pyrophosphate in soil solution.

Also included in Fig. 14.4 is Cu^{2+} maintained by soil-Cu which shows that both $CuHP_2O_7^-$ and $CuH_2P_2O_7^\circ$ contribute significantly to total Cu^{2+} in soil solution, especially at low pH. The species $CuP_2O_7^{2-}$ and $Cu_2P_2O_7^\circ$ do not contribute significantly to total soluble copper.

The instability of $Cu_2P_2O_7(c)$ in soils is demonstrated by the dashed line indicating the $P_2O_7^{4-}$ activity that would be necessary for $Cu_2P_2O_7(c)$ to form when Cu^{2+} is controlled by soil-Cu. Only as this line meets the horizontal $P_2O_7^{4-}$ line representing equilibrium with β-$Ca_2P_2O_7(c)$ can $Cu_2P_2O_7(c)$ form. The pH of this intersection is obtained as follows:

			$\log K^\circ$
$Cu_2P_2O_7(c)$	$\rightleftharpoons$	$2Cu^{2+} + P_2O_7^{4-}$	-15.22
$2Cu^{2+}$	$\rightleftharpoons$	$2\,\text{soil-Cu} + 4H^+$	$2(-2.80)$
$2Ca^{2+} + P_2O_7^{4-}$	$\rightleftharpoons$	$\beta\text{-}Ca_2P_2O_7(c)$	14.70
$2\,\text{soil-Ca}$	$\rightleftharpoons$	$2Ca^{2+}$	$2(-2.50)$
$Cu_2P_2O_7(c) + 2\,\text{soil-Ca}$	$\rightleftharpoons$	$\beta\text{-}Ca_2P_2O_7(c) + 4H^+ + 2\,\text{soil-Cu}$	-11.12

(4.3)

from which

$$pH = 2.78 \tag{14.4}$$

Thus $Cu_2P_2O_7(c)$ is metastable to β-$Ca_2P_2O_7(c)$ above pH 2.78 and cannot form. Below this pH, $Cu_2P_2O_7(c)$ can form. Figure 14.4 is drawn to represent the conditions in which more than enough pyrophosphate is present to transform all of the soil-Cu into $Cu_2P_2O_7(c)$. In this case both pyrophosphate minerals would coexist, and Cu^{2+} activity would be governed by $Cu_2P_2O_7(c)$ through β-$Ca_2P_2O_7(c)$ and soil-Ca. If there were more than enough soil-Cu to dissolve all of the β-$Ca_2P_2O_7$, the soil-Cu line would continue upward below pH 2.78 and the $P_2O_7^{4-}$ line would be depressed. Since very few soils are as acid as pH 2.78, the possibility of $Cu_2P_2O_7(c)$ forming in soils is very unlikely.

14.2 HYDROLYSIS AND SOLUTION COMPLEXES OF Cu(II)

The hydrolysis reactions of Cu^{2+} are given by Reactions 16 through 20 of Table 14.1 and are plotted in Fig. 14.5. The predominant ion below pH 6.9

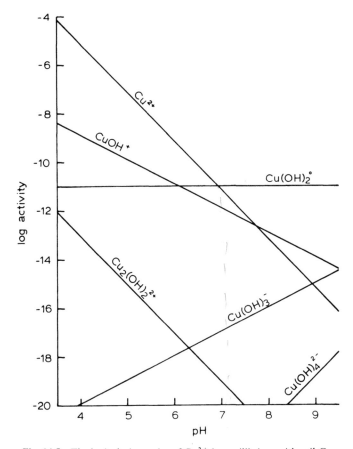

Fig. 14.5 The hydrolysis species of Cu^{2+} in equilibrium with soil-Cu.

is Cu^{2+} while $Cu(OH)_2^\circ$ is the major solution species above this pH. The hydrolysis species $CuOH^+$ becomes slightly significant near pH 7 whereas $Cu(OH)_3^-$, $Cu(OH)_4^{2-}$, and $Cu_2(OH)_2^{2+}$ are insignificant in soils.

Several Cu^{2+} complexes are given by Reactions 21 through 35 of Table 14.1. The pyrophosphate complexes were considered in the previous section and the remainder of these complexes are shown in Fig. 14.6 in equilibrium with soil-Cu.

The complexes of great importance in soils are $CuSO_4^\circ$ and $CuCO_3^\circ$. At $10^{-2.36}$ M SO_4^{2-}, $CuSO_4^\circ$ and Cu^{2+} activities are equal, and $CuSO_4^\circ$ changes 10-fold for each 10-fold change in SO_4^{2-}. At $10^{-2.35}$ atm of CO_2 the activities of $CuCO_3^\circ$ and $Cu(OH)_2^\circ$ are equal. Tenfold changes in CO_2 change $CuCO_3^\circ$ accordingly. Other complexes that may contribute more than

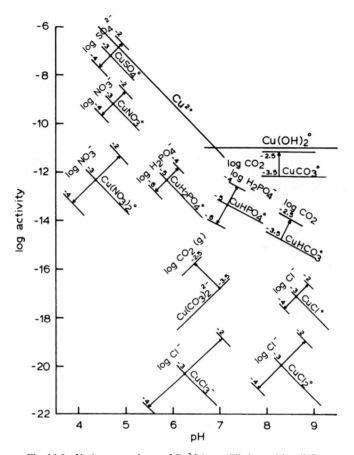

Fig. 14.6 Various complexes of Cu^{2+} in equilibrium with soil-Cu.

1% to total copper in solution include $CuHCO_3^+$ at high CO_2 and neutral pH, $CuHPO_4^o$ at high phosphate and neutral pH, and $Cu(CO_3)_2^{2-}$ at high CO_2 and high pH. Complexes such as $CuNO_3^+$, $Cu(NO_3)_2^o$, $CuH_2PO_4^+$, $Cu(CO_3)_2^{2-}$, $CuCl^+$, $CuCl_2^o$, and $CuCl_3^-$ do not contribute significantly to total copper at the level of anions normally found in soils.

Total inorganic copper in soil solution $[Cu_{inorg}]$ can be fairly well represented by the equation

$$[Cu_{inorg}] = [Cu^{2+}] + [CuSO_4^o] + [Cu(OH)_2^o] + [CuCO_3^o] \quad (14.5)$$

Substituting activities gives

$$[Cu_{inorg}] = \frac{(Cu^{2+})}{\gamma_{Cu^{2+}}} + (CuSO_4^o) + (Cu(OH)_2^o) + (CuCO_3^o) \quad (14.6)$$

EFFECT OF REDOX ON COPPER

When each term on the right side of this equation is expressed in terms of its equilibrium constant given in Table 14.1, Eq. 14.6 can be rearranged to give:

$$(Cu^{2+}) = \frac{[Cu_{inorg}]}{\dfrac{1}{\gamma_{Cu^{2+}}} + 10^{2.36}(SO_4^{2-}) + \dfrac{10^{-13.78}}{(H^+)^2} + \dfrac{10^{-11.43}CO_2(g)}{(H^+)^2}} \quad (14.7)$$

Thus the activity of Cu^{2+} can be estimated from this equation with measurements of (1) total inorganic copper in solution, (2) pH, (3) ionic strength, (4) the activity of SO_4^{2-}, and (5) the partial pressure of $CO_2(g)$. If conditions are such that other inorganic copper complexes contribute significantly to soluble copper, they must also be included in Eq. 14.5 and reflected in Eq. 14.7.

Copper is normally present in soil solution as both organic and inorganic complexes (Hodgson et al., 1966). For this reason total soluble copper is expressed as

$$[\text{Total soluble copper}] = [Cu_{inorg}] + [Cu_{org}] \quad (14.8)$$

and some independent means must be used to distinguish between the two forms of copper. As indicated in Section 13.2, soils can be extracted in the presence and absence of carbon black which absorbs many of the soluble organic constituents. In this way Cu^{2+} activities can be estimated and used to test the solubility relationships depicted in Fig. 14.1.

14.3 EFFECT OF REDOX ON COPPER

Copper equilibria which are affected by redox are given by Reactions 36 through 38 of Table 14.1. According to Reaction 36:

$$Cu^{2+} + e^- \rightleftharpoons Cu^+ \quad \log K° = 2.62 \quad (14.9)$$

$$\log \frac{Cu^+}{Cu^{2+}} = 2.62 - pe \quad (14.10)$$

Thus Cu^+ and Cu^{2+} activities are equal at $pe = 2.62$, and their ratio changes 10-fold for each unit change in pe.

What are the stable copper minerals in soils which become reduced? Figure 14.7 was developed to answer this question. Soil-Cu defined by Reaction 14.1 is stable above $pe + pH$ of 14.89. Below this redox $Cu_2Fe_2O_4$ (cuprous ferrite) in equilibria with soil-Fe becomes the stable copper mineral.

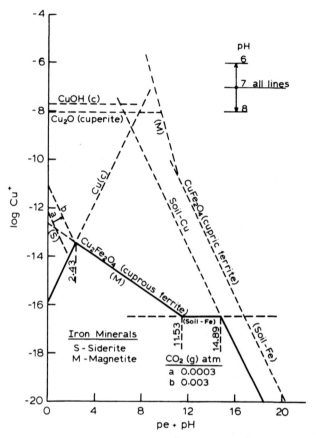

Fig. 14.7 The effect of redox on the solubility of Cu^+ and the stability of copper minerals at pH 7 showing shifts for other pH values.

Under these conditions the solubility of copper can be depicted as follows:

	log $K°$
$\alpha\text{-}Cu_2Fe_2O_4$(cuprous ferrite) $+ 8H^+ \rightleftharpoons 2Cu^+ + 2Fe^{3+} + 4H_2O$	-13.52
$2Fe^{3+} + 6H_2O \rightleftharpoons 2Fe(OH)_3$(soil-Fe) $+ 6H^+$	$2(-2.70)$
$\alpha\text{-}Cu_2Fe_2O_4 + 2H^+ + 2H_2O \rightleftharpoons 2Fe(OH)_3$(soil-Fe) $+ 2Cu^+$	-18.93

(14.11)

EFFECT OF REDOX ON COPPER

for which
$$\log Cu^+ = -9.47 - pH \qquad (14.12)$$

As $pe + pH$ drops below 11.53 magnetite becomes more stable than soil-Fe and therefore controls iron solubility (Chapter 10). In this redox range the solubility of copper is controlled by the reactions:

		$\log K^\circ$
$3\alpha\text{-}Cu_2Fe_2O_4\text{(cuprous ferrite)} + 24H^+ \rightleftharpoons 6Cu^+ + 6Fe^{3+} + 12H_2O$		$3(-13.53)$
$6Fe^{3+} + 2e^- + 8H_2O \rightleftharpoons 2Fe_3O_4\text{(magnetite)} + 16H^+$		$2(3.42)$
$3\alpha\text{-}Cu_2Fe_2O_4(c) + 2e^- + 8H^+ \rightleftharpoons 2Fe_3O_4(c) + 6Cu^+ + 4H_2O$		-33.75

$$(14.13)$$

for which
$$\log Cu^+ = -5.63 - \tfrac{1}{3}(pe + pH) - pH \qquad (14.14)$$

Equation 14.14 is plotted in Fig. 14.7 as the $Cu_2Fe_2O_4$(cuprous ferrite) line in equilibrium with magnetite (M). These equilibrium relationships allow Cu^+ to increase $\tfrac{1}{3}$ log unit for each unit decrease in $pe + pH$. Cuprous ferrite is stable until $pe + pH$ drops to 2.43 below which Cu(c) controls copper solubility. The latter transition occurs before $FeCO_3$(siderite) can form to control iron solubility (Fig. 14.7). Thus $\log Cu^+$ normally does not exceed $10^{-13.4}$ M reached at $pe + pH$ of 2.43 where cuprous ferrite, magnetite, and Cu(c) coexist. This equilibrium redox can be obtained as follows:

		$\log K^\circ$
$3\alpha\text{-}Cu_2Fe_2O_4\text{(cuprous ferrite)} + 24H^+ \rightleftharpoons 6Cu^+ + 6Fe^{3+} + 12H_2O$		$3(-13.53)$
$6Fe^{3+} + 2e^- + 8H_2O \rightleftharpoons 2Fe_3O_4\text{(magnetite)} + 16H^+$		$2(3.42)$
$6Cu^+ + 6e^- \rightleftharpoons 6Cu(c)$		$6(8.87)$
$3\alpha\text{-}Cu_2Fe_2O_4(c) + 8e^- + 8H^+ \rightleftharpoons 3Fe_3O_4(c) + 6Cu(c) + 4H_2O$		19.47

$$(14.15)$$

for which

$$pe + pH = 2.43 \quad (14.16)$$

The dashed lines in Fig. 14.7 represent metastable equilibria. For example, $CuFe_2O_4$(cupric ferrite) in equilibrium with soil-Fe is always metastable with respect to soil-Cu. Below $pe + pH$ of 11.53 where magnetite controls iron solubility cupric ferrite becomes even more soluble. The mineral CuOH(c) is always metastable to Cu_2O(cuperite), and both are metastable to cuprous ferrite. The solubility of Cu^+ in soils is expected to follow the solid lines shown in Fig. 14.7.

Although Fig. 14.7 was constructed for pH 7.0, it can be easily interpreted for any pH. All lines in this figure move upward by one log unit of Cu^+ for each unit decrease in pH, but their relative positions remain unchanged with changes in pH.

14.4 COMPLEXES OF Cu(I)

The chloride complexes of Cu(I) are given by Reactions 43 through 46 of Table 14.1. These complexes are plotted in Fig. 14.8 in equilibrium with cuprous ferrite at $pe + pH$ 11.53. This is the redox at which soil-Fe and magnetite can coexist. The activities of Cu^+ and Cu^{2+} are included for comparison. At this selected redox $CuCl°$ and $CuCl_2^-$ complexes approach that of Cu^+ if Cl^- is near 10^{-3} M. The $CuCl_3^{2-}$ complex requires $> 10^{-2}$ M Cl^- to equal Cu^+. At $pe + pH$ 11.53 the sum of all Cu(I) species is less than that of Cu^{2+}. As redox rises, the Cu^{2+} line rises (equilibrium with cuprous ferrite and soil-Fe). Above $pe + pH$ 14.89 copper is controlled by soil-Cu and Cu^{2+} activity is no longer redox dependent. As redox drops below $pe + pH$ of 11.53, Cu^+ and its accompanying Cl^- complexes (in equilibrium with cuprous ferrite and magnetite) increase while Cu^{2+} decreases. Thus the Cu(I) complexes in solution can exceed the Cu(II) species as redox drops.

Figures 14.7 and 14.6 are useful in selecting the solution species of copper that contribute significantly to total inorganic copper in solution. The $Cu_2Cl_4^{2-}$ species is not sufficiently stable at the low Cu^+ level permitted by cuprous ferrite to appear in Fig. 14.8. As $pe + pH$ drops below 12 and Cl^- rises above 10^{-3} M, it is necessary to add $[Cu^+]$, $[CuCl°]$, and $[CuCl_2^-]$ to the right hand of Eq. 14.5 to represent additional solution species that contribute significantly to total inorganic copper. Equation 14.5 can then be modified as follows to represent total inorganic copper in solution:

$$[Cu_{inorg}] = [Cu^{2+}] + [CuSO_4°] + [Cu(OH)_2°] + [CuCO_3°] + [Cu^+] \\ + [CuCl°] + [Cu(Cl_2)^-] \quad (14.17)$$

REFERENCES

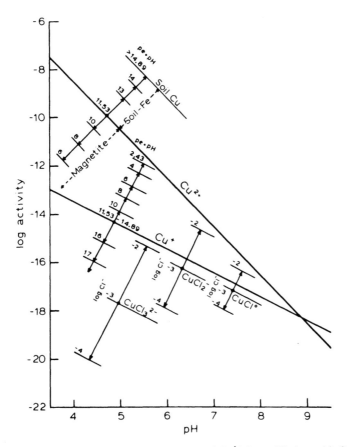

Fig. 14.8 The activities of Cu^+, $CuCl$ complexes, and Cu^{2+} in equilibrium with $Cu_2Fe_2O_4$ (cuprous ferrite) at $pe + pH$ 11.53 where soil-Fe and magnetite coexist. Shifts with changing redox are also included.

Expressing each of the terms of Eq. 14.17 in terms of stability constants, Cl^- activity, pH, and redox gives the appropriate equation corresponding to Eq. 14.7. If additional copper complexes are formed, they too must be included. The effect of redox on copper equilibria are developed further in Chapter 17 (See Fig. 17.5).

REFERENCES

Hodgson, J. F., W. L. Lindsay, and J. F. Trierweiler. 1966. Micronutrient cation complexing in soil solution: II. Complexing of zinc and copper in displaced solution from calcareous soils. Soil Sci. Soc. Am. Proc. 30:723–726.

Norvell, W. A. and W. L. Lindsay. 1969. Reactions of EDTA complexes of Fe, Zn, Mn, and Cu with soils. Soil Sci. Soc. Am. Proc. 33:86–91.

Norvell, W. A. and W. L. Lindsay. 1972. Reactions of DTPA chelates of iron, zinc, copper, and manganese with soils. Soil Sci. Soc. Am. Proc. 36:778–783.

PROBLEMS

14.1 Discuss the possibility that $CuFe_2O_4$(cupric ferrite) could be the mineral referred to as soil-Cu. Outline experiments that could be used to test this hypothesis.

14.2 How useful would $Cu_3(PO_4)_2 \cdot 2H_2O$(c) be as either a copper or a phosphate fertilizer in soils? Explain.

14.3 How would the $Cu_3(PO_4)_2 \cdot 2H_2O$(c) solubility line in Fig. 14.3 change if phosphate were controlled by octacalcium phosphate in neutral and calcareous soils?

14.4 From Fig. 14.4 discuss the following:
   a. The condition in which $Cu_2P_2O_7$(c) can form in soils.
   b. The conditions in which both $Cu_2P_2O_7$(c) and β-$Ca_2P_2O_7$(c) can coexist in soils.
   c. What controls Cu^{2+} activity if both $Cu_2P_2O_7$(c) and β-$Ca_2P_2O_7$(c) coexist in soils.
   d. How this figure would change if there were more than enough soil-Cu to precipitate all of the pyrophosphate as $Cu_2P_2O_7$(c).

14.5 Under what conditions can $CuCO_3^0$ and $CuHCO_3^+$ have equal activities?

14.6 A soil shows the following:
   a. pH = 7.
   b. $CO_2(g) = 10^{-2.52}$ atm.
   c. SO_4^{2-} activity and $H_2PO_4^-$ activity both equal to 10^{-3} M.
   d. Equilibrium with soil-Cu is maintained.
   e. The ratio of soluble $Cu_{organic}/Cu_{inorganic} = 25$.
   f. Ionic strength = 0.005.

Develop the necessary equation and calculate the mole fraction contribution of $CuHPO_4^0$ to total soluble copper in this soil.

14.7
   a. Discuss the significance of Fig. 14.7 as it relates to copper solubility in soils.
   b. If the log $K^°$ value for soil-Cu given by Reaction 7 of Table 14.1 were increased from 2.8 to 3.8, how would this affect the redox at which $Cu_2Fe_2O_4$(cuprous ferrite) could form?

14.8 A soil has the following characteristics:

$$\text{pH} = 7.0, \, pe = 3.0, \, \text{Cl}^- = 10^{-2.5} \, M, \, \text{SO}_4^{2-} = 10^{-4} \, M,$$
$$\text{H}_2\text{PO}_4^- = 10^{-6} \, M, \text{ and } \text{CO}_2(g) = 0.003 \text{ atm.}$$

From Fig. 14.5, 14.6, and 14.8 estimate the mole fraction contribution of $\Sigma\text{Cu(I)}$ to total inorganic copper assuming activities are unity.

FIFTEEN

CHELATE EQUILIBRIA

Metal chelation is important in soils because it increases the solubility of metal ions and affects many important physical, chemical, and biological processes. During podzolization, for example, organic acids generated near the soil surface chelate Fe^{3+} and Al^{3+}, and the chelated metals are carried by percolating water into the lower horizons where equilibrium relationships are sufficiently different that the metal ions separate from the organic chelates and precipitate again.

Chelation affects the availability and movement of many plant nutrients. For example, when micronutrient cations are absorbed by plant roots, the activity of these cations in the immediate vicinity of roots drops, metal chelates dissociate, and diffusion gradients are established enabling metal chelates to become the major vehicle of transport (Lindsay, 1974).

Chelates also are used as micronutrient fertilizers to supply plants with Fe, Zn, Mn, and Cu. Such chelates differ widely in their ability to maintain soluble micronutrient cations in soils of different characteristics. The fact that a metal chelate is used to correct a given nutrient deficiency is no guarantee that it will be effective. Other cations in the soil may displace the added metal and render it ineffective.

Chelating agents also are used to extract available nutrients from soils. Selection of effective extractants for this purpose is often frustrating unless the principles governing their behavior in soils are understood.

The purpose of this chapter is to present a systematic approach for predicting metal chelate equilibria in soils. Stability-pH diagrams are developed to show that metal ions in soils tend to approach equilibrium with both chelating agents and solid phases controlling the metal ion activities. Examples are given to show that the predicted relationships are valid and have wide application to soils and plant nutrition.

15.1 METAL CHELATES AND THEIR STABILITY CONSTANTS

Most metal chelates consist of rather complex organic anions with two or more functional groups that are capable of sharing pairs of electrons with a centrally located metal ion. These pairs of electrons enter the vacant electron shells of the enclosed metal ion. In this way a chelate contains one or more ring structures that include the metal ion and the arms of the chelating ligand. In this respect chelates differ from complexes that have only a single bond between the metal ion and the electron donor.

Let us consider the chelating agent ethylenediaminetetraacetic acid (EDTA),

$$\begin{array}{c} HOOC-CH_2 CH_2-COOH \\ \diagdown \diagup \\ N-CH_2-CH_2-N \\ \diagup \diagdown \\ HOOC-CH_2 CH_2-COOH \end{array}$$

EDTA $H_4L°$

As pH increases one or more H^+ ions dissociate from the four carboxyl groups giving the series of protonated EDTA species in solution shown in Fig. 15.1. The higher the pH the greater is the concentration and charge of the dissociated species. When all four H^+ ions have been removed, the free ligand, L^{4-}, remains

$$^-:OOC-CH_2 \qquad\qquad CH_2-COO:^-$$
$$\diagdown\qquad\qquad\diagup$$
$$\ddot{N}-CH_2-CH_2-\ddot{N}$$
$$\diagup\qquad\qquad\diagdown$$
$$^-:OOC-CH_2 \qquad\qquad CH_2-COO:^-$$

EDTA L^{4-}

This ligand has six pairs of electrons that can be shared with a metal ion to form a metal chelate. The EDTA ligand combines with Fe^{3+} to give $FeEDTA^-$ as depicted in Fig. 15.2. The centrally located Fe^{3+} is surrounded by the four carboxyls and two nitrogen donor groups. Similarly EDTA combines with several different metal cations to form stable metal chelates.

The FeL^- chelate has a charge of -1 and behaves as a monovalent anion. It is free to move in the soil solution or to be absorbed onto solid surfaces. The strength of the chemical bond between a chelating ligand and a metal ion differs with different chelating agents and metal ions. The most convenient

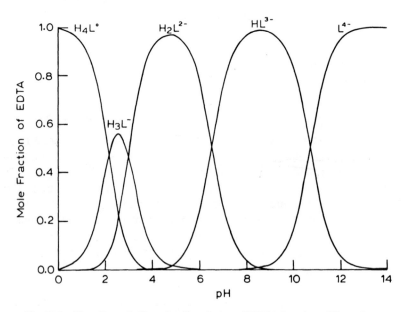

Fig. 15.1 The effect of pH on the dissociation of EDTA based on $K_{0.01}^m$ values.

METAL CHELATES AND THEIR STABILITY CONSTANTS

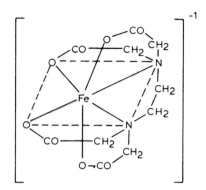

Fig. 15.2 A three-dimensional sketch of FeEDTA$^-$.

way of expressing this bonding strength is with a formation constant. For Fe^{3+} and EDTA the reaction can be written

$$Fe^{3+} + L^{4-} \rightleftharpoons FeL^- \qquad \log K^m_{0.01} = 26.50 \qquad (15.1)$$

In soils it is convenient to use $K^m_{0.01}$, which is a mixed equilibrium constant corresponding to an ionic strength of 0.01 M. This means that all reactants and products are expressed in terms of concentrations except H^+, OH^-, and e^-, which are expressed as activities (Chapter 2). The equilibrium constant for Reaction 15.1 is expressed as

$$K_{FeL^-} = \frac{[FeL^-]}{[Fe^{3+}][L^{4-}]} = 10^{26.50} \qquad (15.2)$$

in which K_{FeL^-} is the formation constant of the FeL$^-$ species. The greater the formation constant, the greater is the stability of the metal chelate and the more likely it will remain intact.

A summary of the equilibrium constants for several metal chelates is given in Table 15.1. Because of the complexity of metal chelate reactions, equilibrium relationships can be greatly simplified by using mixed concentration constants (K^m) rather than activity constants ($K°$). An ionic strength of 0.01 M corresponds to a 0.0033 M CaCl$_2$ solution which approximates that of many well-drained soils. The $K^c_{0.1}$ values in Table 15.1 were taken largely from Martell and Smith (1974). The $K^m_{0.01}$ values were calculated from the $K^c_{0.1}$ values using the Davies' equation (Eq. 2.13). In this compilation charges on the reacting species were omitted to simplify tabulation of the data for ligands of different charge. It is understood that the charges are

TABLE 15.1 STABILITY CONSTANTS ($\log K_{0,1}^c$) at 25°C FOR METAL-LIGAND REACTIONS SELECTED FROM MARTELL AND SMITH (1974) AND CORRECTED TO MIXED CONSTANTS ($\log K_{0.01}^m$) USING DAVIES' EQUATION (Eq. 2.13)

Reaction	HEEDTA		EGTA		EDTA		DTPA		CDTA		EDDHA	
	$K_{0.1}^c$	$K_{0.01}^m$	$K_{0.1}^c$	$K_{0.01}^m$	$K_{0.1}^c$	$K_{0.01}^m$	$K_{0.1}^c$	$K_{0.01}^m$	$K_{0.1}^c$	$K_{0.01}^m$	$K_{0.1}^c$	$K_{0.01}^m$
H + L ⇌ HL	9.81	10.23	9.40	9.95	10.17	10.72	10.45	11.13	12.3	12.85	11.68	12.23
2H + L ⇌ H$_2$L	15.18	15.89	18.18	19.15	16.28	17.25	18.98	20.21	18.42	19.39	21.92	22.89
3H + L ⇌ H$_3$L	17.78	18.66	20.84	22.10	18.96	20.22	23.26	24.91	21.95*	23.21	30.56	31.82
4H + L ⇌ H$_4$L	—	—	22.84	24.27	20.96	22.39	25.91	27.85	24.37*	25.80	36.88	38.31
5H + L ⇌ H$_5$L	—	—	—	—	—	—	27.73	29.84	—	—	—	—
Ca + L ⇌ CaL	8.2	8.95	10.86	11.86	10.61	11.61	10.75	12.02	13.15	14.15	7.20	8.20
Ca + H + L ⇌ CaHL	—	—	14.65*	15.95	13.79*	15.09	16.86	18.55	—	—	16.48	17.78
2Ca + L ⇌ Ca$_2$L	—	—	—	—	—	—	11.81	13.83	—	—	—	—
Mg + L ⇌ MgL	7.0	7.75	5.28	6.28	8.83	9.83	9.34	10.61	11.07	12.07	8.00	9.00
Mg + H + L ⇌ MgHL	—	—	12.90*	14.20	12.68*	13.98	16.19	17.88	—	—	16.88	18.18
Fe(II) + L ⇌ Fe(II)L	12.2	12.95	11.80	12.80	14.27	15.27	16.40	17.67	18.90	19.90	14.30	15.30
Fe(II) + H + L ⇌ Fe(II)HL	14.95*	15.87	16.10*	17.40	16.97	18.27	21.70	23.39	21.70	23.00	—	—
Fe(II) + L ⇌ Fe(II)OHL + H	3.23	3.68	—	—	5.20*	5.79	10.99	11.71	—	—	—	—
2Fe(II) + L ⇌ Fe(II)$_2$L	—	—	—	—	—	—	19.38*	21.40	—	—	—	—
Fe(III) + L ⇌ FeL	19.8	20.92	20.5	22.00	25.0	26.50	27.3‡	29.19	30.0	31.50	33.90	35.40
Fe(III) + H + L ⇌ FeHL	—	—	—	—	26.3	27.97	30.86	33.04	—	—	—	—
Fe(III) + L ⇌ FeOHL + H	18.82	19.77	—	—	17.15*	18.36	16.86	18.33	20.4*	21.61	—	—
Fe(III) + L ⇌ Fe(OH)$_2$L + 2H	9.80	10.46	—	—	8.10*	8.89	—	—	—	—	—	—
Zn + L ⇌ ZnL	14.60	15.35	12.60	13.60	16.44	17.44	18.29	19.56	19.35	20.35	16.80	17.80
Zn + H + L ⇌ ZnHL	—	—	17.56*	18.86	19.44	20.74	23.89	25.58	22.25*	23.55	24.54	25.84
2Zn + L ⇌ Zn$_2$L	—	—	15.9*	17.40	—	—	22.67	24.69	—	—	—	—
Cu + L ⇌ CuL	17.50	18.25	17.57	18.57	18.70	19.70	21.38	22.65	21.92	22.92	23.90¶	24.90

Reaction												
Cu + H + L ⇌ CuHL	19.92	20.84	21.85	23.15	21.70	23.00	26.19	27.88	25.02*	26.32	31.94	33.24
2Cu + L ⇌ Cu$_2$L	—	—	21.88	23.38	—	—	28.17	30.19	—	—	—	—
Mn + L ⇌ MnL	10.80	11.55	12.18	13.18	13.81	14.81	15.51	16.78	17.43	18.43	—	—
Mn + H + L ⇌ MnHL	—	—	16.28*	17.58	16.91	18.21	19.91	21.60	20.23	21.53	—	—
2Mn + L ⇌ Mn$_2$L	—	—	—	—	—	—	17.60*	19.62	—	—	—	—
Cd + L ⇌ CdL	13.10	13.85	16.5	17.50	16.36	17.36	19.00	20.27	19.84	20.84	13.13	14.13
Cd + H + L ⇌ CdHL	—	—	19.97*	21.27	19.26	20.56	23.17	24.86	22.84*	24.14	21.83	23.13
2Cd + L ⇌ Cd$_2$L	—	—	—	—	—	—	21.30	23.32	—	—	—	—
Pb + L ⇌ PbL	15.50	16.25	14.54	15.54	17.88	18.88	18.66	19.93	20.24	21.24	—	—
Pb + H + L ⇌ PbHL	—	—	19.70†	21.00	20.68*	21.98	23.18	24.87	23.04*	24.34	—	—
2Pb + L ⇌ Pb$_2$L'	—	—	19.14*	20.64	—	—	22.07	24.09	—	—	—	—
Ni + L ⇌ NiL	17.10	17.85	13.50	14.50	18.52	19.52	20.17	21.44	20.2	21.20	19.66	20.66
Ni + H + L ⇌ NiHL	20.14§	21.06	18.60*	19.90	21.72	23.02	25.84	27.53	23.18†	24.48	27.29	28.59
2Ni + L ⇌ Ni$_2$L	—	—	18.40†	19.90	—	—	25.76	27.78	—	—	—	—
Co(II) + L ⇌ CoL	14.50	15.25	12.35	13.35	16.26	17.26	19.15	20.42	19.58	20.58	—	—
Co(II) + H + L ⇌ CoHL	—	—	17.25	18.55	19.26	20.56	24.09	25.78	22.48*	23.78	—	—
2Co(II) + L ⇌ Co$_2$L	—	—	15.65*	17.15	—	—	22.89	24.91	22.07*	23.57	—	—
Hg + L ⇌ HgL	20.05	20.80	22.90	23.90	21.50	22.50	26.40	27.67	24.79	25.79	—	—
Hg + H + L ⇌ HgHL	—	—	25.96	27.26	24.60	25.90	30.64*	32.33	27.89	29.19	—	—
Al + L ⇌ AlL	14.40	15.52	14.49†	15.99	16.50	18.00	18.70	20.59	19.60	21.10	—	—
Al + H + L ⇌ AlHL	16.56†	17.73	18.52†	20.19	19.00	20.67	23.00	25.18	21.89	23.56	—	—
Al + L ⇌ AlOHL + H	9.45†	10.40	9.19†	10.40	10.67	11.88	11.30	12.77	11.68	12.89	—	—

* Stability constant for a contributing reaction (e.g., ML + H ⇌ MHL) was obtained at 20°C.
† Stability constant for a contributing reaction (e.g., ML + H ⇌ MHL) was determined at $\mu = 0.2$ and then corrected with Davies' equation to $\mu = 0.1$ to yield the value shown.
‡ $Fe^{3+} + L^{5-} \rightleftharpoons FeL$ stability constant at 20°C from Bottari and Anderegg, (1967).
§ Same as † but orginal data obtained at $\mu = 1.0$.
¶ This value was calculated from the data of Frost et al. (1958).

present and they are supplied in the detailed development of the text. The chelating agents in Table 15.1 include the following:

EDTA   ethylenediaminetetraacetic acid or ethylendiamine-N,N,N',N'-tetraacetic acid, $C_{10}H_{16}O_8N_2$

HEEDTA hydroxyethylethylenediaminetriacetic acid or N'-(2-hydroxyethyl)ethylenediamine-N,N,N'-triacetic acid, $C_{10}H_{18}O_7N_2$

EGTA   ethyleneglycolbis(ethylamine)tetraacetic acid or 2,2'-ethylene-dioxybis[ethyliminodi(acetic acid)] $C_{14}O_{24}O_{10}N_2$

DTPA   diethylenetriaminepentaacetic acid or diethylenetriamine-N,N,N',N',N'-pentaacetic acid, $C_{14}H_{23}O_{10}N_3$

CDTA   cyclohexane-1,2-diaminetetraacetic acid or *trans*-cyclohexane-1,2-diamine-N,N,N',N'-tetraacetic acid, $C_{14}H_{22}O_8N_2$

EDDHA   ethylenediamine di(*o*-hydroxyphenylacetic acid) or N,N'-ethylenebis 2-(*o*-hydroxylphenyl)glycine, $C_{18}H_{20}O_6N_2$

Before the stability constants of metal chelates have any real meaning in soils, the many competing equilibria must be considered. Initially Lindsay et al. (1967) showed how to develop chelate stability diagrams to predict which metal ions are held by various chelating agents in soils. Their initial developments considered competition among Fe^{3+}, Ca^{2+}, and H^+; later studies included Zn^{2+} (Lindsay and Norvell, 1969) and other metal ions (Norvell, 1972). Their experimental findings confirm that reactions of EDTA and DTPA in soils can be predicted largely from properly developed theoretical relationships (Norvell and Lindsay, 1969, 1972). Thus the use of stability-pH diagrams to describe metal chelate reactions in soils is well documented.

15.2 DEVELOPMENT OF STABILITY-pH DIAGRAMS FOR CHELATES

When a chelating agent is added to soil, it reacts with cations in the soil solution to form various metal chelates. These reactions cause solid phases to dissolve and exchangeable ions to dissociate to replenish the ions that are chelated. The numerous interrelated equilibria involved in these reactions are depicted diagrammatically in Fig. 15.3.

The chelating ligand L is shown at the center of this diagram. Solid phases and exchange sites supplying the metal ions are shown in the outer Region I,

DEVELOPMENT OF STABILITY-pH DIAGRAMS FOR CHELATES

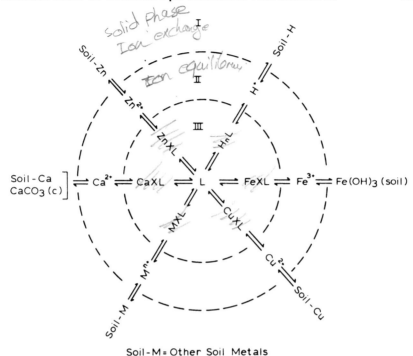

Fig. 15.3 Diagrammatical representation of the equilibrium reactions of metal ions and metal chelates in soils.

and the cations in equilibrium with them are shown in Region II. The equilibrium reactions controlling the concentrations of metal ions in soils are summarized in Table 15.2. The chelated metals are shown as various MXL species in Region III. In these expressions X indicates that some of the chelated species contain either H^+ or OH^- ions. For example, FeXL for EDTA includes the following significant species: FeL^-, $FeHL°$, $FeOHL^{2-}$, and $Fe(OH)_2L^{3-}$. The amount of any chelated species that forms depends upon all competing reactions. How can all of these competing reactions shown in Fig. 15.3 be systematized in a meaningful way?

The distribution of H^+, Fe^{3+}, Ca^{2+}, and Zn^{2+} on the EDTA ligand in soils can be obtained as follows: the total ligand concentration (L_t) in the soil solution shown in Region III of Fig. 15.3 can be expressed by the equation:

$$L_t = L + \Sigma H_n L + \Sigma FeXL + \Sigma CaXL + \Sigma ZnXL + \Sigma MXL \quad (15.3)$$

where L is the concentration of free ligand; $\Sigma H_n L$ is the sum of all protonated ligand species; $\Sigma FeXL$, $\Sigma CaXL$, and $\Sigma ZnXL$ are the respective sums of all

TABLE 15.2 EQUILIBRIUM REACTIONS CONTROLLING THE SOLUBILITY OF METAL CATIONS IN SOILS

Reaction No.	Equilibrium Reaction	$\log K°$*	$\log K_{0.01}^m$†
1	$Fe(OH)_3(amorp) + 3H^+ \rightleftharpoons Fe^{3+} + 3H_2O$	3.54	3.94
2	$Fe(OH)_3(soil) + 3H^+ \rightleftharpoons Fe^{3+} + 3H_2O$	2.70	3.10
3	$Fe(OH)_3(soil) + 3H^+ + e^- \rightleftharpoons Fe^{2+} + 3H_2O$	15.74	15.92
4	$Fe_3O_4(magnetite) + 8H^+ \rightleftharpoons 3Fe^{3+} + e^- + 4H_2O$	−3.42	−2.21
5	$Fe_3O_4(magnetite) + 8H^+ + 2e^- \rightleftharpoons 3Fe^{2+} + 4H_2O$	35.69	36.23
6	$FeCO_3(siderite) + 2H^+ \rightleftharpoons Fe^{2+} + CO_2(g) + H_2O$	7.92	8.10
7	$Fe^{3+} + e^- \rightleftharpoons Fe^{2+}$	13.04	12.82
8	soil-Ca $\rightleftharpoons Ca^{2+}$	−2.50	−2.32
9	$CaCO_3(calcite) + 2H^+ \rightleftharpoons Ca^{2+} + CO_2(g) + H_2O$	9.74	9.92
10	soil-Zn + $2H^+ \rightleftharpoons Zn^{2+}$	5.80	5.98
11	soil-Cu + $2H^+ \rightleftharpoons Cu^{2+}$	2.80	2.98
12	$MnO_2(pyrolusite) + 4H^+ + 2e^- \rightleftharpoons Mn^{2+} + 2H_2O$	41.89	42.07
13	$\gamma\text{-}MnOOH(manganite) + 3H^+ + e^- \rightleftharpoons Mn^{2+} + 2H_2O$	25.27	25.45
14	$MnCO_3(rhodochrosite) + 2H^+ \rightleftharpoons Mn^{2+} + CO_2(g) + H_2O$	8.08	8.26
15	soil-Mg $\rightleftharpoons Mg^{2+}$	−3.00	−2.82
16	$MgCa(CO_3)_2(dolomite) + 2H^+ \rightleftharpoons Mg^{2+} + CO_2(g) + H_2O + CaCO_3(calcite)$	8.72	8.90

* Calculated from $\Delta G_f°$ values in the appendix or taken from the first table in each chapter.
† Calculated from $\log K°$ using Davies' equation (Eq. 2.13) as shown in Section 2.7.

Fe^{3+}, Ca^{2+}, and Zn^{2+} ligand species; and MXL includes all other metal ligands.

For EDTA the L_t term of Eq. 15.3 includes

$$L_t = FeL^- + FeHL^\circ + FeOHL^{2-} + Fe(OH)_2L^{3-} + ZnL^{2-} \\ + ZnHL^- + HL^{3-} + CaL^{2-} + CaHL^- + \Sigma MXL \quad (15.4)$$

Each term in Eq. 15.4 with the exception of ΣMXL can be expressed as a function of the free ligand concentration, the H^+ activity, and the respective formation constants of the complexes and solid phases involved. For example, the FeL^- term of Eq. 15.4 can be obtained by combining the following equilibrium reactions:

$\log K^m_{0.01}$

$Fe^{3+} + L^{4-}$	$\rightleftharpoons$	FeL^-	26.50
$Fe(OH)_3(\text{soil}) + 3H^+$	$\rightleftharpoons$	$Fe^{3+} + 3H_2O$	3.10
$Fe(OH)_3(\text{soil}) + 3H^+ + L^{4-}$	$\rightleftharpoons$	$FeL^- + 3H_2O$	29.60

(15.5)

From this equilibrium reaction

$$[FeL^-] = 10^{29.60}(H^+)^3[L^{4-}] \quad (15.6)$$

The ZnL^{2-} term of Eq. 15.4 can be obtained from the following equilibria:

$\log K^m_{0.01}$

soil-Zn + $2H^+$	$\rightleftharpoons$	Zn^{2+}	5.98
$Zn^{2+} + L^{4-}$	$\rightleftharpoons$	ZnL^{2-}	17.44
soil-Zn + $2H^+ + L^{4-}$	$\rightleftharpoons$	ZnL^{2-}	23.42

(15.7)

from which

$$[ZnL^{2-}] = 10^{23.42}(H^+)^2[L^{4-}] \quad (15.8)$$

In neutral and acid soils, Ca^{2+} is fairly well buffered by the soil exchange sites and usually lies between 1 and 5×10^{-3} M. The selected reference level of $10^{-2.5}$ M (Ca^{2+}) (Chapter 7) corresponds to $10^{-2.32}$ M at 0.01 ionic strength. Thus

$\log K^m_{0.01}$

$Ca^{2+} + L^{4-}$	$\rightleftharpoons$	CaL^{2-}	11.61
soil-Ca	$\rightleftharpoons$	Ca^{2+}	-2.32
soil-Ca + L^{4-}	$\rightleftharpoons$	CaL^{2-}	9.29

(15.9)

from which

$$[CaL^{2-}] = 10^{9.29}[L^{4-}] \tag{15.10}$$

For calcareous soils the following equilibria apply

			$\log K^m_{0.01}$
$Ca^{2+} + L^{4-}$	$\rightleftharpoons$	CaL^{2-}	11.61
$CaCO_3(\text{calcite}) + 2H^+$	$\rightleftharpoons$	$Ca^{2+} + CO_2(g) + H_2O$	9.92
$CaCO_3(c) + L^{4-} + 2H^+$	$\rightleftharpoons$	$CaL^{2-} + CO_2(g) + H_2O$	21.53

$$\tag{15.11}$$

from which

$$[CaL^{2-}] = \frac{10^{21.53}(H^+)^2[L^{4-}]}{CO_2(g)} \tag{15.12}$$

At a $CO_2(g)$ of 0.003 atm

$$[CaL^{2-}] = 10^{24.05}(H^+)^2[L^{4-}] \tag{15.13}$$

Equations similar to 15.6, 15.8, 15.10, and 15.13 can be developed for each term in Eq. 15.4 except for ΣMXL. The latter term presents certain problems in that it includes all possible metal-ligand species not included in the other terms. Without a knowledge of the solid phases governing the concentrations of these other ions, no specific mathematical statement can be developed for them.

There are two ways of testing whether or not a suspected ΣMXL complex may contribute significantly to L_t. The first is to examine a list of formation constants and develop ratios such as the following:

$$\frac{[ML^{4-n}]}{[FeL^-]} = \frac{K_{ML}[M^{n+}][L^{4-}]}{K_{FeL}[Fe^{3+}][L^{4-}]} \tag{15.14}$$

In this expression, the $[L^{4-}]$ factor is common to both numerator and denominator and can be cancelled. The $[Fe^{3+}]$ can be estimated from the solubility of soil-Fe (Reaction 2 of Table 15.2). Thus an estimate of $[M^{n+}]$ permits estimation of the ratio $[ML^{4-n}]/[FeL^-]$.

A second method of estimating the contribution of any MXL term is to measure it experimentally. The concentration of all suspected metals in solution are measured in the presence and absence of the chelating agent, and the difference is attributed to chelation. In this way the ΣMXL term can be estimated experimentally. This method is particularly useful in testing whether or not a suspected metal chelate contributes significantly to L_t even when there is no way of independently estimating the concentration of that metal ion in the soil solution. Data obtained by Norvell and Lindsay (1969) show that for the soils they used, the ΣMXL term contributed less than 5% of L_t in the case of EDTA.

When each term in Eq. 15.4 is expressed either as a measured parameter or in terms of $[L^{4-}]$, (H^+), and the appropriate formation constants, it is possible to calculate the mole fraction of each of the chelated species present, i.e., FeL^-/L_t or ZnL^{2-}/L_t, or any combination of chelated species such as $\Sigma FeXL/L_t$ or $\Sigma ZnXL/L_t$, as a function of pH. This is possible since the free ligand $[L^{4-}]$ in both numerator and denominator can be cancelled.

The distribution of chelated species of EDTA in soils when H^+, Ca^{2+}, Fe^{3+}, Zn^{2+} and Mg^{2+} are the competing ions is shown in Fig. 15.4. This figure shows that FeL^- is the predominant species at low pH, ZnL^{2-} from pH 6 to 7, and CaL^{2-} at higher pH values. In calcareous soils at 0.003 atm of CO_2, ZnL^{2-} is present at a constant 0.19 mole fraction. The chelated species remain fixed because the concentrations of both Ca^{2+} and Zn^{2+},

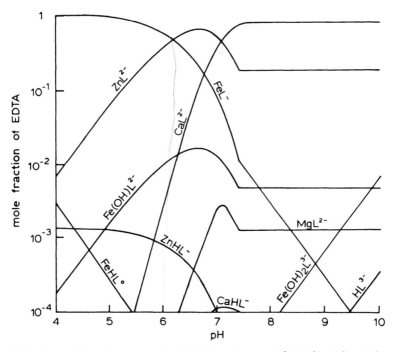

Fig. 15.4 The mole fraction diagram for EDTA in soils when Zn^{2+}, Fe^{3+}, Ca^{2+}, Mg^{2+}, and H^+ are competing metal ions at 0.003 atm of $CO_2(g)$.

which are the major competing ions, decrease at the same rate as pH increases. The relationships depicted in this figure are those expected in soils to which ZnEDTA is added or in which sufficient labile zinc is present to attain equilibrium with EDTA.

If metal ions other than Zn^{2+}, Fe^{3+}, Ca^{2+}, Mg^{2+}, and H^+ are complexed by EDTA to a significant extent, the mole fraction scale represented in Fig. 15.4 becomes the fraction of EDTA in solution not complexed by those metal ions, $[L_t - MXL]$. For EDTA to be an effective micronutrient fertilizer it is necessary that a significant fraction of chelate remains complexed with the micronutrient to increase its mobility in soils. In Fig. 15.4, ZnL^{2-} reaches a maximum at pH 6.7 where it attains a mole fraction of 0.67. Experimental studies of Norvell and Lindsay (1969) confirmed this predicted behavior of ZnEDTA in soils of different pH when ZnEDTA was reacted with soils for 30 days.

Mole fraction calculations of DTPA using modifications of Eq. 15.4 appropriate for this chelate have also been made. The results of these calculations are shown in Fig. 15.5. This figure represents the mole fraction of

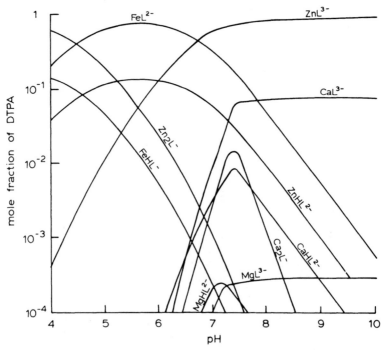

Fig. 15.5 The mole fraction diagram of DTPA in soils when Fe^{3+}, Ca^{2+}, Mg^{2+}, Zn^{2+}, and H^+ are competing metal ions at 0.003 atm $CO_2(g)$.

DTPA in soil solution when the competing metal ions are Zn^{2+}, Fe^{3+}, Ca^{2+}, Mg^{2+}, and H^+. The species Zn_2L^- predominates below pH 4.5, FeL^{2-} in the pH range of 4.5 to 6.5, and ZnL^{3-} above pH 6.5. These equilibrium conditions might be expected to occur following the addition of ZnDTPA to soils, or in soils containing sufficient labile zinc to saturate the added chelate. If metal ions other than Zn^{2+}, Fe^{3+}, Ca^{2+}, Mg^{2+}, and H^+ from the soil are complexed by DTPA to a significant extent, the mole fraction represented on the ordinate refers to $[L_t - \Sigma MXL]$ rather than L_t.

Stability diagrams also can be developed to show various mole fraction combinations of specific metals. For example, the $\Sigma ZnXL$ fractions chelated by EDTA and DTPA are given in Fig. 15.6 when the competing metal ions are again Zn^{2+}, Fe^{3+}, Ca^{2+}, Mg^{2+}, and H^+. The ability of EDTA to chelate zinc is greatest near pH 6.6. At lower pH, Fe^{3+} competes more favorably whereas at higher pH, Ca^{2+} is favored. With DTPA, maximum chelation of Zn^{2+} occurs below pH 4 and near pH 7.4 depending upon $CO_2(g)$. At low pH the chelation of zinc results from the formation of Zn_2L^- which is equivalent to a tetravalent cation. In calcareous soils increasing CO_2

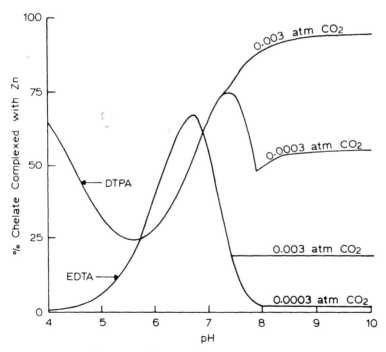

Fig. 15.6 A summary diagram showing ZnEDTA and ZnDTPA complexes in soils as affected by pH and $CO_2(g)$ when Zn^{2+}, Fe^{3+}, Ca^{2+}, Mg^{2+}, and H^+ are the competing metal ions.

depresses Ca^{2+} and enables Zn^{2+} to compete more favorably for the DTPA ligand.

15.3 EFFECT OF REDOX ON METAL CHELATE STABILITY

The chelate stability diagrams developed so far apply to well-oxidized soils where the activities of Fe^{2+} and Mn^{2+} are too low to compete successfully with Fe^{3+}, Zn^{2+}, Ca^{2+}, and Mg^{2+} for the chelating ligands. When soils are reduced, however, the concentration of Mn^{2+} increases according to Reactions 12, 13, 14 of Table 15.2 depending upon whether pyrolusite, manganite, or rhodochrosite is the stable manganese mineral (Fig. 11.2). In addition, Fe^{2+} and Fe^{3+} are controlled by Reactions 2 through 7 of Table 15.2 depending on whether soil-Fe, magnetite, or siderite controls iron solubility (Fig. 10.8). These controls for iron and manganese can be used to show the effect of $pe + pH$ on the distribution of cations or chelating ligands.

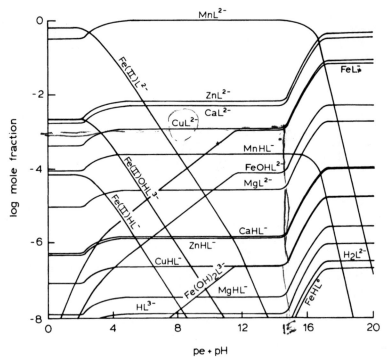

Fig. 15.7 Effect of redox on the equilibrium relationships of EDTA in soils at pH 7.0 when $CO_2(g)$ is 0.003 atm.

For EDTA under reduced conditions the L_t term of Eq. 15.3 includes

$$L_t = FeL^- + FeHL^° + FeOHL^{2-} + Fe(OH)_2L^{3-} + Fe(II)L^{2-}$$
$$+ Fe(II)HL^- + Fe(II)OHL^{3-} + ZnL^{2-} + ZnHL^- + CaL^{2-}$$
$$+ CaHL^- + MnL^{2-} + MnHL^- + \Sigma MXL \quad (15.15)$$

Expressing each term in Eq. 15.15 by the appropriate equilibrium constants and other parameters in the equilibrium reactions of Table 15.1 and 15.2 permits calculation of the mole fraction distribution of metals on EDTA. Sommers and Lindsay (1979), Abuzkhar (1978), and Abuzkhar and Lindsay (1981) have developed mole fraction diagrams for several chelating agents as a function of $pe + pH$. Only a few such diagrams are given here as examples of how Mn^{2+} and Fe^{2+} displace other metal ions.

Figure 15.7 shows the effect of redox on the equilibrium relationships of EDTA in a neutral soil of pH 7.0 when CO_2 is 0.003 atm. According to this diagram ZnL^{2-} and CaL^{2-} are the most stable species at $pe + pH > 16$. The next most stable species are FeL^- and CuL^{2-}, which reflect equal chelating ability of EDTA for Cu^{2+} and Fe^{3+} under these conditions.

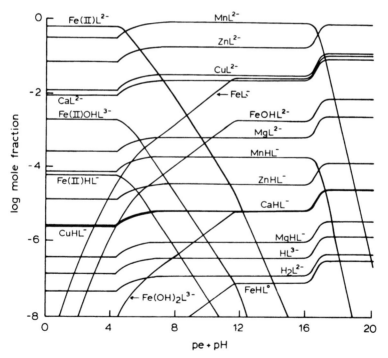

Fig. 15.8 Effect of redox on the equilibrium relationships of EDTA in soils at pH 7.0 when $CO_2(g)$ is 0.1 atm.

In the $pe + pH$ range of 4 to 16, MnL^{2-} is the major chelated species, and it displaces Zn^{2+}, Cu^{2+}, Fe^{3+}, and Ca^{2+} causing them to decrease by approximately 100-fold. Only as $pe + pH$ drops below 4 does Fe^{2+} occupy a significant fraction of EDTA.

Increasing CO_2 to 0.1 atm (Fig. 15.8) reduces the competition from Mn^{2+} in the $pe + pH$ range of 4 to 16 and allows the other cations to compete more favorably. This displacement results from the fact that Mn^{2+} activity is limited by the formation of $MnCO_3$(rhodocrosite). Changes in slope of the Fe(III) species at $pe + pH < 11.53$ reflect the lowering of Fe^{3+} activity as controlled by magnetite, while changes between $pe + pH$ of 4 and 5 reflect the control of iron solubility by $FeCO_3$(siderite).

Increasing soil pH to 8.0 while CO_2 is held at 0.003 atm (Fig. 15.9) lowers the effectiveness of EDTA to retain Fe(III) and increases its ability to chelate Ca^{2+}. Magnesium also forms a significant species, MgL^{2-}. Again in the $pe + pH$ range of 4 to 16, Mn^{2+} occupies a dominant role, and at $pe + pH < 4$ the Fe(II) species becomes important and displaces some of all the other cations.

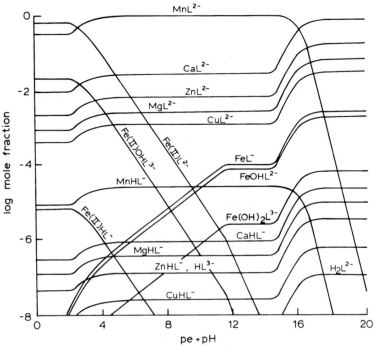

Fig. 15.9 Effect of redox on the equilibrium relationships of EDTA in soils at pH 8.0 when $CO_2(g)$ is 0.003 atm.

Similar mole fraction calculations were made for DTPA using appropriate modifications of Eq. 15.15. Equilibrium relationships for DTPA at pH 7.0 and CO_2 at 0.003 atm are shown in Fig. 15.10. At $pe + pH$ values > 16 Cu^{2+}, Zn^{2+}, and Fe^{3+} compete favorably for DTPA to give CuL^{3-}, ZnL^{3-}, and FeL^{2-}, respectively. In the $pe + pH$ range of 4 to 16 Mn^{2+} displaces a portion of all other cations by approximately 100-fold. Two manganese species are involved, MnL^{3-} and Mn_2L^-. At $pe + pH < 6$, four ferrous species, $Fe(II)OHL^{4-}$, $Fe_2(II)L^-$, $Fe(II)L^{3-}$, and $Fe(II)HL^{2-}$, become important and displace some of all other metal cations.

Reddy and Patrick (1977) recently showed that reducing soils to which Cu^{2+} and Zn^{2+} were added as chelates of EDTA and DTPA showed a loss of these elements from solution. Figures 15.7 to 15.10 predict such displacement and show that Mn^{2+} is largely responsible for this displacement. Abuzkhar (1978) and Abuzkhar and Lindsay (1981) gave further experimental evidence that chelated metals such as EDTA and DTPA are affected by reduction and their behavior can be predicted by properly developed

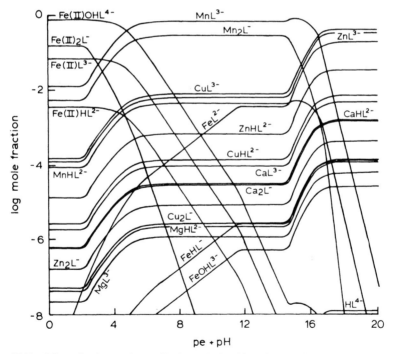

Fig. 15.10 Effect of redox on the equilibrium relationships of DTPA in soils at pH 7.0 when $CO_2(g)$ is 0.003 atm.

theoretical relationships and verified by properly controlled experimental studies.

Metal chelates may still be effective in supplying available zinc, copper, and iron in reduced soils so long as a small fraction of the added metal remains chelated (Fig. 15.7 through 15.10). This hypothesis needs further examination.

15.4 CHELATION IN HYDROPONICS

Halvorson and Lindsay (1972) developed computer programs to calculate the equilibrium levels of chelated metals in nutrient solutions. In hydroponic solutions solid phases are generally not present, except for iron oxides; therefore it is necessary to estimate the activity of a metal ion through an

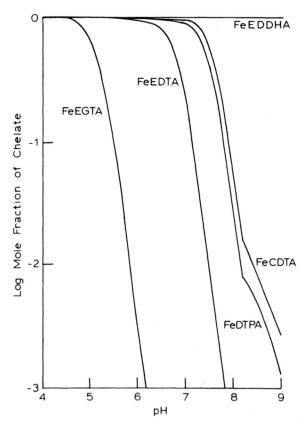

Fig. 15.11 Effectiveness of various metal chelating agents to hold iron in a modified Hoagland hydroponic solution (adapted from Halvorson and Lindsay, 1972).

iterative process beginning with the total metal in solution and subtracting out that which becomes chelated.

Figure 15.11 shows the ability of five synthetic chelating agents to chelate Fe^{3+} in a well-aerated modified Hoagland solution defined in Table 15.3 from Halvorson and Lindsay (1972). As pH rises, FeEGTA becomes unstable above pH 5, FeEDTA above pH 6.5, FeDTPA and FeCDTA above pH 7.2, whereas FeEDDHA remains stable throughout the entire pH range of soils. This means that no cations normally encountered in well-aerated hydroponic solutions are capable of displacing Fe^{3+} from FeEDDHA.

For hydroponic solutions of different compositions the equilibrium relationships shown in Fig. 15.11 are shifted slightly. Increasing the concentration of any competing cation tends to displace some of all the remaining cations from the chelate. Equilibrium conditions can be calculated from the initial concentrations of all components in solution and the competing equilibrium reactions involved.

Halvorson and Lindsay (1977) grew plants in hydroponic solutions containing either FeEDTA or FeDTPA at pH 5.2 and 7.5.

TABLE 15.3 COMPOSITION OF THE MODIFIED HOAGLAND NUTRIENT SOLUTION USED TO DEVELOP FIG. 15.11

Nutrient Ion Added	Concentration (μM)
NH_4^+	400
K^+	2,400
Ca^{2+}	1,600
Mg^{2+}	800
$H_2PO_4^-$	400
NO_3^-	5,600
SO_4^{2-}	800
Fe^{3+}	100
Mn^{2+}	4.56
Zn^{2+}	1.54
Cu^{2+}	0.315
$H_3BO_3^0$	23.1
MoO_4^{2-}	0.104
Cl^-	9.12
Chelating agent	100

Source: Halvorson and Lindsay (1972).

Equilibrium relationships of these chelates are given in Fig. 15.11. The plants grew well at low pH, but when iron was supplied as FeDTPA, the plants at pH 7.5 became severely zinc deficient. Calculation of the free Zn^{2+} concentration in these nutrient solutions (Fig. 15.12) helps to explain why zinc deficiency appeared. At pH 7.5 EDTA depressed Zn^{2+} to approximately $10^{-10.2}$ M, whereas DTPA depressed it to approximately $10^{-11.6}$ M. By repeating these growth studies at different levels of added zinc, they found that whenever free Zn^{2+} dropped below $10^{-10.6}$ M, corn plants showed zinc deficiency. They further concluded that chelated zinc was not absorbed by the corn roots, but that chelated zinc provides a buffered reserve of Zn^{2+}

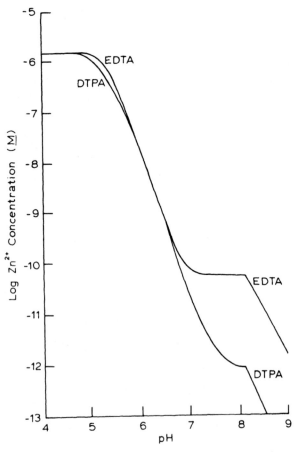

Fig. 15.12 The free Zn^{2+} remaining in a modified Hoagland nutrient solution to which $100\,\mu M$ FeEDTA or FeDTPA and $1.54\,\mu M$ total zinc were added (adapted from Halvorson and Lindsay, 1977).

next to the absorbing cell membrane by increasing the diffusion gradients of zinc through the apparent free space of roots. This study is an example of how chemical equilibria can be used to resolve many seemingly complex problems involving chelating agents even in dynamic systems.

15.5 USE OF CHELATES TO ESTIMATE METAL ION ACTIVITIES IN SOILS

Chelating agents can be used to estimate the activity of cations in soils even though the free metal ions are below the normal analytical detection limits or where unknown complexes preclude the estimation of metal ion activity from total metal concentrations in soil solution. Studies relating to the development of these techniques include Lindsay et al., 1967; Lindsay and Norvell, 1969; Norvell and Lindsay, 1969; 1972, 1981; Norvell, 1971; Vlek et al., 1974.

An example of how EDTA was used to obtain a solubility product of $Fe(OH)_3$(soil) is demonstrated by the investigations of Norvell and Lindsay (1981). In their study ^{14}C-EDTA at 10^{-4} M containing initial mole fractions of Fe^{3+} and Ca^{2+} that varied from 0 to 1 was added to soils. These suspensions were shaken for 7 to 11 days, centrifuged, and filtered, and the clear solutions were analyzed for Fe, Zn, Cu, Ca, Mg, Na, K, and ^{14}C. Labeled ^{14}C was used to determine how much of the added L_t was lost from solution due to adsorption onto the soil. From these measurements it was possible to calculate the activity of Fe^{3+}. The theoretical development for EDTA that permits these calculations includes

$$L_t = [FeL^-] + [Fe(OH)L^{2-}] + [CaL^{2-}] + [MgL^{2-}] \\ + [CuL^{2-}] + [ZnL^{2-}] \tag{15.16}$$

The solubilities of iron, copper, and zinc in soils near pH 7 are normally so low that measurable concentrations of these metals are $< 10^{-6}$ M. In the presence of EDTA the measured concentration can be taken as

$$Fe_t = [FeL^-] + [Fe(OH)L^{2-}]$$
$$Zn_t = [ZnL^{2-}]$$
$$Cu_t = [CuL^{2-}]$$

For calcium, both $[Ca^{2+}]$ and $[CaL^{2-}]$ contribute significantly to solution Ca_t, whereas for magnesium only $[Mg^{2+}]$ is important because of the low stability of $[MgL^{2-}]$. Thus

$$Ca_t = [Ca^{2+}] + [CaL^{2-}]$$
$$Mg_t = [Mg^{2+}]$$

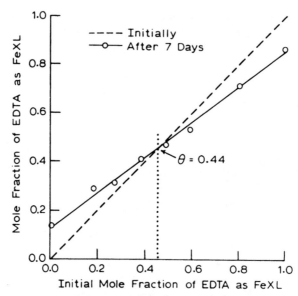

Fig. 15.13 Changes in the mole fraction of EDTA present as FeXL in a Monticello loam during a 7-day equilibration period used to estimate the Fe^{3+} activity in this soil (from Norvell and Lindsay, 1981).

Combining these expressions with the appropriate formation constants gives an expression for $[Fe^{3+}]$

$$[Fe^{3+}] = \frac{[FeL^-] + [FeOHL^{2-}]}{[CaL^{2-}]} \cdot \frac{[Ca^{2+}]K_{CaL}}{K_{FeL} + (OH^-)K_{FeOHL}} \quad (15.17)$$

The $[Fe^{3+}]$ of a soil can be estimated from Eq. 15.17 providing the first factor on the right side of this equation representing the ratio of chelated iron to chelated calcium can be measured under equilibrium conditions. To do this Norvell and Lindsay (1981) equilibrated soil with 10^{-4} M EDTA varying in Fe^{3+} from 0 to 100% using Ca^{2+} as the counter ion. The iron remaining in solution after 8 days was measured.

Typical results are plotted in Fig. 15.13 for the Monticello loam. A slight adjustment, explained by Norvell and Lindsay (1981), was necessary to account for the differential adsorption of the different chelated metals onto soil. For the Monticello loam, a unique mole fraction of 0.44 of chelated iron showed that no iron was gained or lost from the chelate as a result of competitive equilibria between Ca^{2+} and Fe^{3+}. Substituting 0.44/0.56 into Eq. 15.17 for the expression:

$$\frac{[FeL^-] + [FeOHL^{2-}]}{[CaL^{2-}]},$$

along with measurements or estimates for $[Ca^{2+}]$ and pH gives a value for $[Fe^{3+}]$ which, in turn, can be converted to (Fe^{3+}) from the relationship

$$(Fe^{3+}) = [Fe^{3+}]\gamma_{Fe^{3+}} \tag{15.18}$$

For the soils they used an average solubility value of

$$(Fe^{3+})(OH^-)^3 = 10^{-39.30 \pm 0.23} \tag{15.19}$$

was obtained for the solubility of Fe(III) in soils. This value has been used throughout this text as a reference solubility of iron in soils and corresponds to Reaction 2 of Table 10.1 and to Reaction 2 of Table 15.2. The significance of soil-Fe and its relationships to other iron oxides is discussed in Section 10.1.

Similar studies have been made to determine the solubilities of soil-Zn and soil-Cu (Norvell and Lindsay, 1969, 1972; Lindsay and Norvell, 1969). Further investigations of metal ion activities using these techniques are to be encouraged.

15.6 USE OF CHELATING AGENTS AS SOIL TESTS

Chelating agents are often used as soil test extractants to estimate the plant availability of various elements in soils. The DTPA micronutrient soil test was developed from an understanding of the chemical equilibrium relationships of chelating agents in soils (Lindsay and Norvell, 1978) and is used here as an example of how such principles can be applied.

In the DTPA soil test 10 g of air-dry soil is shaken with 20 ml of extractant containing 0.005 M DTPA, 0.01 M $CaCl_2$, and 0.1 M triethanolamine adjusted to pH 7.30. After 2 hr the suspension is filtered and zinc, iron, manganese, and copper are determined in the filtrate by atomic absorption spectrophotometry.

The soil test successfully separated 77 Colorado soils into deficiency and nondeficient categories based on crop responses to micronutrient fertilizers. The critical levels of DTPA-extractable nutrients above which fertilizer responses were not obtained were: 0.8 ppm for zinc, 4.5 ppm for iron, 1 ppm for manganese, and 0.2 ppm for copper.

The theoretical basis for the DTPA soil test is illustrated in Fig. 15.14 and is based on an ionic strength of 0.1 (Lindsay and Norvell, 1978). The percent of DTPA present that can potentially become saturated with each of the micronutrient cations in equilibrium with solid phases normally encountered in soils are shown as a function of pH.

Approximately 90% of the chelate can become saturated with Zn^{2+} or Cu^{2+} if equilibria were attained with soil-Zn or soil-Cu, respectively,

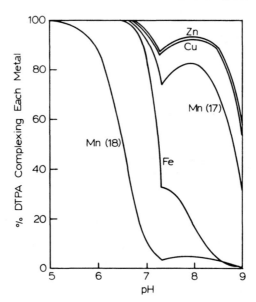

Fig. 15.14 The equilibrium levels of Mn, Fe, Zn, or Cu extracted by DTPA in competition with Ca^{2+} commonly found during extraction of soils with the DTPA soil test. The numbers in parentheses refer to $pe + pH$ values (Lindsay and Norvell, 1978).

and approximately 32% of the chelate can be saturated with Fe^{3+} in equilibrium with soil-Fe. The amount of Mn^{2+} that can be extracted is highly redox dependent. At $pe + pH$ of 17 approximately 85% of the chelate could be saturated with Mn^{2+} in equilibrium with MnO_2(pyrolusite) as shown in Fig. 11.2, whereas at $pe + pH$ of 18 this drops to 4%.

Selection of pH 7.30 permits the extracting solution to attain equilibrium with $CaCO_3$ and $CO_2(g)$ at about 10 times the CO_2 level of the atmosphere. This avoids the dissolution of $CaCO_3$ with the exposure and release of occluded micronutrients that are normally not accessible to plant roots. Since soils normally contain large quantities of iron and manganese relative to zinc and copper (Table 1.1), the soil test was designed to avoid excessive extraction of iron and manganese. The chelating agent DTPA was used because it provides the most suitable combination of equilibrium constants for the simultaneous extraction of zinc, iron, manganese, and copper.

Triethanolamine is added at 0.1 M to provide a pH buffer because of the great sensitivity of iron and manganese to pH. The 0.005 M level of DTPA provides ample chelating capacity to avoid interferences among nutrients. This quantity of DTPA is sufficient to extract 10 times the molecular weight of the micronutrient (ppm dry weight basis). For zinc this would be $65 \times 10 =$

650 ppm. For a critical level of zinc of 0.5 ppm, this is only $\frac{1}{812}$ of the chelating capacity. Obviously equilibrium is not attained during a typical soil extraction.

The 0.01 M CaCl$_2$ is sufficient to convert the chelate to 0.005 M CaDTPA^{3-} and provide an additional 0.005 M Ca^{2+}, which tends to equalize Ca^{2+} for most soils. In actual operation protonated triethanolamine displaces additional Ca^{2+} from the soil exchange and raises free Ca^{2+} to approximately 0.01 M.

The parameters mentioned here combine to give a micronutrient soil test that has been surprisingly successful, as shown by its 10-year track record (Lindsay and Norvell, 1978).

More recently Sommers and Lindsay (1979) developed stability diagrams describing the equilibrium relationships of seven chelating agents with eleven metal ions in soils. From these diagrams it is easy to see which chelating agents can be used to extract various metals depending on soil characteristics and particularly pH. Thus theoretical equilibrium relationships of metal ions and chelating agents provide useful guidelines for developing soil test extractants for various purposes.

15.7 NATURAL CHELATES IN SOILS

So far in this chapter only synthetic chelating agents have been discussed. This is not to infer that natural chelates are not important. To the contrary, natural chelates are even more important because they are present in most soils. Major problems in dealing with natural chelates include (1) identifying the chelates that are present in soils, (2) determining how long they persist, and (3) obtaining meaningful equilibrium constants for reactions in which they participate. As soon as these relationships can be determined, natural chelates can be treated similarly to synthetic chelating agents discussed herein.

Numerous examples could be cited of studies that demonstrate the presence of natural chelates in soils and the role they play in the chelation of metals, but only a few will be mentioned (Schnitzer and Khan, 1972; Mortensen, 1963; Stevenson and Ardakani, 1972). Norvell (1972) included citrate and oxalate in treatment of chelates in soils and showed that aluminum, calcium, and magnesium are major cations under the condition he considered. Recent investigation (Graustein et al., 1977) shows the occurrence of calcium oxalate in many soils as the result of fungi activity. The reported concentrations are sufficient to cause the solubilization of metals such as Fe^{3+}. Further investigations of metal complexing by soil organic matter that leads to the identification of specific complexes or chelates and the thermodynamic equilibrium constants controlling their reactions are to be encouraged.

The models presented in this chapter help to demonstrate the principles involved in metal chelation in soils and some of the parameters that must be included in the theoretical treatment of metal chelates.

REFERENCES

Abuzkhar, A. A. 1978. Effect of redox on the stability of metal chelates in soils. Ph.D. Dissertation, Colorado State University, Fort Collins, Colorado.

Abuzkhar, A. A. and W. L. Lindsay. 1981. Effect of redox on the equilibrium relationships of EDTA in soils. Soil Sci. Soc. Am. J. 45: (in press).

Bottari, E. and G. Anderegg. 1967. Komplexone. XLII. Die untersuchung der 1:1 complexes von einigen drei- und vierwertigen metallionen mit polyaminocarboxylaten mittels redoxmessungen. Helv. Chim. Acta 50:2349–2356.

Frost, A. E., H. H. Freedman, S. K. Westerback, and A. E. Martell. 1958. Chelating tendencies of N,N'-ethylenebis-2-(o-hydroxyphenyl)-glycine. J. Am. Chem. Soc. 80:530–536.

Graustein, W. C., K. Cromack, Jr., and P. Sollins. 1977. Calcium oxalate: occurrence in soils effect on nutrient and geochemical cycles. Science 198:1252–1254.

Halvorson, A. D. and W. L. Lindsay. 1972. Equilibrium relationships of metal chelates in hydroponic solutions. Soil Sci. Soc. Am. Proc. 36:755–761.

Halvorson, A. D. and W. L. Lindsay. 1977. The critical Zn^{2+} concentration for corn and the nonabsorption of chelated zinc. Soil Sci. Soc. Am. J. 41:531–534.

Lindsay, W. L. 1974. Role of chelation in micronutrient availability. In E. W. Carson, Ed., The Plant Root and Its Environment, pp. 507–524. University of Virginia, Charlottesville.

Lindsay, W. L., J. F. Hodgson, and W. A. Norvell. 1967. The physicochemical equilibrium of metal chelates in soils and their influence on the availability of micronutrient cations, pp. 305–316. Trans. Comm. II & IV Int. Soc. Soil Sci., Aberdeen.

Lindsay, W. L. and W. A. Norvell. 1969. Equilibrium relationships of Zn^{2+}, Fe^{3+}, Ca^{2+}, and H^+ with EDTA and DTPA in soils. Soil Sci. Soc. Am. Proc. 33:62–68.

Lindsay, W. L. and W. A. Norvell. 1978. Development of a DTPA soil test for zinc, iron, manganese, and copper. Soil Sci. Soc. Am. J. 42:421–428.

Martell, A. L. and R. M. Smith. 1974. Critical Stability Constants. Vol. 1. Amino Acids. Plenum Press, New York.

Mortensen, J. L. 1963. Complexing of metals by soil organic matter. Soil Sci. Soc. Am. Proc. 27:179–186.

Norvell, W. A. 1971. Solubility of Fe^{3+} in soils. Ph.D. Thesis, Colorado State University. Diss. Abstr. 5111-B.

Norvell, W. A. 1972. Equilibria of metal chelates in soil solution. In J. J. Mortvedt, P. M. Giordano, and W. L. Lindsay, Eds., Micronutrients in Agriculture, pp. 115–138. Soil Science Society of America, Madison, Wi.

Norvell, W. A. and W. L. Lindsay. 1969. Reactions of EDTA complexes of Fe, Zn, Mn, and Cu with soils. Soil Sci. Soc. Am. Proc. 33:86–91.

Norvell, W. A. and W. L. Lindsay. 1972. Reactions of DTPA chelates of Fe, Zn, Cu, and Mn with soils. Soil Sci. Soc. Am. Proc. 36:755–761.

Norvell, W. A. and W. L. Lindsay. 1981. Estimation of iron(III) solubility from EDTA chelate equilibria in soils. Soil Sci. Soc. J. 44:(in press).

Reddy, C. N. and W. H. Patrick, Jr. 1977. Effect of redox potential on the stability of Zn and Cu chelates in flooded soils. Soil Sci. Soc. Am. J. 41:429–432.

Schnitzer, M. and S. U. Khan. 1972. Humic Substances in the Environment. Marcell Dekker, New York.

Sommers, L. E. and W. L. Lindsay. 1979. Effect of pH and redox on predicted heavy metal-chelate equilibria in soils. Soil Sci. Soc. Am. J. 43:39–47.

Stevenson, F. J. and M. S. Ardakani. 1972. Organic matter reactions involving micronutrients in soils. In J. J. Mortvedt, P. M. Giordano, and W. L. Lindsay, Eds., Micronutrients in Agriculture, pp. 79–114. Soil Science Society of America, Madison, Wi.

Vlek, P. L. G., T. J. M. Blom, J. Beek, and W. L. Lindsay. 1974. Determination of the solubility product of various iron hydroxides and jarosite by the chelation method. Soil Sci. Soc. Am. Proc. 38:429–432.

PROBLEMS

15.1 Calculate the necessary equations and plot the mole fraction of total EDTA that is complexed with iron over the pH range of 4 to 10 when the competing cations are Ca^{2+}, Fe^{3+}, and H^+. Consider Ca^{2+} to be at $10^{-2.5} M$ or in equilibrium with $CaCO_3$, which ever is smaller. Plot curves for CO_2 both at 0.0003 atm and 0.03 atm. Consider Fe^{3+} as being in equilibrium with $Fe(OH)_3$(soil).

15.2 Calculate the activity of Zn^{2+} in the hydroponic nutrient study of Halvorson and Lindsay (1977) at pH 7.5 when FeEDTA and when FeDTPA are used to furnish iron.

15.3 Calculate the activity of Fe^{3+} in a hydroponic nutrient solution at pH 7.5 similar to that of Halvorson and Lindsay (1972) when FeEDDHA at $10^{-4}\ M$ is used to supply iron and 1 % of the iron has been removed by plant roots.

15.4 Soils of different pH were shaken with $10^{-4}\ M$ ZnEDTA for 30 days. Analyses of the solution phase at the end of this period were as follows:

pH	Total Ca (μ mol)	Total EDTA (μ mol)	Total Fe (μ mol)	Total Zn (μ mol)	Total Cu (μ mol)
5.70	2500	93.5	87	6.0	0.5
6.10	2000	84.6	38	38	1.0
6.75	2400	88.3	7.0	57	1.0
7.30	1500	93.4	0.5	51	1.0
7.85	850	92.5	0.0	23	0.5

Develop the necessary equations to calculate (Zn^{2+}) and develop an equilibrium constant ($\log K°$) for the reaction:

$$\text{soil-Zn} + 2H^+ \rightleftharpoons Zn^{2+}$$

15.5 Calculate the DTPA soil test levels of Fe and Zn extracted by the DTPA soil test when simultaneous equilibrium is obtained with the extractant, and soil-Fe and soil-Zn govern the activities of these metal ions. For this purpose consider that only negligible quantities of Cu and Mn or cations (other than Ca) are complexed. What is the capacity of the extracting solution for each of the micronutrient cations if any one of them were to completely saturate the chelate? The extracting solution consists of 0.01 M $CaCl_2$ and 0.005 M DTPA buffered at pH 7.30.

SIXTEEN

NITROGEN

There may be some deficiencies.

Inorganic chemistry.

s⁺ close to
p electron
d
f

Chelates

(E)

Left
less complex

Right
more complex elements.

— charge is strong, magnetic attr, south & north pole, chelate got electron, fail and become parts of above shell, chelate have sharing.

The nitrogen content of most soils ranges from 200 to 4000 ppm with an average of approximately 1400 ppm (Table 1.1). A large portion of the nitrogen in soils is present as organic matter.

Nitrogen is an unusual element in that it slowly escapes from soils to the atmosphere, which contains approximately 78% N_2(g). This form of nitrogen is unavailable to plants and must be either oxidized or reduced before it can be utilized. Since nitrogen is the nutrient element that is most often deficient for plants, much attention has been given to it. Nevertheless, many of the chemical and biological transformations of nitrogen in the environment are still poorly understood.

The objective of this chapter is to present a thermodynamic development that depicts the overall driving forces that ultimately govern chemical transformations of the various nitrogen species. The development given here is based largely on a recent paper (Lindsay et al., 1981) and touches briefly on earlier developments of Stumm and Morgan (1970).

16.1 OXIDATION STATES OF NITROGEN

Nitrogen exists in many different oxidation states ranging from $+6$ to -3. Some of the common inorganic forms of nitrogen that exist in each of these oxidation states are summarized in Table 16.1. Obviously redox relationships are necessary in any development that attempts to determine which nitrogen species are stable and the conditions that affect their stability.

In the past most nitrogen transformations have been attributed to microorganisms. There are several reasons for this: (1) most of the nitrogen in soils is present as organic nitrogen and must be mineralized before it is free to participate in other chemical reactions; (2) microorganisms are the main source of electrons enabling soils to become reduced so that reduced nitrogen species can form; and (3) some microorganisms can utilize atmospheric N_2 and with the help of special enzymes catalyze its reduction in isolated regions of living cells. For these and other reasons, nitrogen transformations become rather complex.

Dinitrogen contains a triple bond (N≡N), which apparently requires a high activation energy for N_2(g) to be reactive even though its reactions may be thermodynamically feasible. In spite of these complexities overall thermodynamic relationships are useful in distinguishing between possible and impossible reactions.

The equilibrium relationships among various nitrogen species are summarized in Table 16.2. The equilibrium constants given here were calculated from the ΔG_f° values documented in the Appendix. Recently Van Cleemput

TABLE 16.1 INORGANIC NITROGEN SPECIES AND THEIR OXIDATION STATES

Nitrogen Species	Name	Oxidation State of Nitrogen
$NO_3(g)$	Nitrogen Trioxide	$+6$
$N_2O_5(g)$	Dinitrogen pentoxide	$+5$
NO_3^-	Nitrate ion	$+5$
$NO_2(g)$	Nitrogen dioxide	$+4$
$N_2O_4(g)$	Dinitrogen tetroxide	$+4$
$N_2O_3(g)$	Dinitrogen trioxide	$+3$
NO_2^-	Nitrite ion	$+3$
$NO(g), NO°$	Nitric oxide	$+2$
$N_2O(g), N_2O°$	Nitrous oxide	$+1$
$N_2(g)$	Dinitrogen	0
N_3^-	Azide	$-\frac{1}{3}$
$NH_2OH°$	Hydroxylamine	-1
$NH(g)$	Imidogen	-1
$NH_2(g)$	Amidogen	-2
$N_2H_4(g), N_2H_4°$	Hydrazine	-2
$N_2H_5^+$	Protonated hydrazine	-2
$NH_3(g), NH_3°$	Ammonia	-3
NH_4^+	Ammonium ion	-3

and Baert (1974) summarized their own compilation of free energy values for biological environments. Nitrogen equilibria can be combined in many different ways to show the equilibrium relationships among the various nitrogen species. Many such equilibria will be demonstrated in the sections that follow.

16.2 EQUILIBRIUM BETWEEN ATMOSPHERIC N_2 AND O_2

The activities of various inorganic nitrogen species in equilibrium with atmospheric $N_2(g)$ and $O_2(g)$ are shown in Fig. 16.1. Oxygen at 0.2 atm corresponds to an equilibrium $pe + pH$ of 20.61 (Chapter 2). Since redox reactions seldomly attain equilibrium with atmospheric $O_2(g)$, this graph depicts potential equilibrium conditions more nearly than it does actual conditions.

TABLE 16.2 EQUILIBRIUM CONSTANTS FOR NITROGEN REACTIONS USED IN THIS TEXT

Reaction No.	Equilibrium Reaction	log $K°$
1	$NO_3^- \rightleftharpoons NO_3(g) + e^-$	-39.87
2	$NO_3^- + H^+ \rightleftharpoons \frac{1}{2}N_2O_5(g) + \frac{1}{2}H_2O$	-9.08
3	$NO_3^- + 2H^+ + e^- \rightleftharpoons NO_2(g) + H_2O$	13.03
4	$NO_3^- + 2H^+ + 2e^- \rightleftharpoons NO_2^- + H_2O$	28.64
5	$NO_3^- + 3H^+ + 2e^- \rightleftharpoons \frac{1}{2}N_2O_3(g) + \frac{3}{2}H_2O$	30.62
6	$NO_3^- + 4H^+ + 3e^- \rightleftharpoons NO(g) + 2H_2O$	48.41
7	$NO_3^- + 5H^+ + 4e^- \rightleftharpoons \frac{1}{2}N_2O° + \frac{5}{2}H_2O$	75.52
8	$NO_3^- + 6H^+ + 5e^- \rightleftharpoons \frac{1}{2}N_2(g) + 3H_2O$	105.15
9	$NO_3^- + 6H^+ + \frac{16}{3}e^- \rightleftharpoons \frac{1}{3}N_3^- + 3H_2O$	84.80
10	$NO_3^- + 7H^+ + 6e^- \rightleftharpoons NH_2OH° + 2H_2O$	67.69
11	$NO_3^- + 7H^+ + 6e^- \rightleftharpoons NH(g) + 3H_2O$	46.81
12	$NO_3^- + 8H^+ + 7e^- \rightleftharpoons NH_2(g) + 3H_2O$	74.02
13	$NO_3^- + 10H^+ + 8e^- \rightleftharpoons NH_4^+ + 3H_2O$	119.07
	Other Reactions	
14	$NO_2(g) \rightleftharpoons \frac{1}{2}N_2O_4(g)$	0.42
15	$HNO_2° \rightleftharpoons H^+ + NO_2^-$	-3.15
16	$NO(g) \rightleftharpoons NO°$	-2.73
17	$N_2O(g) \rightleftharpoons N_2O°$	0.54
18	$NH_2(g) \rightleftharpoons \frac{1}{2}N_2H_4(g)$	17.18
19	$\frac{1}{2}N_2H_4(g) \rightleftharpoons \frac{1}{2}N_2H_4°$	2.74
20	$\frac{1}{2}N_2H_4° + \frac{1}{2}H^+ \rightleftharpoons \frac{1}{2}N_2H_5^+$	3.98
21	$NH_3(g) \rightleftharpoons NH_3°$	1.76
22	$NH_3° + H^+ \rightleftharpoons NH_4^+$	9.28
23	$NH_4^+ + H_2O \rightleftharpoons NH_4OH° + H^+$	-9.25
24	$H^+ + e^- + \frac{1}{4}O_2(g) \rightleftharpoons \frac{1}{2}H_2O$	20.78

An example of how the lines in Fig. 16.1 were developed is shown for NO_3^- by combining Reactions 8 and 24 of Table 16.2 to give

$$\begin{array}{rcl} & & \log K° \\ \frac{1}{2}N_2(g) + 3H_2O & \rightleftharpoons NO_3^- + 6H^+ + 5e^- & -105.15 \\ 5H^+ + 5e^- + \frac{5}{4}O_2(g) & \rightleftharpoons \frac{5}{2}H_2O & 5(20.78) \\ \hline \frac{1}{2}N_2(g) + \frac{5}{4}O_2(g) + \frac{1}{2}H_2O & \rightleftharpoons NO_3^- + H^+ & -1.25 \end{array}$$

(16.1)

EQUILIBRIUM BETWEEN ATMOSPHERIC N_2 AND O_2

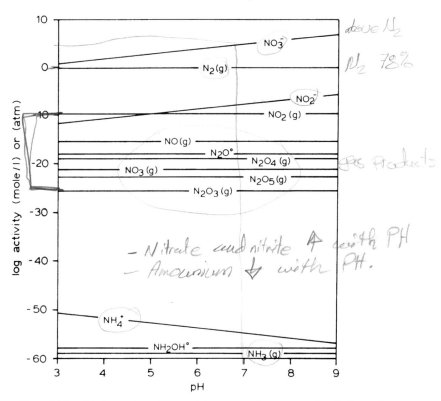

Fig. 16.1 The activities of ionic species and the partial pressures of gaseous species of nitrogen in equilibrium with $N_2(g)$ and $O_2(g)$ of the air at 0.78 and 0.20 atm. respectively.

which reduces to

$$\log NO_3^- = -1.25 + \tfrac{1}{2}\log N_2(g) + \tfrac{5}{4}\log O_2(g) + pH \quad (16.2)$$

When $N_2(g) = 0.78$ atm and $O_2(g) = 0.20$ atm

$$\log NO_3^- = -2.18 + pH \quad (16.3)$$

Equation 16.3 is shown in Fig. 16.1 as the NO_3^- line. This line represents the activity of NO_3^- required in soils if equilibrium were attained with atmospheric $N_2(g)$ and $O_2(g)$.

Figure 16.1 shows that NO_3^- is the most stable form of nitrogen in equilibrium with $N_2(g)$ and $O_2(g)$ of the air. The other nitrogen species are present at only trace amounts. For example, nitrite (NO_2^-) at pH 7 is present at only 10^{-8} M, whereas NH_4^+ shows an equilibrium level of only 10^{-55} M. Most of the gaseous oxides of nitrogen lie in the range of 10^{-10}

to 10^{-25} atm. The species NH(g), NH$_2$(g), N$_2$H$_2$(g), N$_2$H$_4^o$, N$_2$H$_5^+$, and N$_3^-$ all fall below the 10^{-60} atm or molar level shown here.

Another conclusion that can be drawn from Fig. 16.1 is that N$_2$(g) and O$_2$(g) in the air are thermodynamically unstable and potentially can form NO$_3^-$. At pH 7 the concentration of NO$_3^-$ that would be required for equilibrium with atmospheric N$_2$(g) is approximately 10^5 M, which is an impossibly high concentration. Thus N$_2$(g) and O$_2$(g) in the air have a great potential to form NO$_3^-$ providing the right catalyst is present. Such thermodynamic relationships offer a real challenge to chemical engineers to produce nitrogen fertilizers from air without the expenditure of energy. Fortunately such catalysts are not generally present in nature, or the earth would be deluged with nitric acid and O$_2$(g) in the air would be drastically reduced.

16.3 EFFECT OF REDOX ON NITROGEN STABILITY

Soils are nearly always more reduced than pe + pH of 20.61 which corresponds to 0.20 atm of O$_2$(g). The effect of redox on the equilibrium relationships of various nitrogen species at pH 7.0 were calculated from the reactions in Table 16.2 and are shown in Fig. 16.2. In this plot NO$_3^-$ was arbitrarily set at 10^{-3} M or 14 ppm nitrogen, which is near that found in many soils. The procedure used to establish the lines in this diagram is illustrated for N$_2$(g). From Reaction 8 of Table 16.2

$$NO_3^- + 6H^+ + 5e^- \rightleftharpoons \tfrac{1}{2}N_2(g) + 3H_2O \qquad \log K^\circ = 105.15$$

$$\frac{N_2^{1/2}(g)}{(NO_3^-)(H^+)(H^+)^5(e^-)^5} = 10^{105.15}$$

from which

$$\tfrac{1}{2} \log N_2(g) = 105.15 + \log NO_3^- - \text{pH} - 5(pe + \text{pH}) \qquad (16.4)$$
$$= 105.15 + (-3) - (7) - 5(pe + \text{pH})$$
$$= 95.15 - 5(pe + \text{pH})$$

$$\log N_2(g) = 190.30 - 10(pe + \text{pH}) \qquad (16.5)$$

A plot of Eq. 16.5 shows the log of N$_2$(g) in equilibrium with 10^{-3} M NO$_3^-$ and is represented by the N$_2$(g) line in Fig. 16.2. Other lines in this figure were obtained similarly using the appropriate equations from Table 16.2.

The equilibrium relationships in Fig. 16.2 show that N$_2$(g) rapidly becomes the stable nitrogen species in slightly reduced environments and equilibrium would require impossibly high pressures of N$_2$(g). The lines in this figure show the relative stabilities of the different nitrogen species compared to NO$_3^-$ at 10^{-3} M as pe + pH changes from 0 to 21.

EFFECT OF REDOX ON NITROGEN STABILITY

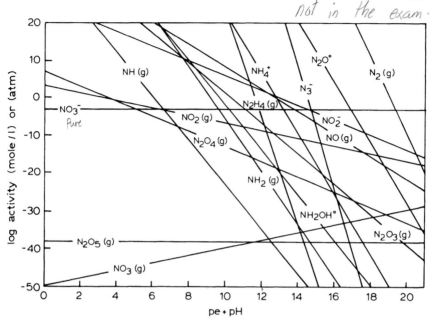

Fig. 16.2 The effect of redox on the equilibrium levels of various nitrogen species in equilibrium with 10^{-3} M NO_3^- at pH 7.0.

Nitrogen species that could be present in significant quantities in the redox range of $pe + pH$ of 14 to 19 include: $N_2(g)$, $N_2O°$, NO_2^-, N_3^-, and $NO(g)$. Some of these species that may be involved as intermediates in the reactions leading to denitrification will be discussed later.

In the $pe + pH$ range of 12 to 14, NH_4^+, $NH_3(g)$, $N_2H_5^+$, $N_2H_2(g)$, and $N_2H_4°$ rapidly become more stable than NO_3^-. Although $NH_3(g)$, $N_2H_5^+$, and $N_2H_4°$ were omitted from Fig. 16.2 to avoid further cluttering of the diagram, their locations can be readily obtained from Eq. 20 through 23 of Table 16.2. For example, combining Reactions 21 and 22 gives

$$NH_4^+ \rightleftharpoons NH_3(g) + H^+ \qquad \log K° = -11.04 \qquad (16.6)$$

which at pH 7 reduces to

$$NH_4^+ \rightleftharpoons NH_3(g) \qquad \log K° = -4.04 \qquad (16.7)$$

Equation 16.7 indicates that the $NH_3(g)$ line in Fig. 16.2 would be parallel to the NH_4^+ line and lie -4.04 log units below it. The lines for $N_2H_5^+$ and $N_2H_4°$ can be obtained similarly.

In the $pe + pH$ range of 9 to 12 the species $N_2O_3(g)$, $NH_2OH°$, and $NH_2(g)$ become significant. In the $pe + pH$ range of 4 to 9 $NO_2(g)$, $NH(g)$, and $N_2O_4(g)$ become significant. The highly oxidized species $N_2O_5(g)$ and $NO_3(g)$ are not significant in natural soil environments.

Equilibrium relationships among the various nitrogen species are depicted slightly differently in Fig. 16.3. In this plot total nitrogen in solution is limited at 10^{-3} M or by equilibrium with $N_2(g)$ of the air (0.78 atm). Several interesting conclusions can be drawn from this figure. In highly oxidized soils ($pe + \text{pH} > 19$), NO_3^- is the stable form of nitrogen, and $N_2(g)$ of the air could oxidize to NO_3^-. In the $pe + \text{pH}$ range of 3.5 to 19, $N_2(g)$ is the stable form of nitrogen and all other nitrogen species tend to change to $N_2(g)$ and escape to the atmosphere. Only under highly reducing environments ($pe + \text{pH} < 3.5$) is NH_4^+ the stable species. Under such low redox conditions it is thermodynamically possible for $N_2(g)$ from the air to reduce to NH_4^+ and accumulate in soils. Similarly, such reduced environments are undoubtedly maintained in the organelles of certain microorganisms where nitrogen fixation occurs. The lines in this figure are drawn for pH 7.0, but changes that would occur at pH 5 and 8 are shown by arrows and lines labeled 5 and 8, respectively.

The important conclusion to be drawn from Fig. 16.3 is that $N_2(g)$ is the only stable nitrogen species in the $pe + \text{pH}$ range of 3.5 to 19.0. All other

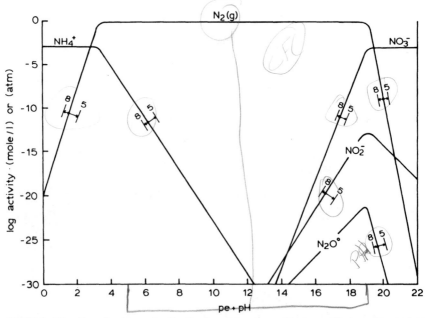

Fig. 16.3 The effect of redox on the distribution of nitrogen species in equilibrium with 10^{-3} M total nitrogen or atmospheric $N_2(g)$ at 0.78 atm.

EFFECT OF REDOX ON NITROGEN STABILITY

forms of nitrogen are unstable. Since most agricultural soils lie in this redox range, equilibrium conditions would require that most nitrogen be released to the atmosphere as $N_2(g)$. The broad range of stability of $N_2(g)$ no doubt accounts for the fact that nitrogen in soils is slowly and continually lost to the atmosphere as $N_2(g)$. It is well recognized that NO_3^-, NH_4^+, and traces of NO_2^- are commonly present in soils. The fact that these forms of inorganic nitrogen are present in soils suggests that equilibrium relationships with $N_2(g)$ are not readily attained.

Nonequilibrium with $N_2(g)$

The $N \equiv N$ bond of $N_2(g)$ makes it one of the most inert of all diatomic molecules. Let us recognize this fact and look at the equilibrium relationships among the remaining nitrogen species when $N_2(g)$ is considered to be unreactive. This condition is depicted in Fig. 16.4 where total nitrogen in solution is fixed at 10^{-3} M.

From Fig. 16.4 it can be seen that NO_3^- is the most stable species above $pe + pH$ of 16.7, $N_2O°$ is most stable between 9.5 and 16.7, and NH_4^+ is most stable below $pe + pH$ 9.5. Under these conditions NO_2^- reaches a maximum

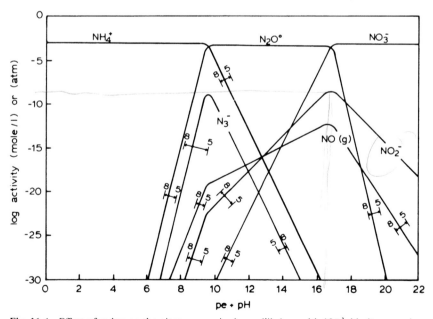

Fig. 16.4 Effect of redox on the nitrogen species in equilibrium with 10^{-3} M nitrogen when nonequilibrium with $N_2(g)$ is recognized. Drawn at pH 7 with modifications for pH 5 and 8.

level of about 10^{-8} M at $pe + pH$ of 16.8 and decreases at both higher and lower redox. Also NO(g) attains 10^{-12} atm at this same redox. Similarly, N_3^- (azide) reaches a maximum of 10^{-9} M at $pe + pH$ of 9.5. The partial pressure of $N_2O(g)$ in equilibrium with $N_2O°$ is given by Reaction 17 of Table 16.2. If $N_2O(g)$ were plotted, its line would be 0.54 log unit below the $N_2O°$ line and lie at $10^{-4.38}$ M in this intermediate redox range. If the equilibria depicted in Fig. 16.4 were attained, $N_2O(g)$ would be lost from soils in the $pe + pH$ range of 10 to 17 either from the reduction of NO_3^- or the oxidation of NH_4^+.

Since $N_2O(g)$ is not an abundant nitrogen species in the atmosphere (~300 ppb), nor has $N_2O°$ been recognized as a major solution species in soils, the equilibria depicted in Fig. 16.4 are apparently not achieved. If they were, it would be impossible to prevent the loss of NH_4^+ and NO_3^- from soils in the $pe + pH$ range of 10 to 16.

Nonequilibrium with N_2 and N_2O

If both $N_2(g)$ and $N_2O°$ are sufficiently unreactive that equilibrium between them and other nitrogen species is not attained, the equilibrium relationships

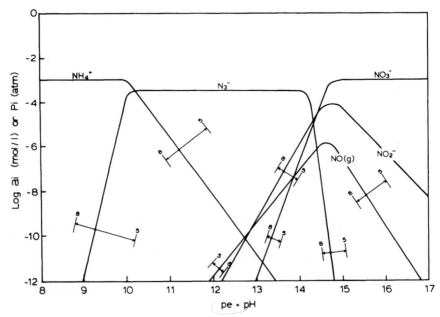

Fig. 16.5 Effect of redox on nitrogen species in equilibrium with total nitrogen at 10^{-3} M when nonequilibrium with $N_2(g)$ and $N_2O°$ are considered. Drawn at pH 7 with modifications for pH 5 and 8.

EFFECT OF REDOX ON NITROGEN STABILITY

among the remaining nitrogen species would be those depicted in Fig. 16.5. In this diagram total nitrogen in solution was arbitrarily set at 10^{-3} M. Under these conditions, NO_3^- is stable above $pe + pH$ of about 15, N_3^- (azide) is stable between 10 and 14.2, and NH_4^+ is stable below 10.2. The stability of NH_4^+ is highly pH dependent and at pH 5 would increase to about 11.2 as can be extrapolated from Fig. 16.5. Both NO_2^- and $NO(g)$ become significant species at $pe + pH$ around 14.5, attaining approximately 10^{-4} M and 10^{-6} atm, respectively.

Based on Fig. 16.5, it is thermodynamically possible for N_3^- to form in soils at ordinary levels of NO_3^- and NH_4^+ in the redox range of $pe + pH$ 10.2 to 14.3. However, nothing has been found in the literature to suggest that N_3^- may be involved in denitrification for soils, nor have significant quantities of N_3^- been reported in soils. One possibility is that N_3^- forms very slowly, and since it is unstable relative to $N_2O°$ or N_2, it rapidly transforms to these species. A second possibility is that even though N_3^- is thermodynamically stable, it does not form. Very few studies have been made to determine the $G_f°$ for N_3^-, so the present value documented in the Appendix should be accepted with some reservation. The possible involvement of N_3^- in nitrogen transformation in the redox range of $pe + pH$ of 10 to 14.5 needs further examination.

Nonequilibrium with N_2, N_2O, and N_3^-

If the species $N_2(g)$, $N_2O°$, and N_3^- are all considered to be relatively unreactive, that is, they do not readily attain equilibrium with other nitrogen species, then the equilibrium relationships among the remaining nitrogen species are those shown in Fig. 16.6. Again total nitrogen in solution was arbitrarily set at 10^{-3} M. Under these conditions NO_3^- is the most stable species above $pe + pH$ of 14.3, NO_2^- is most stable in the range of 12.7 to 14.3, and NH_4^+ is most stable below 12.7.

The equilibria shown in Fig. 16.6 reflect more nearly the conditions generally observed in soils, namely, NH_4^+ is stable in highly reduced soils, NO_3^- in oxidized soils, and frequently traces of NO_2^- are found in soils of intermediate redox. Many interesting hypotheses regarding possible nitrogen transformations and denitrification reactions can be developed from Fig. 16.6.

Of particular interest is the fact that $NO(g)$ reaches a maximum partial pressure at the same $pe + pH$ range that NO_2^- becomes a stable species. The stability of $NO(g)$ is highly pH-dependent and increases at low pH. The stability range of NO_2^- narrows from the NH_4^+ side but remains fixed on the NO_3^- side as pH lowers. In contrast the stability of $NO(g)$ increases

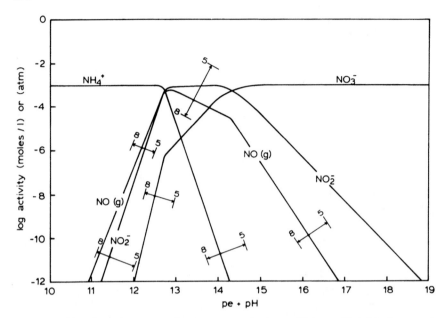

Fig. 16.6 The effect of redox on the nitrogen species in equilibrium with total nitrogen at 10^{-3} M when nonequilibrium with $N_2(g)$, $N_2O°$ and N_3^- are recognized. Drawn at pH 7 with modifications for pH 5 and 8.

greatly and shifts to the right. To show these changes the equilibrium relationships in Fig. 16.6 are redrawn in Fig. 16.7 for pH 4. At this pH the equilibrium partial pressure of NO(g) peaks at $pe + pH$ of 13.9 where it reaches approximately $10^{-1.2}$ atm and the solution species NO° reaches 10^{-4} M.

The hypothesis proposed herein is that NO (nitric oxide) may play the key role in the denitrification process by providing a kinetically feasible intermediate reaction product. Once formed, NO may transform to N_2O and/or $N_2(g)$ since both are thermodynamically more stable than NO at this redox. In this way NO(g) need not accumulate nor even attain the equilibrium levels depicted in Fig. 16.6 and 16.7.

If NO is an intermediate reaction product in the denitrification process, chemical denitrification would be largely restricted to the narrow redox range of $pe + pH$ 13 to 15. Chemical denitrification would also be aided by decreasing pH. Fluctuating redox conditions that bring soils into the intermediate $pe + pH$ of 13 to 15 would result in greater denitrification than would either higher or lower redox conditions that avoid this critical redox range. As shown in Fig. 16.6 and 16.7, NO(g) or NO° can only form in this narrow redox range. These stability relationships may help to explain

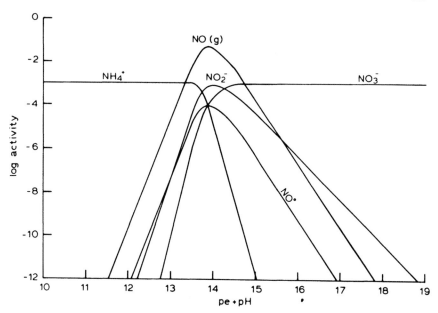

Fig. 16.7 Nitrogen species at pH 4.0 in equilibrium with total N of 10^{-3} M when N_2, N_2O and N_3^- are considered to be inert.

why denitrification studies often give conflicting results when redox is neither controlled nor monitored.

Natural processes like lightning are known to oxidize $N_2(g)$ and give small quantities of various oxides of nitrogen. These oxides are scrubbed from the atmosphere by rain and provide some of the fixed nitrogen used by plants. Such fixation of nitrogen along with microbiological fixation of $N_2(g)$ helps to replace nitrogen in soils that is lost through denitrification.

Further investigations are needed to examine the hypothesis proposed in this chapter, namely that NO may be the kinetically feasible intermediate reaction product in the chemical denitrification processes that occur in soils. Increased interest in denitrification in recent years reflects the importance of this process in the environment. Hopefully the development presented herein will help to show how thermodynamic relationships can be used to distinguish between possible and impossible reactions and suggest possible intermediates in the transformation of nitrogen species.

REFERENCES

Lindsay, W. L., M. Sadiq, and L. K. Porter. 1981. Thermodynamics of inorganic nitrogen transformations. Soil Sci. Soc. Am. J. 45: (in press).

Stumm, W. and J. J. Morgan. 1970. Aquatic Chemistry. Wiley-Interscience, New York, pp. 330-331.

Van Cleemput, O. and L. Baert. 1974. Gibbs standard free energy changes of different nitrogen reactions as corrected for biological real pH and O_2 activities. Overdruk Uit:Meded. Fak. Land. Gent 39:1-28.

PROBLEMS

16.1 From the reactions in Table 16.2 develop the equation for plotting the NO_2^- line in Fig. 16.1. Explain the significance of this figure.

16.2 From the reactions in Table 16.2 develop the necessary equation to plot the line for $N_2H_5^+$ in Fig. 16.2.

16.3 From Fig. 16.2 discuss the possible reactions that could occur in a 10^{-3} M solution of NO_3^- when:
  a. $pe + pH$ is maintained at 20.6
  b. $pe + pH$ is maintained at 18.0
  c. $pe + pH$ is maintained at 14.5, but N_2O, N_2, and N_3^- do not attain equilibrium with other nitrogen species.

16.4 Develop equations for the NO_2^- lines in Fig. 16.3. Prepare a plot of this species at pH 3, 5, 7, and 9, and discuss the effect of pH on the stability of nitrite.

16.5 Develop equations for the $NO°$ species and plot it in Fig. 16.4. Under what conditions is $NO°$ more stable than $N_2O°$?

16.6 Discuss the theoretical possibility that N_3^- may be an intermediate reaction product in denitrification in soils. Is there experimental evidence available to support or refute this role?

16.7 Develop an equation for the $NO(g)$ line in Fig. 16.7. Prepare a plot showing log $NO(g)$ versus $pe + pH$ for pH 2, 4, 6, and 8. Discuss the effect of redox and pH on chemical denitrification if nitric oxide is the key intermediate reaction product in this process.

16.8 What partial pressures of $N_2(g)$ and $N_2O(g)$ would be required for equilibrium with $10^{-1.3}$ atm $NO(g)$ as shown for $pe + pH$ 13.9 in Fig. 16.7? How does this relate to the possible role of NO in the denitrification process?

SEVENTEEN

SULFUR

The sulfur content of the lithosphere is estimated at 600 ppm. Sulfur in soils is highly variable and generally ranges from 30 to 10,000 ppm with an estimated average of 700 ppm. Since sulfur is essential for all living organisms, it has been widely studied. Many of the reactions of sulfur in soils are closely associated with organic matter and the activity of microorganisms.

The most common form of sulfur in soils is SO_4^{2-}. Since sulfur occurs in many different oxidation states, redox relationships must be considered in dealing with its overall stability and mineral transformations.

In this chapter the solubility relationships of the various sulfur species and the sequence of metal sulfide precipitations will be critically examined. These developments are of particular interest in submerged soils and in soils where the cycling of heavy metals is exposed to fluctuating redox conditions. Complexes of sulfate are included in the separate chapters dealing with the respective cations.

17.1 EFFECT OF REDOX ON SULFUR SPECIATION

The oxidation states of sulfur range from $+6$ to -2 (Table 17.1). Included among these species are solids, solution species, and gases. When soils are reduced, the lower oxidation states become more significant.

Equilibrium relationships among the various sulfur species are summarized in Table 17.2. Sulfate is recognized as the stable species in all but highly

TABLE 17.1 SULFUR SPECIES AND THEIR OXIDATION STATES

Chemical Formula	Name	Oxidation State
SO_4^{2-}, HSO_4^-, $H_2SO_4^\circ$	Sulfate	$+6$
$SO_3(g)$	Sulfur trioxide	$+6$
SO_3^{2-}, HSO_3^-, $H_2SO_3^\circ$	Sulfite	$+4$
SO_2°, $SO_2(g)$	Sulfur dioxide	$+4$
$S_2O_4^{2-}$, $HS_2O_4^-$, $H_2S_2O_4^\circ$	Dithionite	$+3$
$S_2O_3^{2-}$, $HS_2O_3^-$, $H_2S_2O_3^\circ$	Thiosulfate	$+2$
$SO(g)$	Sulfur monoxide	$+2$
$S_2O(g)$	Disulfur oxide	$+1$
S(rhombic)	Sulfur (rhombic)	0
S_3^{2-}, S_4^{2-}, S_5^{2-}, etc.	Polysulfides	$-\frac{2}{3}, -\frac{1}{2}, -\frac{2}{5}$
S_2^{2-}, $HS(g)$	Bisulfide	-1
$HS(g)$		-1
S^{2-}, HS^-, H_2S°, $H_2S(g)$	Sulfide	-2

EFFECT OF REDOX ON SULFUR SPECIATION

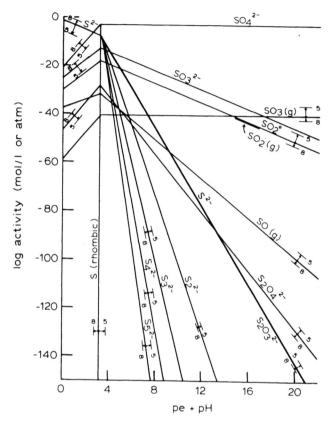

Fig. 17.1 The effect of redox on the sulfur species in equilibrium with 10^{-3} M SO_4^{2-} or rhombic sulfur. Drawn for pH 7 with shifts for pH 5 and 8.

reduced soils. An activity level of 10^{-3} M SO_4^{2-} is used in this text as a reference level of SO_4^{2-}. When necessary this level can be easily adjusted to correspond to that of any given soil. The solubility of SO_4^{2-} in soils is limited by the solubility of $CaSO_4 \cdot 2H_2O$(gypsum) given by Reaction 54 of Table 17.2. For soils in equilibrium with soil-Ca at $10^{-2.5}$ M Ca^{2+}, SO_4^{2-} would be limited at $10^{-2.14}$ M. When Ca^{2+} activity is depressed by the formation of $CaCO_3$(calcite), the SO_4^{2-} activity can rise accordingly.

Reactions 1 through 15 of Table 17.2 show the equilibrium relationships between SO_4^{2-} and each of the deprotonated sulfur species. These relationships are plotted in Fig. 17.1 as a function of redox. An example of how these lines were obtained is given for S^{2-} beginning with Reaction 14 of Table 17.2:

$$SO_4^{2-} + 8e^- + 8H^+ \rightleftharpoons S^{2-} + 4H_2O \qquad \log K° = 20.74$$

TABLE 17.2 EQUILIBRIUM REACTIONS OF SULFUR SPECIES AT 25°C

Reaction No.	Equilibrium Reaction	log $K°$
	Redox Reactions	
1	$SO_4^{2-} + 2H^+ \rightleftharpoons SO_3(g) + H_2O$	−23.87
2	$SO_4^{2-} + 2e^- + 2H^+ \rightleftharpoons SO_3^{2-} + H_2O$	−3.73
3	$SO_4^{2-} + 2e^- + 4H^+ \rightleftharpoons SO_2(g) + 2H_2O$	5.04
4	$SO_4^{2-} + 2e^- + 4H^+ \rightleftharpoons SO_2^\circ + 2H_2O$	5.35
5	$SO_4^{2-} + 3e^- + 4H^+ \rightleftharpoons \frac{1}{2}S_2O_4^{2-} + 2H_2O$	5.26
6	$SO_4^{2-} + 4e^- + 6H^+ \rightleftharpoons SO(g) + 3H_2O$	−2.03
7	$SO_4^{2-} + 4e^- + 5H^+ \rightleftharpoons \frac{1}{2}S_2O_3^{2-} + \frac{5}{2}H_2O$	18.68
8	$SO_4^{2-} + 5e^- + 7H^+ \rightleftharpoons \frac{1}{2}S_2O(g) + \frac{3}{2}H_2O$	−61.44
9	$SO_4^{2-} + 6e^- + 8H^+ \rightleftharpoons S(rhombic) + 4H_2O$	35.78
10	$SO_4^{2-} + \frac{20}{3}e^- + 8H^+ \rightleftharpoons \frac{1}{3}S_3^{2-} + 4H_2O$	31.38
11	$SO_4^{2-} + \frac{13}{2}e^- + 8H^+ \rightleftharpoons \frac{1}{2}S_4^{2-} + 4H_2O$	32.73
12	$SO_4^{2-} + \frac{32}{5}e^- + 8H^+ \rightleftharpoons \frac{1}{5}S_5^{2-} + 4H_2O$	33.48
13	$SO_4^{2-} + 7e^- + 8H^+ \rightleftharpoons \frac{1}{2}S_2^{2-} + 4H_2O$	28.54
14	$SO_4^{2-} + 8e^- + 8H^+ \rightleftharpoons S^{2-} + 4H_2O$	20.74
15	$\frac{1}{2}S_2^{2-} + e^- \rightleftharpoons S^{2-}$	−7.80
	Acids	
16	$H_2SO_4^\circ \rightleftharpoons H^+ + HSO_4^-$	1.98
17	$HSO_4^- \rightleftharpoons H^+ + SO_4^{2-}$	−1.98
18	$H_2SO_3^\circ \rightleftharpoons H^+ + HSO_3^-$	−1.91
19	$HSO_3^- \rightleftharpoons H^+ + SO_3^{2-}$	−7.18
20	$H_2S_2O_4^\circ \rightleftharpoons H^+ + HS_2O_4^-$	−0.37
21	$HS_2O_4^- \rightleftharpoons H^+ + S_2O_4^{2-}$	−2.49
22	$H_2S_2O_3^\circ \rightleftharpoons H^+ + HS_2O_3^-$	−0.60
23	$HS_2O_3^- \rightleftharpoons H^+ + S_2O_3^{2-}$	−1.61
24	$HS(g) \rightleftharpoons H^+ + \frac{1}{2}S_2^{2-}$	12.98
25	$H_2S(g) \rightleftharpoons H_2S^\circ$	−0.99
26	$H_2S^\circ \rightleftharpoons H^+ + HS^-$	−7.02
27	$HS^- \rightleftharpoons H^+ + S^{2-}$	−12.90
28	$H_2S(g) \rightleftharpoons 2H^+ + S^{2-}$	−20.92
	Metal Sulfides	
29	$\alpha\text{-}Ag_2S(c) \rightleftharpoons 2Ag^+ + S^{2-}$	−49.02
30	$\beta\text{-}Ag_2S(c) \rightleftharpoons 2Ag^+ + S^{2-}$	−48.97
31	$CaS(oldhamite) \rightleftharpoons Ca^{2+} + S^{2-}$	−0.78
32	$CdS(greennokite) \rightleftharpoons Cd^{2+} + S^{2-}$	−27.07
33	$Cu_2S(chalcocite) \rightleftharpoons 2Cu^+ + S^{2-}$	−48.54
34	$CuS(covellite) \rightleftharpoons Cu^{2+} + S^{2-}$	−36.10
35	$\alpha\text{-}Fe_{0.95}S(pyrrohotite) \rightleftharpoons 0.85Fe^{2+} + 0.10Fe^{3+} + S^{2-}$	−18.74

TABLE 17.2 (*Continued*)

Reaction No.	Equilibrium Reaction	log $K°$
36	α-FeS(trolite) $\rightleftharpoons Fe^{2+} + S^{2-}$	−16.21
37	FeS_2(pyrite) $\rightleftharpoons Fe^{2+} + S_2^{2-}$	−26.93
38	FeS_2(pyrite) + 2e$^- \rightleftharpoons Fe^{2+} + 2S^{2-}$	−42.52
39	FeS_2(markasite) $\rightleftharpoons Fe^{2+} + S_2^{2-}$	−26.23
40	Fe_2S_3(c) $\rightleftharpoons 2Fe^{3+} + 3S^{2-}$	−88.00
41	α-HgS(cinnibar, red) $\rightleftharpoons Hg^{2+} + S^{2-}$	−52.03
42	β-HgS(metacinnabar, black) $\rightleftharpoons Hg^{2+} + S^{2-}$	−51.66
43	Hg_2S(c) $\rightleftharpoons Hg_2^{2+} + S^{2-}$	−54.77
44	MgS(c) $\rightleftharpoons Mg^{2+} + S^{2-}$	5.59
45	MnS(green) $\rightleftharpoons Mn^{2+} + S^{2-}$	−11.67
46	MnS(alabanite) $\rightleftharpoons Mn^{2+} + S^{2-}$	−12.84
47	MnS_2(hauerite) $\rightleftharpoons Mn^{2+} + S_2^{2-}$	−14.79
48	MnS_2(hauerite) + 2e$^- \rightleftharpoons Mn^{2+} + 2S^{2-}$	−30.38
49	MoS_2(molybdenite) + $4H_2O \rightleftharpoons MoO_4^{2-} + 2S^{2-} + 2e^- + 8H^+$	−96.49
50	PbS(galena) $\rightleftharpoons Pb^{2+} + S^{2-}$	−27.51
51	α-ZnS(sphalerite) $\rightleftharpoons Zn^{2+} + S^{2-}$	−24.70
52	β-ZnS(wurtzite) $\rightleftharpoons Zn^{2+} + S^{2-}$	−22.50

Soil Cation Reactions

53	Ag(c) $\rightleftharpoons Ag^+ + e^-$	−13.50
54	$CaSO_4 \cdot 2H_2O$(gypsum) $\rightleftharpoons Ca^{2+} + SO_4^{2-} + 2H_2O$	−4.64
55	soil-Ca $\rightleftharpoons Ca^{2+}$	−2.50
56	$CaCO_3$(calcite) + $2H^+ \rightleftharpoons Ca^{2+} + CO_2(g) + H_2O$	9.74
57	soil-Cd $\rightleftharpoons Cd^{2+}$	−7.00
58	$CdCO_3$(otavite) + $2H^+ \rightleftharpoons Cd^{2+} + CO_2(g) + H_2O$	6.16
59	α-$Cu_2Fe_2O_4$(cuprous ferrite) + $8H^+ \rightleftharpoons 2Cu^+ + 2Fe^{3+} + 4H_2O$	−13.53
60	$Cu^{2+} + e^- \rightleftharpoons Cu^+$	2.62
61	$Cu^{2+} + 2e^- \rightleftharpoons Cu(c)$	11.49
62	soil-Fe + $3H^+ \rightleftharpoons Fe^{3+}$	2.70
63	Fe_3O_4(magnetite) + $8H^+ + 2e^- \rightleftharpoons 3Fe^{2+} + 4H_2O$	35.69
64	$FeCO_3$(siderite) + $2H^+ \rightleftharpoons Fe^{2+} + CO_2(g) + H_2O$	7.92
65	$Fe^{3+} + e^- \rightleftharpoons Fe^{2+}$	13.04
66	Hg(l) $\rightleftharpoons \frac{1}{2}Hg_2^{2+} + e^-$	−13.46
67	Hg(l) $\rightleftharpoons Hg^{2+} + 2e^-$	−28.86
68	soil-Mg $\rightleftharpoons Mg^{2+}$	−3.00
69	$MgCa(CO_3)_2$(dolomite) + $2H^+ \rightleftharpoons Mg^{2+} + CO_2(g) + H_2O + CaCO_3(c)$	8.72
70	$MnCO_3$(rodochrosite) + $2H^+ \rightleftharpoons Mn^{2+} + CO_2(g) + H_2O$	8.08
71	soil-Mo $\rightleftharpoons MoO_4^{2-} + 0.8H^+$	−12.40*
72	soil-Pb $\rightleftharpoons Pb^{2+}$	−8.00
73	$PbCO_3$(cerussite) + $2H^+ \rightleftharpoons Pb^{2+} + CO_2(g) + H_2O$	4.65
74	soil-Zn + $2H^+ \rightleftharpoons Zn^{2+}$	5.80

* Vlek and Lindsay (1977)

which can be expressed as

$$\frac{(S^{2-})}{(SO_4^{2-})(e^-)^8(H^+)^8} = 10^{20.74} \qquad (17.1)$$

or

$$\log S^{2-} = 20.74 + \log SO_4^{2-} - 8(pe + pH) \qquad (17.2)$$

When SO_4^{2-} is 10^{-3} M, this equation reduces to

$$\log S^{2-} = 17.74 - 8(pe + pH) \qquad (17.3)$$

A plot of Eq. 17.3 is shown as the S^{2-} line in Fig. 17.1. The other lines in this figure were obtained similarly using the appropriate reactions from Table 17.2.

Lines in Fig. 17.1 shift with pH when the number of electrons and protons in the reaction are unequal. These lines are drawn at pH 7 with shifts for pH 5 and 8 indicated by short lines and arrows. The pH shifts are minor compared to the overall redox dependence of the various sulfur species. It is apparent from Fig. 17.1 that SO_4^{2-} is by far the most stable sulfur species in the redox range above $pe + pH$ of 4. The species of intermediate stability include: SO_3^{2-}, SO_2°, $SO_2(g)$, $SO_3(g)$, which generally are present at less than 10^{-20} atm or M. They can be discounted as being prevalent in soils.

The next most stable species include: $SO(g)$, S^{2-}, $S_2O_3^{2-}$, and $S_2O_4^{2-}$. Of these, S^{2-} is most significant because of the insolubility of many metal sulfides. The polyanions of sulfur include S_2^{2-}, S_3^{2-}, S_4^{3-}, S_5^{3-}, etc., which increase rapidly with decreasing redox. These species become significant as $pe + pH$ drops near 4. The vertical line at $pe + pH$ 3.13 shows the redox at which elemental sulfur (rhombic S) can form. The equation for this line is obtained from Reaction 9 of Table 17.2

$$pe + pH = 5.96 + \tfrac{1}{6}\log SO_4^{2-} - \tfrac{1}{3}pH \qquad (17.4)$$

When SO_4^{2-} is 10^{-3} M,

$$pe + pH = 5.46 - \tfrac{1}{3}pH \qquad (17.5)$$

and at pH 7

$$pe + pH = 3.13 \qquad (17.6)$$

Rhombic S can be considered to be an infinite assemblage of sulfur atoms that approach the oxidation state of zero. The lines to the left of the rhombic S line in Fig. 17.1 are shown in equilibrium with S (rhombic). Before the redox can drop below $pe + pH$ of 3.13, SO_4^{2-} activity must decrease below 10^{-3} M or pH must rise above 7. Both conditions are favored by conversion of SO_4^{2-} to S(rhombic) or by the precipitation of metal sulfides. As SO_4^{2-} drops, so

17.2 DISSOCIATION OF SULFUR ACIDS

The dissociation constants of the various sulfur acids are given by Reactions 16 through 28 of Table 17.2. These acids are discussed in the order in which they appear in the table.

Sulfuric acid ($H_2SO_4^\circ$) is a strong acid and is essentially fully dissociated to SO_4^{2-} throughout the pH range of soils. Reaction 17 of Table 17.2 indicates that at pH 1.98, 50% of the total sulfate would be present as HSO_4^-, and this percentage decreases by a factor of 10 for each unit increase in pH.

Sulfurous acid ($H_2SO_3^\circ$) is much weaker. At pH 7.18 equal amounts of HSO_3^- and SO_3^{2-} are present and their ratio changes by a factor of 10 for each unit change in pH. Although sulfurous acid may be added to soils, it is unstable and soon disappears. In most soils it oxidizes to SO_4^{2-}, whereas in highly reduced soils, it may be reduced to sulfides. Because of the instability of SO_3^{2-}, metal sulfites are generally too soluble to persist in soils.

Both dithionitic acid ($H_2S_2O_4^\circ$) and thiosulfuric acid ($H_2S_2O_3^\circ$) are strong acids (Reactions 20 through 23 of Table 17.2) and are essentially dissociated in the pH range of most soils. These anions are also unstable in soils (Fig. 17.1) and do not persist. In most soils they oxidize to SO_4^{2-}, whereas in highly reduced soils they reduce to sulfides.

In highly reduced soils, hydrogen sulfide (H_2S°) becomes an important solution species. Under these conditions it maintains a relatively high partial pressure of $H_2S(g)$, sufficient to escape from soils. According to Reaction 25 of Table 17.2, a 0.1 M solution of H_2S° has a $H_2S(g)$ partial pressure near one atmosphere. Hydrogen sulfide is also a very weak acid and is predominantly present as H_2S° below pH 7.02 (Reaction 26 of Table 17.2). Between pH 7.02 and 12.90 HS^- is the major solution species while S^{2-} is the predominant solution species above pH 12.90 (Reaction 27 of Table 17.2).

Equilibrium relationships of important sulfide and bisulfide species in equilibrium with 10^{-3} M SO_4^{2-} are shown in Fig. 17.2. In neutral soils of pe + pH of 4, S^{2-} approaches 10^{-14} M, H_2S° and HS^- approach 10^{-8} M, and $H_2S(g)$ approaches 10^{-7} atm. As pH drops to 5, $H_2S(g)$ increases to 10^{-3} atm, a level that is easily detected by odor and lost to the atmosphere. The partial pressure of $H_2S(g)$ that can occur in soils is highly redox dependent and increases 8 log units for each unit decrease in pe + pH (Eq. 17.). At pe + pH of 3.13, where S(rhombic) can form, $H_2S(g)$ approaches 1 atm

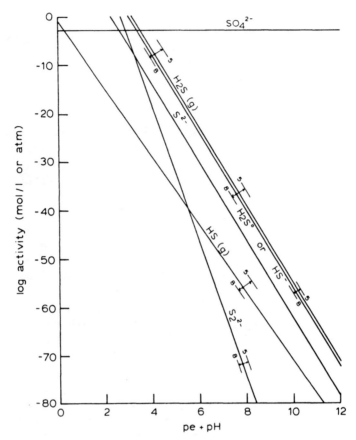

Fig. 17.2 The effect of redox on the sulfide and bisulfide species in equilibrium with 10^{-3} M SO_4^{2-}. Drawn for pH 7 with shifts for pH 5 and 8.

(Fig. 17.2) and can escape from soils. Increasing pH lowers the partial pressure of $H_2S(g)$ at this redox (Fig. 17.2) and lowers the redox required for the precipitation of S(rhombic) (Fig. 17.1). Thus elemental sulfur is not likely to form in soils that are open to the atmosphere.

17.3 FORMATION OF ELEMENTAL SULFUR IN SOILS

Sometimes it is important to know whether or not elemental S(rhombic) can form in soils, and if so, under what conditions. Figure 17.3 was developed to answer these questions.

FORMATION OF ELEMENTAL SULFUR IN SOILS

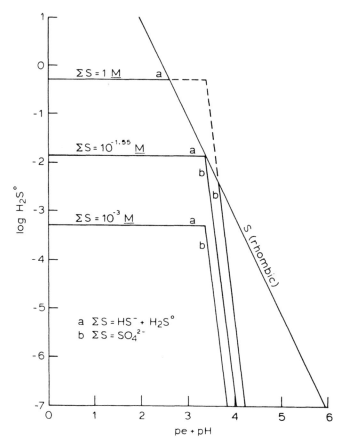

Fig. 17.3 Total sulfur and redox range necessary for the formation of S(rhombic) at pH 7.0.

The S(rhombic) line in Fig. 17.3 shows the minimum activity of $H_2S°$ required for the precipitation of S(rhombic). This line corresponds to the equilibrium reaction:

$$S(rhombic) + 2e^- + 2H^+ \rightleftharpoons H_2S° \qquad \log K° = 4.89 \quad (17.7)$$

from which

$$\log H_2S° = 4.89 - 2(pe + pH) \qquad (17.8)$$

The ΣS lines correspond to the indicated levels of total sulfur in solution. The a branch of these lines corresponds to the redox region where $\Sigma S = H_2S° + HS^-$. At pH 7 these species are approximately equal (Reaction 26,

Table 17.2) so the indicated $H_2S°$ activity is approximately half the total sulfur. The *b* branch of these lines corresponds to the redox region where SO_4^{2-} is stable, or $\Sigma S = SO_4^{2-}$. The equilibrium reaction used to construct the *b* lines is

$$SO_4^{2-} + 8e^- + 10H^+ \rightleftharpoons H_2S° + 4H_2O \quad \log K° = 40.67 \quad (17.9)$$

from which

$$\log H_2S° = 40.67 + \log SO_4^{2-} - 2pH - 8(pe + pH) \quad (17.10)$$

For pH 7.0 and 10^{-3} M SO_4^{2-}, Eq. 17.10 simplifies to

$$\log H_2S° = 23.67 - 8(pe + pH) \quad (17.11)$$

From Fig. 17.3 it is apparent that S(rhombic) cannot form in soils of pH 7.0 having only 10^{-3} M total sulfur in solution because such soils would always be undersaturated with respect to S(rhombic). Only when total sulfur in solution exceeds $10^{-1.55}$ M is it possible for S(rhombic) to form. Higher levels of soluble sulfur can lead to the precipitation of elemental sulfur. For example, a 1 M solution of sulfur can precipitate S(rhombic) in the $pe + pH$ range of 2.60 to 3.63 as indicated by the dashed lines that rise above the S(rhombic) line in Fig. 17.3.

As shown in Section 17.2, $H_2S(g)$ can be lost to the atmosphere at low redox. The partial pressure of $H_2S(g)$ is obtained from the relationship

$$H_2S° \rightleftharpoons H_2S(g) \quad \log K° = 0.99$$

Thus the partial pressure of $H_2S(g)$ in atmospheres is equal to 10 times the molar activity of $H_2S°$. Thus the $H_2S(g)$ lines in Fig. 17.3 would lie one log unit above the $H_2S°$ lines.

The formation of elemental sulfur in soils is highly unlikely for several reasons: (1) the $pe + pH$ must be between 2.5 and 4.0, (2) total sulfur in solution must exceed $10^{-1.55}$ M, and (3) the system must be confined, otherwise $H_2S(g)$ will escape. Furthermore, the formation of S(rhombic) apparently requires a considerably activation energy because the reverse reaction, the oxidation of elemental sulfur, does not normally occur unless sulfur oxidizing bacteria are present.

17.4 FORMATION OF METAL SULFIDES

The solubilities of several metal sulfides are given by Reactions 29 through 52 of Table 17.2. The $pe + pH$ to which soils must be lowered before each of the metal sulfides can form in soils is shown in Fig. 17.4. An example of how

FORMATION OF METAL SULFIDES

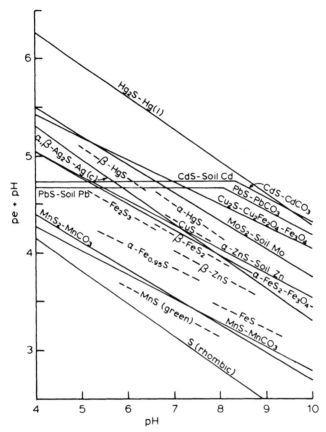

Fig. 17.4 The redox at which various metal sulfides precipitate in soils when SO_4^{2-} is 10^{-3} M and the cations are controlled as indicated by Reactions 53 through 70 of Table 17.2.

the lines in this figure were developed is shown for $Hg_2S(c)$ by combining the following equilibrium reactions:

$$\begin{array}{rcll}
 & & & \log K^\circ \\
Hg_2^{2+} + S^{2-} & \rightleftharpoons & Hg_2S(c) & 54.77 \\
2\,Hg(l) & \rightleftharpoons & Hg_2^{2+} + 2e^- & 2(-13.46) \\
SO_4^{2-} + 8e^- + 8H^+ & \rightleftharpoons & S^{2-} + 4H_2O & 20.74 \\
\hline
2\,Hg(l) + SO_4^{2-} + 6e^- + 8H^+ & \rightleftharpoons & Hg_2S(c) + 4H_2O & 48.59
\end{array}$$

(17.12)

At equilibrium,

$$\frac{1}{(SO_4^{2-})(e^-)^6(H^+)^6(H^+)^2} = 10^{48.59} \qquad (17.13)$$

In log form

$$6(pe + \text{pH}) = 48.59 + \log SO_4^{2-} - 2\text{pH} \quad (17.14)$$

or

$$pe + \text{pH} = 8.10 + \tfrac{1}{6}\log SO_4^{2-} - \tfrac{1}{3}\text{pH} \quad (17.15)$$

When SO_4^{2-} is 10^{-3} M,

$$pe + \text{pH} = 7.60 - \tfrac{1}{3}\text{pH} \quad (17.16)$$

Equation 17.16 is plotted as the Hg_2S-$Hg(l)$ line in Fig. 17.4. This line represents the $pe + \text{pH}$ below which $Hg(l)$ is transformed to $Hg_2S(c)$. The pH dependency of this transformation is apparent. The line was developed for SO_4^{2-} at 10^{-3} M, but it can be readily adjusted for any SO_4^{2-} level by using Eq. 17.15. The two phases, $Hg(l)$ and $Hg_2S(c)$, can only coexist at equilibrium for the conditions defined by this line. Above the line only $Hg(l)$ is stable, whereas below the line only $Hg_2S(c)$ is stable. If the redox of a soil is lowered sufficiently slow that near equilibrium conditions are maintained, $Hg(l)$ will disappear and $Hg_2S(c)$ will form and become the controlling phase of mercury solubility (See Section 21.4). The other lines in this diagram were obtained similarly by combining appropriate reactions from Table 17.2.

In developing Fig. 17.4 it was necessary to select minerals or soil phases that control the various cations in soils under the redox at which the metal sulfides precipitate. These selections are discussed in detail in the separate chapters of this text dealing with each element. These selections are documented in Table 17.3 along with a mathematical expression of the $pe + \text{pH}$ at which each sulfide first precipitates. The sulfide minerals with an asterisk (*) are arranged sequentially in the order in which they can form in soils. Minerals without an asterisk are metastable in soils and normally do not form. The metastable sulfides are shown in Fig. 17.4 by short dashed lines. The metastable lines lie below the stable sulfide for that element. If near equilibrium conditions are maintained, the redox cannot drop below a given line until the transformation occurring at that line is complete. When complete, the initial solid phase is dissolved and the metal sulfide remains to control the solubility relationships of that element. The metastable sulfides are only included to show the redox at which they would have formed had the initial phase governing that cation remained. In some cases the metastable sulfide becomes the stable phase at lower redox (for example see HgS(cinnabar) in Fig. 21.9).

The sequence of metal sulfides that form in soils and the $pe + \text{pH}$ at which they precipitate are summarized in Table 17.4. The first-formed sulfide is $Hg_2S(c)$ and the least likely to form is $MgS(c)$. As pointed out in Section 17.2 and 17.3 S(rhombic) normally does not form in soils because the low redox that is required raises the partial pressure of $H_2S(g)$ sufficiently that it

TABLE 17.3 EXPRESSIONS GIVING THE pe + pH AT WHICH VARIOUS METAL SULFIDES FORM IN SOILS WHEN THE INDICATED SOLID PHASES CONTROL THE METAL CATIONS

No.	Sulfide Mineral	pe + pH	Soil Phases and Reactions† Controlling the Cation Activities
1	Cu_2S(chalcocite)*	$6.70 + 0.12 \log SO_4^{2-} - 0.23$ pH	$Cu_2Fe_2O_4$(c) (59), Fe_3O_4 (63)
2	CuS(covellite)	$6.63 + 0.14 \log SO_4^{2-} - 0.27$ pH	$Cu_2Fe_2O_4$(c) (59), Fe_3O_4 (63)
3	Hg_2S(c)*	$8.10 + 0.17 \log SO_4^{2-} - 0.33$ pH	Hg(l) (66)
4	α-HgS(cinnabar, red)	$7.32 + 0.17 \log SO_4^{2-} - 0.33$ pH	Hg(l) (67)
5	β-HgS(cinnabar, black)	$7.26 + 0.17 \log SO_4^{2-} - 0.33$ pH	Hg(l) (67)
6	CdS(greennokite)*	$5.10 + 0.13 \log SO_4^{2-}$	soil-Cd (57)
7	CdS(greennokite)*	$6.75 + 0.13 \log SO_4^{2-} - 0.13 \log CO_2 - 0.25$ pH	$CdCO_3$(otavite) (58)
8	PbS(galena)*	$5.03 + 0.13 \log SO_4^{2-}$	soil-Pb (72)
9	Pb(galena)	$6.61 + 0.13 \log SO_4^{2-} - 0.13 \log CO_2 - 0.25$ pH	$PbCO_3$(cerussite) (73)
10	MoS_2(molybdenite)*	$6.98 + 0.11 \log SO_4^{2-} - 0.29$ pH	soil-Mo (71)
11	α-Ag_2S(c)	$6.63 - 0.33$ pH	Ag(c) (53)
12	β-Ag_2S(c)	$6.62 - 0.33$ pH	Ag(c) (53)
13	α-ZnS(sphalerite)*	$6.41 + 0.13 \log SO_4^{2-} - 0.25$ pH	soil-Zn (74)
14	β-ZnS(wurtzite)	$6.13 + 0.13 \log SO_4^{2-} - 0.25$ pH	soil-Zn (74)
15	α-FeS_2(pyrite)	$6.54 + 0.14 \log SO_4^{2-} - 0.27$ pH	Fe_3O_4(magnetite) (63)
16	β-FeS_2(markasite)	$6.49 + 0.14 \log SO_4^{2-} - 0.27$ pH	Fe_3O_4(magnetite) (63)
17	Fe_2S_3(c)	$6.34 + 0.13 \log SO_4^{2-} - 0.26$ pH	Fe_3O_4(magnetite) (63)
18	α-$Fe_{0.95}S$(pyrrhotite)	$5.80 + 0.12 \log SO_4^{2-} - 0.23$ pH	Fe_3O_4(magnetite) (63)
19	α-FeS(troilite)	$5.63 + 0.12 \log SO_4^{2-} - 0.23$ pH	Fe_3O_4(magnetite) (63)
20	MnS_2(hauerite)	$5.71 + 0.14 \log SO_4^{2-} - 0.25 \log CO_2 - 0.29$ pH	$MnCO_3$(rodochrosite) (70)
21	MnS(alabanite)*	$5.21 + 0.13 \log SO_4^{2-} - 0.13 \log CO_2 - 0.25$ pH	$MnCO_3$(rodochrosite) (70)
22	MnS(green)	$5.06 + 0.13 \log SO_4^{2-} - 0.13 \log CO_2 - 0.25$ pH	$MnCO_3$(rodochrosite) (70)
23	S(rhombic)	$5.96 + 0.17 \log SO_4^{2-} - 0.33$ pH	
24	CaS(oldhamite)	$2.38 + 0.13 \log SO_4^{2-}$	soil-Ca (55)
25	CaS(oldhamite)	$3.91 + 0.13 \log SO_4^{2-} - 0.13 \log CO_2 - 0.25$ pH	$CaCO_3$(calcite) (56)
26	MgS(c)	$1.52 + 0.13 \log SO_4^{2-}$	soil-Mg (68)
27	MgS(c)	$2.98 + 0.13 \log SO_4^{2-} - 0.13 \log CO_2 - 0.25$ pH	$MgCa(CO_3)_2$(dolomite)-$CaCO_3$(calcite) (69)

†The numbers in ( ) refer to equations in Table 17.2.
*Indicate stable sulfides in highly reduced soils.

escapes (Fig. 17.2). Similarly the likely loss of $H_2S(g)$ makes it virtually impossible for CaS(oldhamite) and MgS to form in soils.

Since α-Ag_2S(c) and β-Ag_2S(c) have very similar solubilities, both may appear as stable phases. Although MnS_2(hauerite) and MnS(alabanite) have similar solubilities at pH 7, MnS_2(hauerite) is more stable at lower pH and MnS(alabanite) is more stable at higher pH.

Four different iron sulfides are included in the present development. Of these FeS_2(pyrite) is the most stable. The mineral, β-FeS_2(markasite), is only slightly less stable, whereas Fe_2S_3(c), α-$Fe_{0.95}S$(pyrrohotite), and FeS(trolite) are less stable. At lower redox some of these stability relationships shift.

Zinc deficiency of rice growing on submerged soils is often attributed to the formation of insoluble zinc sulfides. Both α-ZnS(sphalerite) and α-FeS_2 (pyrite) precipitate at approximately the same redox, pe + pH 4.28 and 4.24, respectively. Thus Zn^{2+} activity cannot be depressed significantly by the formation of ZnS(sphalerite) because the redox will be poised by the transformation of iron oxides into FeS_2(pyrite). Since soils generally contain much more iron than zinc, this transition would require considerable quantities of sulfate. Only after the labile iron oxides have been transformed to FeS_2(pyrite) can pe + pH drop and allow ZnS(sphalerite) to depress

TABLE 17.4 THE pe + pH AT WHICH VARIOUS METAL SULFIDES ARE PRECIPITATED WHEN pH = 7.0, SO_4^{2-} = 10^{-3} M, AND CO_2(g) = 0.0003 atm

pe + pH	Sulfide Mineral
5.27	Hg_2S(c)
4.73	Cu_2S(chalcocite)
4.73	CdS(greennokite)
4.66	PbS(galena)
4.61	MoS_2(molybdenite)
4.30	α-Ag_2S(c)
4.29	β-Ag_2S(c)
4.28	α-ZnS(sphalerite)
4.24	α-FeS_2(pyrite)
3.53	MnS_2(hauerite)
3.52	MnS(alabanite)
3.13	S(rhombic)
2.00	CaS(oldhamite)
1.14	MgS(c)

the activity of Zn^{2+}. It seems unlikely that zinc deficiencies are caused by precipitation of zinc sulfides except temporarily when reduction occurs so rapidly that the iron oxides are unable to respond to redox changes. These interesting solubility relationships need further investigation.

For manganese sulfides to form, very low redox levels are required. For a soil of pH 7, $pe + pH$ must drop to 3.5. In highly reduced soils Mn^{2+} solubility is controlled by $MnCO_3$(rodochrosite). Increasing $CO_2(g)$, as generally occurs in submerged soils, shifts the MnS_2-$MnCO_3$ and MnS-$MnCO_3$ lines in Fig. 17.4 downward (Reactions 21 and 22 of Table 17.3), requiring even lower redox levels for manganese sulfides to precipitate.

17.5 EFFECT OF SULFIDES ON METAL SOLUBILITIES

An example of how the formation of metal sulfides in soils can affect the activities of metal cations is demonstrated for copper in Fig. 17.5.

Above $pe + pH$ of 14.89, Cu^{2+} is controlled by soil-Cu and is independent of redox ($pe + pH$), but the activity of Cu^{2+} increases 2 log units for each unit decrease in pH. The activity of Cu^+, on the other hand, increases one log unit for each unit decrease in $pe + pH$ and increases for each unit decrease in pH.

In the redox range of 14.89 to 4.73, copper solubility is controlled by $Cu_2Fe_2O_4$(cuprous ferrite) (See Section 14.3). Between $pe + pH$ 14.89 and 11.53, where soil-Fe controls Fe^{3+} activity, Cu^{2+} decreases one log unit for each unit drop in redox, while Cu^+ remains constant. In the $pe + pH$ range of 11.53 to 4.73 where magnetite controls Fe^{3+} activity, log Cu^{2+} decreases 2/3 log unit for each unit decrease in redox whereas log Cu^+ increases 1/3 log unit.

At $pe + pH$ of 4.73 Cu_2S(chalcocite) becomes the controlling solid phase of copper solubility. This relationship is obtained as follows:

$$
\begin{array}{rcll}
& & & \log K° \\
3\alpha\text{-}Cu_2Fe_2O_4(\text{cuprous ferrite}) + 24H^+ & \rightleftharpoons & 6Cu^+ + 6Fe^{3+} + 12H_2O & 3(-13.53) \\
6Cu^+ + 3S^{2-} & \rightleftharpoons & 3Cu_2S(\text{chalcocite}) & 3(48.54) \\
6Fe^{3+} + 2e^- + 8H_2O & \rightleftharpoons & 2Fe_3O_4(\text{magnetite}) + 16H^+ & 3(3.42) \\
3SO_4^{2-} + 24H^+ + 24e^- & \rightleftharpoons & 3S^{2-} + 12H_2O & 3(20.74) \\
\hline
Cu_2Fe_2O_4(c) + 3SO_4^{2-} + 26e^- + 32H^+ & \rightleftharpoons & 3Cu_2S(c) + 2Fe_3O_4(c) + 16H_2O & 174.09 \\
\end{array}
$$

(17.17)

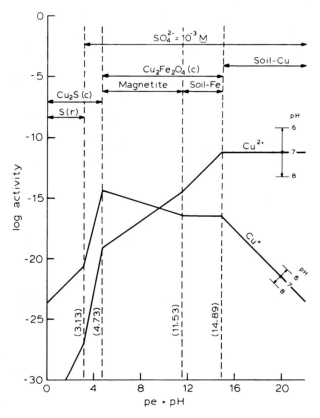

Fig. 17.5 The effect of redox on the stability of copper minerals in soils and the activities of Cu^+ and Cu^{2+} they maintain. Drawn for pH 7 with shifts for pH 6 and 8.

For which

$$\frac{1}{(SO_4^{2-})^3(e^-)^{26}(H^+)^{26}(H^+)^6} = 10^{174.09} \quad (17.18)$$

Thus

$$pe + pH = 6.70 + 0.12 \log SO_4^{2-} - 0.23\, pH \quad (17.19)$$

When $SO_4^{2-} = 10^{-3}\, M$ and pH = 7.0, Eq. 17.19 becomes

$$pe + pH = 4.73 \quad (17.20)$$

Thus at $pe + pH$ of 4.73, $Cu_2Fe_2O_4$(cuprous ferrite) dissolves to precipitate Cu_2S(chalcocite) which depresses the solubilities of both Cu^+ and Cu^{2+}

(Fig. 17.5). Equation 17.19 can be used to show how different levels of SO_4^{2-} and pH affect the pe + pH at which chalcocite can first form. Thus chalcocite can greatly depress the solubility of copper as redox drops below pe + pH of 4.73. Should S(rhombic) form (Section 17.3), copper solubility would continue to be depressed as redox drops below pe + pH of 3.13.

The formation of metal sulfides in soils will depress the activities of the cations involved. Figure 17.4 shows the pe + pH at which the various metal sulfides can first precipitate and below which that cation will be depressed providing there is sufficient sulfur present to precipitate all of the metal cations that are capable of precipitating. These metal sulfides can be very important in highly reduced soils and other environments.

REFERENCES

Vlek, P. L. G. and W. L. Lindsay. 1977. Thermodynamic stability of molybdenum minerals in soils. Soil Sci. Soc. Am. J. 41:42–46.

PROBLEMS

17.1 Develop the equations and plot the lines for SO_3^{2-} and S^{2-} as shown in Fig. 17.1.

17.2 Develop the equations and plot the relationship between $H_2S(g)$ and pe + pH for a soil of pH 6.8 having $10^{-2.5}$ M SO_4^{2-}. Also include a line for S(rhombic). What partial pressure of $H_2S(g)$ is necessary before S(rhombic) can form in this soil?

17.3 Explain how the diagram in Fig. 17.1 would change if 10^{-3} M total soluble sulfur were used to control the sulfur species rather than 10^{-3} M SO_4^{2-} or S(rhombic).

17.4 Develop a plot of log $H_2S°$ and $H_2S(g)$ versus pe + pH in equilibrium with total sulfur of 0.01 M for a soil of pH 7. Also include the solubility of S(rhombic) and discuss the redox range in which this mineral may be stable.

17.5 Develop the equations and plot the line for pyrite in Fig. 17.4. How would this line shift if soil-Fe rather than magnetite controlled iron solubility in this soil?

17.6 Identify a sulfide in Fig. 17.4 other than α-HgS(cinnabar) that changes from a metastable to the stable sulfide as redox is lowered. Write an equation relating the coexistence of the two sulfides.

17.7 Develop the equations and prepare a plot of Zn^{2+} activity versus $pe + pH$ showing the effects of pH, ZnS(sphalerite), and S(rhombic) on Zn^{2+} solubility. Explain why zinc deficiencies in paddy soils cannot be attributed to the formation of insoluble zinc sulfides.

17.8 From Fig. 17.5 explain why copper is not severely deficient in highly reduced soils of $pe + pH$ of 4 even though Cu^+ activity is depressed to approximately 10^{-17} M.

17.9 Discuss the implications of this chapter with regard to the use of soils as a disposal medium for heavy metals and organic wastes.

EIGHTEEN
SILVER

The silver content of the lithosphere is reported at 0.07 ppm, while soils commonly range from 0.01 to 5 with an average of 0.05 ppm. This represents $10^{-5.33}$ M Ag at 10% moisture.

In recent years silver iodide has been used as a nucleating agent for cloud seeding to enhance precipitation, and there is some concern about the fate of silver in the environment (Klein, 1978). A descriptive view of silver in the environment has been given by Carson and Smith (1975) and a more quantitative view by Lindsay and Sadiq (1978) and Jenne et al. (1978). The present treatment is an update of how to treat silver solubility in soils.

18.1 EFFECT OF REDOX ON THE STABILITY OF SILVER MINERALS

The equilibrium reactions of silver minerals and solution complexes are summarized in Table 18.1. Silver exists in four oxidation states: 0, +1, +2, and +3, however, it will be shown herein that only 0 and +1 oxidation states are important in soils.

Figure 18.1 shows that Ag(c) can be stable in the redox range of soils. This diagram was constructed from Reactions 1 through 7 of Table 18.1. An example of how the lines in Fig. 18.1 were developed is given for the coexistence of Ag(c) and $Ag_2O(c)$:

$$
\begin{array}{rcll}
& & & \log K^\circ \\
\hline
Ag^+ + e^- & \rightleftharpoons & Ag(c) & 13.50 \\
\tfrac{1}{2}Ag_2O(c) + H^+ & \rightleftharpoons & Ag^+ + \tfrac{1}{2}H_2O & \tfrac{1}{2}(12.59) \\
\hline
\tfrac{1}{2}Ag_2O(c) + e^- + H^+ & \rightleftharpoons & Ag(c) + \tfrac{1}{2}H_2O & 19.80 \quad (18.1)
\end{array}
$$

from which

$$pe + pH = 19.80 \quad (18.2)$$

Thus at $pe + pH$ of 19.80 both Ag(c) and $Ag_2O(c)$ can coexist. At lower redox Ag(c) is stable whereas at higher redox $Ag_2O(c)$ is stable. The other lines in this figure were obtained similarly.

The minerals $Ag_2O(c)$ and AgOH(c) are stable between $pe + pH$ of 19.80 and 24.14, AgO(c) between 24.14 and 28.98, and $Ag_2O_3(c)$ above $pe + pH$ of 28.98. Since most soils do not attain redox values of $pe + pH > 18$, silver oxides are not expected in soils. Of the minerals represented in Fig. 18.1, only Ag(c) is important.

EFFECT OF REDOX ON THE STABILITY OF SILVER MINERALS

Only in highly reduced soils can silver precipitate as sulfides. The following equilibrium reactions combine to give the line separating Ag(c) from α-Ag$_2$S(c):

			log $K°$
2 Ag(c)	$\rightleftharpoons$	2 Ag$^+$ + 2e$^-$	2(−13.50)
2 Ag$^+$ + S^{2-}	$\rightleftharpoons$	α-Ag$_2$S(c)	49.02
SO$_4^{2-}$ + 8e$^-$ + 8H$^+$	$\rightleftharpoons$	S^{2-} + 4H$_2$O	20.74
SO$_4^{2-}$ + 6e$^-$ + 8H$^+$ + 2 Ag(c)	$\rightleftharpoons$	α-Ag$_2$S(c) + 4H$_2$O	42.76 (18.3)

which reduces to

$$pe + pH = 7.13 + \tfrac{1}{6} \log SO_4^{2-} - \tfrac{1}{3} pH \qquad (18.4)$$

When SO$_4^{2-}$ = 10^{-3} M,

$$pe + pH = 6.63 - \tfrac{1}{3} pH \qquad (18.5)$$

and for pH 7

$$pe + pH = 4.29 \qquad (18.6)$$

Thus the redox at which Ag(c) transforms to α-Ag$_2$S(c) occurs at $pe + pH$ of 4.29 and is affected only slightly by changes in pH and SO$_4^{2-}$ as seen from

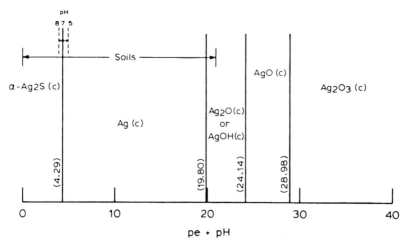

Fig. 18.1 The redox stability range of various silver minerals when SO$_4^{2-}$ is 10^{-3} M.

TABLE 18.1 EQUILIBRIUM REACTIONS OF SILVER MINERALS AND SOLUTION SPECIES

Reaction No.	Chemical Reaction	log $K°$
	Redox Reactions	
1	$Ag^+ + e^- \rightleftharpoons Ag(c)$	13.50
2	$Ag^{2+} + e^- \rightleftharpoons Ag^+$	33.64
	Oxides and Hydroxides	
3	$Ag_2O(c) + 2H^+ \rightleftharpoons 2Ag^+ + H_2O$	12.59
4	$AgOH(c) + H^+ \rightleftharpoons Ag^+ + H_2O$	6.29
5	$AgO(c) + 2H^+ \rightleftharpoons Ag^{2+} + H_2O$	−3.21
6	$Ag(OH)_2(c) + 2H^+ \rightleftharpoons Ag^{2+} + 2H_2O$	2.36
7·	$\frac{1}{2}Ag_2O_3(c) + H^+ + e^- \rightleftharpoons AgO(c) + \frac{1}{2}H_2O$	28.98
	Ag(I) Halides	
8	$AgF(c) \rightleftharpoons Ag^+ + F^-$	−0.38
9	$AgCl(c) \rightleftharpoons Ag^+ + Cl^-$	−9.75
10	$AgBr(c) \rightleftharpoons Ag^+ + Br^-$	−12.31
11	$AgI(c) \rightleftharpoons Ag^+ + I^-$	−16.08
	Ag(I) Sulfides	
12	$\alpha\text{-}Ag_2S(c) \rightleftharpoons 2Ag^+ + S^{2-}$	−49.02
13	$\beta\text{-}Ag_2S(c) \rightleftharpoons 2Ag^+ + S^{2-}$	−48.97
	Other Ag minerals	
14	$Ag_2CO_3(c) + 2H^+ \rightleftharpoons 2Ag^+ + CO_2(g) + H_2O$	7.07
15	$Ag_2MoO_4(c) \rightleftharpoons 2Ag^+ + MoO_4^{2-}$	−11.55
16	$AgNO_3(c) \rightleftharpoons Ag^+ + NO_3^-$	0.19
17	$Ag_3PO_4(c) + 2H^+ \rightleftharpoons 3Ag^+ + H_2PO_4^-$	3.58
18	$Ag_2SO_4(c) \rightleftharpoons 2Ag^+ + SO_4^{2-}$	−4.81
19	$Ag_2SO_3(c) \rightleftharpoons 2Ag^+ + SO_3^{2-}$	−13.81
	Complexes	
20	$Ag^+ + Br^- \rightleftharpoons AgBr°$	4.38
21	$Ag^+ + 2Br^- \rightleftharpoons AgBr_2^-$	7.34
22	$Ag^+ + 3Br^- \rightleftharpoons AgBr_3^{2-}$	8.70
23	$Ag^+ + 4Br^- \rightleftharpoons AgBr_4^{3-}$	9.00
24	$Ag^+ + Cl^- \rightleftharpoons AgCl°$	3.31
25	$Ag^+ + 2Cl^- \rightleftharpoons AgCl_2^-$	5.26

TABLE 18.1 (Continued)

Reaction No.	Chemical Reaction	log $K°$
26	$Ag^+ + 3Cl^- \rightleftharpoons AgCl_3^{2-}$	6.40
27	$Ag^+ + 4Cl^- \rightleftharpoons AgCl_4^{3-}$	6.10
28	$Ag^+ + F^- \rightleftharpoons AgF°$	−0.38
29	$Ag^+ + I^- \rightleftharpoons AgI°$	6.58
30	$Ag^+ + 2I^- \rightleftharpoons AgI_2^-$	11.70
31	$Ag^+ + 3I^- \rightleftharpoons AgI_3^{2-}$	13.60
32	$Ag^+ + 4I^- \rightleftharpoons AgI_4^{3-}$	14.40
33	$Ag^+ + H_2O \rightleftharpoons AgOH° + H^+$	−12.04
34	$Ag^+ + 2H_2O \rightleftharpoons Ag(OH)_2^- + 2H^+$	−24.01
35	$Ag^+ + 3H_2O \rightleftharpoons Ag(OH)_3^{2-} + 3H^+$	−48.19
36	$3Ag^+ + 4H_2O \rightleftharpoons Ag_3(OH)_4^- + 4H^+$	−37.18
37	$Ag^+ + NO_3^- \rightleftharpoons AgNO_3°$	−0.20
38	$Ag^+ + NH_4^+ \rightleftharpoons AgNH_3^+ + H^+$	−5.91
39	$Ag^+ + 2NH_4^+ \rightleftharpoons Ag(NH_3)_2^+ + 2H^+$	−11.36
40	$Ag^+ + SO_4^{2-} \rightleftharpoons AgSO_4^-$	1.30
41	$2Ag^+ + SO_4^{2-} \rightleftharpoons Ag_2SO_4°$	0.00
42	$Ag^+ + SO_3^{2-} \rightleftharpoons AgSO_3^-$	5.60
43	$2Ag^+ + SO_3^{2-} \rightleftharpoons Ag_2SO_3°$	8.68
44	$Ag^+ + H^+ + S^{2-} \rightleftharpoons AgHS°$	26.94
45	$Ag^+ + 2H^+ + 2S^{2-} \rightleftharpoons Ag(HS)_2^-$	45.84
	Other Reactions	
46	$SO_4^{2-} + 8e^- + 8H^+ \rightleftharpoons S^{2-} + 4H_2O$	20.74
47	$SO_4^{2-} + 2e^- + 2H^+ \rightleftharpoons SO_3^{2-} + H_2O$	−3.73
48	soil-Mo $\rightleftharpoons MoO_4^{2-} + 0.8H^+$	−12.40

Eq. 18.4. The mineral β-Ag$_2$S(c) is only slightly more soluble than α-Ag$_2$S(c) (Reactions 12 and 13 of Table 18.1). Because of their very similar solubilities, both sulfides may be present below pe + pH of 4.29.

What is the predominant silver ion in soils and natural aqueous environments? From Reaction 2 of Table 18.1

$$Ag^{2+} + e^- \rightleftharpoons Ag^+ \qquad \begin{array}{c}\log K°\\ \hline 33.64\end{array} \qquad (18.7)$$

This reaction indicates that the ratio of Ag^+/Ag^{2+} is unity at $pe = 33.60$. In moderately well-oxidized soil where pe is approximately 11, the ratio of

Ag^+/Ag^{2+} would be $10^{33.6-11.0} = 10^{22.6}$. Thus in the redox range of soils Ag^+ is the predominant silver ion with very little Ag^{2+} and even less Ag^{3+}. Silver chemistry in soils and natural aqueous environments is largely limited to Ag(c) and to Ag^+ with its minerals and complexes.

18.2 SOLUBILITY OF SILVER HALIDES AND SULFIDES

The Ag^+ ion combines with the halide anions: Br^-, Cl^-, F^-, and I^- to form relatively insoluble minerals. The solubility relationships of these minerals are shown in Fig. 18.2. Superimposed in this diagram is the activity of Ag^+ in equilibrium with Ag(c) shown at various values of *pe*. These lines are readily obtained from the equilibrium reaction:

$$Ag(c) \rightleftharpoons Ag^+ + e^- \qquad \log K° = -13.50$$

which gives:

$$\log Ag^+ = -13.50 + pe \qquad (18.8)$$

When $pe = 0$, $\log Ag^+$ in equilibrium with Ag(c) is -13.50 and increases one unit for each unit increase in *pe*.

From Fig. 18.2 it is possible to predict which silver minerals are stable depending upon redox and halide activities. The mineral that supports the lowest Ag^+ activity is the one that is most stable. All silver halide minerals except AgF(c) are sufficiently insoluble that they must be considered as possible stable minerals.

As examples, let us consider the four soils described in Table 18.2. Based on Fig. 18.2 AgCl(c) can maintain the lowest activity in Soil A ($10^{-8.8}$ M) and would be the stable silver mineral. In Soil B, AgBr(c) is the most stable

TABLE 18.2 EXAMPLE SOILS FOR DISCUSSION OF SILVER MINERALS AND COMPLEXES

Soil	pH	*pe*	Log Activity (mole/liter)		
			Cl^-	Br^-	I^-
A	6.5	11.5	-1	-5	< -8
B	7.5	7.5	-3	-3.5	-8
C	7.0	6.0	-3	-5	-6
D	7.0	1.0	-3	-5	-6

SOLUBILITY OF SILVER HALIDES AND SULFIDES

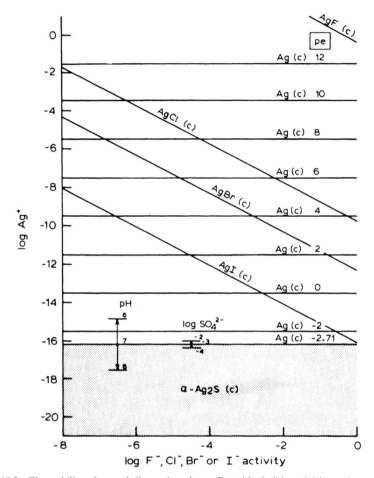

Fig. 18.2 The stability of several silver minerals as affected by halide activities, redox and pH when SO_4^{2-} is 10^{-3} M.

mineral and would maintain Ag^+ at $10^{-8.8}$ M. In Soil C, AgI(c) is the most stable silver mineral and maintains Ag^+ at $10^{-10.1}$ M. The redox of Soils A, B, and C is sufficiently high that Ag(c) cannot form. In Soil D Ag(c) is the stable silver mineral and maintains Ag^+ activity at $10^{-12.5}$ M. None of the silver halides are stable in Soil D, as seen from Fig. 18.2.

The mineral α-Ag_2S(c) is also included in Fig. 18.2, and its equilibrium reactions with Ag(c) and SO_4^{2-} are summarized in Reaction 18.3 which gives

$$pe = 7.13 + \tfrac{1}{6} \log SO_4^{2-} - \tfrac{4}{3}pH \qquad (18.9)$$

When SO_4^{2-} is 10^{-3} M and pH is 7,

$$pe = -2.71 \tag{18.10}$$

Thus when *pe* drops below -2.71, $\alpha\text{-}Ag_2S(c)$ becomes the stable mineral. The *pe* at which this transition occurs is affected slightly by pH and SO_4^{2-} as shown by Eq. 18.9 and as depicted in Fig. 18.2.

Several generalizations can be made from Fig. 18.2. In highly oxidized soils Ag(c) cannot exist because it is too soluble. Instead one of the halide minerals will most likely control Ag^+ solubility. Even though AgI(c) appears to be the most insoluble silver mineral, the activities of Cl^-, Br^-, and I^- must be known before the stable halide mineral can be designated.

In Chapter 17 Ag(c) was used as the solid phase controlling Ag^+ activity to determine the *pe* + pH at which $\alpha\text{-}Ag_2S(c)$ forms. Of the minerals depicted in Fig. 18.2 it becomes apparent that Ag(c) is the appropriate choice for the redox range in which silver sulfides form.

18.3 STABILITY OF OTHER SILVER MINERALS

The solubilities of several additional silver minerals are given by Reactions 14 through 19 of Table 18.1 and are plotted in Fig. 18.3. For this development the anionic activities were selected to represent the levels that are typically found in soils.

The solubility of $AgNO_3(c)$ depicted in Fig. 18.3 corresponds to NO_3^- levels between 10^{-4} and 10^{-2} M. The activity of Ag^+ would have to be >1 M for $AgNO_3(c)$ to form in soils.

The solubility line for $Ag_2SO_4(c)$ shown in Fig. 18.3 was drawn for SO_4^{2-} activities between 10^{-4} and 10^{-2} M. For this mineral to be stable in soils, Ag^+ would have to be near 10^{-1} M.

The solubility of $Ag_2CO_3(c)$ shown in Fig. 18.3 was drawn on the basis that $CO_2(g)$ ranges from 0.0003 to 0.03 atm, the lower value being the CO_2 of normal air. At pH 8 the Ag^+ activity in equilibrium with this mineral is $10^{-3.2}$ M and increases at lower pH values. Obviously $Ag_2CO_3(c)$ is far too soluble to persist in soils.

The solubility of $Ag_3PO_4(c)$ is also represented in Fig. 18.3. At low pH the reference phosphate level was taken as that controlled by $AlPO_4 \cdot 2H_2O$ (variscite) and $Al(OH)_3$(gibbsite). Accordingly, $Ag_3PO_4(c)$ supports a Ag^+ activity identical to that of $Ag_2CO_3(c)$ at 0.003 atm CO_2. Above pH 6.8 $Ca_4H(PO_4)_3 \cdot \frac{5}{2}H_2O$ (octacalcium phosphate) was selected as a representative control of phosphate activity. In this case $Ag_3PO_4(c)$ controls Ag^+ at about 10^{-2} M. Again $Ag_3PO_4(c)$ is too soluble to be of importance in soils. Lower phosphate levels make this mineral even more soluble.

STABILITY OF OTHER SILVER MINERALS

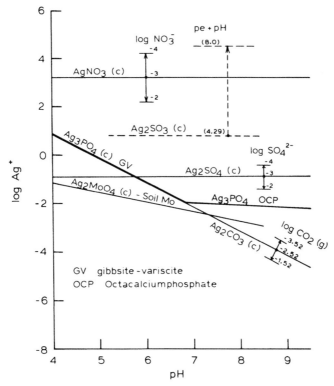

Fig. 18.3 The solubility relationships of various silver minerals at ionic activities commonly found in soils.

The solubility of $Ag_2SO_3(c)$ is highly redox dependent as shown by the reactions

$$\begin{array}{lr} & \log K° \\ Ag_2SO_3(c) \rightleftharpoons 2Ag^+ + SO_3^{2-} & -13.81 \\ SO_3^{2-} + H_2O \rightleftharpoons SO_4^{2-} + 2e^- + 2H^+ & 3.73 \\ \hline Ag_2SO_3(c) + H_2O \rightleftharpoons 2Ag^+ + SO_4^{2-} + 2e^- + 2H^+ & -10.03 \end{array}$$

(18.11)

from which

$$\log Ag^+ = -5.04 - \tfrac{1}{2}\log SO_4^{2-} + (pe + pH) \quad (18.12)$$

A plot of Eq. 18.12 is shown in Fig. 18.3 for $pe + pH$ at 4.29 and 8.00 shows the instability of $\alpha\text{-}Ag_2SO_3(c)$.

A plot of $Ag_2MoO_4(c)$ in equilibrium with soil-Mo shown in Fig. 18.3 shows this mineral to be unstable in soils, because greater than 10^{-3} M Ag^+ would be required for its formation.

18.4 STABILITY OF SILVER HALIDE COMPLEXES

As indicated earlier Ag^+ is the predominant ionic form of silver in the redox range of soils, but how important are other solution complexes of silver? Figure 18.4 was constructed to show the important halide complexes of

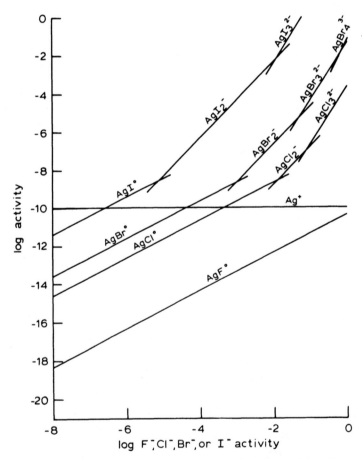

Fig. 18.4 The activity of silver halide complexes in equilibrium with 10^{-10} M Ag^+ as a function of halide activities.

STABILITY OF SILVER HALIDE COMPLEXES

Ag^+. For this comparison an arbitrary Ag^+ activity of 10^{-10} M was chosen. Carson and Smith (1975) reported an average silver content for freshwater of 0.2 ppb, which corresponds to $10^{-8.73}$ M total silver. Since some of this silver is very likely to be complexed, the selection of 10^{-10} M for Ag^+ activity seems appropriate.

From Fig. 18.4 the activities of the various silver halide complexes can be readily determined as a function of the halide ion activities. Only the predominant region of each complex is shown. These lines can be easily extended to show the activities of the complex ions beyond their region of dominance. Halide complexes included in Table 18.1 that do not appear in Fig. 18.4 require greater than 1 M halide activities to become major species. Since such levels are impractical for soils, the graph was not extended to include them.

Even though Fig. 18.4 was constructed with Ag^+ at 10^{-10} M, its use is not restricted to this Ag^+ activity. For example, if Ag^+ activity of a given soil were 10^{-8} M, the activities of the complexes read from Fig. 18.4 need only be multiplied by $10^{-8}/10^{-10}$ or 10^2. Thus the graph can be used to obtain the activity of silver complexes corresponding to any Ag^+ activity.

For example, let us consider Soil B of Table 18.2 and add up the significant silver halide complexes in solution. As shown in Section 18.2 $AgBr(c)$ controls Ag^+ activity in this soil at $10^{-8.8}$ M. Activities of the silver complexes read from Fig. 18.4 need only be multiplied by the factor $10^{-8.8}/10^{-10} = 10^{1.2}$. If activity coefficients can be considered as unity, total soluble silver for Soil B would include the following significant species as obtained from Fig. 18.4:

$$[\text{Total Ag}] = 10^{1.2}[Ag^+ + AgCl^\circ + AgCl_2^- + AgBr^\circ + AgBr_2^- + AgI^\circ]$$
$$= 10^{1.2}[10^{-10} + 10^{-9.7} + 10^{-10.8} + 10^{-9.1} + 10^{-9.7} + {}^{-11.4}]$$
$$= 10^{-7.68} M$$

(18.13)

Thus Fig. 18.2 permits a ready diagnosis of the silver minerals that can exist in soils and the solubility of Ag^+ they support. Figure 18.4 can then be used to estimate the activity of each of the halide complexes in solution to give the total silver in solution. In contrast, measurements of total silver in solution and the activities of the halide anions permits an estimate of the equilibrium Ag^+ activity in soil solution. Activity coefficients can also be calculated and used to convert concentrations to activities when it is necessary to do so. Equation 18.13 must be modified to include the appropriate halide complexes depending upon the activities of halide anions. Other silver complexes, if they are significant, must also be included.

18.5 HYDROLYSIS SPECIES AND OTHER SILVER COMPLEXES

The hydrolysis species of Ag^+ defined by Reactions 33 through 36 of Table 18.1 are plotted in Fig. 18.5. These species do not contribute significantly to total silver in solution. Also Ag^{2+} does not contribute significantly to total silver in solution throughout the redox range of soils.

Silver forms complexes with several other anions, and these complexes will be examined to determine their importance in soils. The complexes are considered in the order in which they appear in Table 18.2.

From Reaction 37 of Table 18.2 comes the equation

$$\log AgNO_3^0 = -0.20 + \log NO_3^- + \log Ag^+ \qquad (18.14)$$

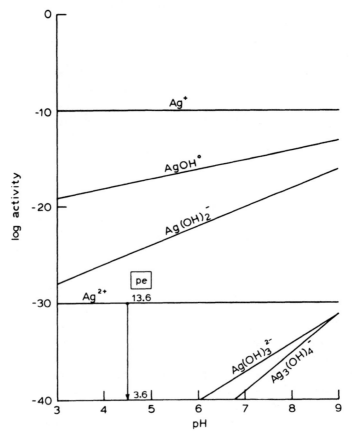

Fig. 18.5 The hydrolysis species of Ag^+ and the effect of pe on the activity of Ag^{2+} in equilibrium with 10^{-10} M Ag^+.

HYDROLYSIS SPECIES AND OTHER SILVER COMPLEXES

For a typical NO_3^- level of 10^{-3} M in soils this equation becomes

$$\log AgNO_3^\circ = -3.20 + \log Ag^+ \quad (18.15)$$

which corresponds to $AgNO_3^\circ/Ag^+$ ratio of $10^{-3.20}$. This complex increases the solubility of silver in soils only very slightly.

From Reaction 38 of Table 18.2 comes the relationship

$$\log AgNH_3^+ = -5.91 + pH + \log NH_4^+ + \log Ag^+ \quad (18.16)$$

In well-aerated soils the thermodynamic stability of NH_4^+ would be less than 10^{-15} M at $pe + pH$ values below 13 (Chapter 16). Thus the $AgNH_3^+$ complex would not be important. Even if NH_4^+ accumulates metastably to a level of 10^{-3} M, as may result temporarily from the mineralization of organic matter or application of ammonium fertilizer, Eq. 18.16 for a soil of pH 7 becomes

$$\log AgNH_3^+ = -1.91 + \log Ag^+ \quad (18.17)$$

Under these conditions, the ratio of $AgNH_3^+/Ag^+ = 10^{-1.91}$, showing that $AgNH_3^+$ becomes only slightly significant as NH_4^+ approaches 10^{-3} M.

From Reaction 39 of Table 18.1 comes the equation

$$\log Ag(NH_3)_2^+ = -11.36 + 2\log NH_4^+ + 2pH + \log Ag^+ \quad (18.18)$$

If NH_4^+ approached 10^{-3} M for a soil at pH 7, Eq. 18.12 becomes

$$\log Ag(NH_3)_2^+ = -3.36 + \log Ag^+ \quad (18.19)$$

which corresponds to a $Ag(NH_3)_2^+/Ag^+$ ratio of $10^{-3.36}$ M. The importance of this species can be fairly well discounted if $NH_4^+ < 10^{-3}$ M.

From Reaction 40 of Table 18.1 comes the equation

$$\log AgSO_4^- = 1.30 + \log SO_4^{2-} + \log Ag^+ \quad (18.20)$$

For a sulfate level of 10^{-3} M this equation becomes

$$\log AgSO_4^- = -1.70 + \log Ag^+ \quad (18.21)$$

which gives a ratio of $AgSO_4^-/Ag^+$ of 1/50, which would increase Ag solubility by only 2%.

For the $Ag_2SO_4^\circ$ complex, when SO_4^{2-} is 10^{-3} M, Reaction 41 of Table 18.1 gives

$$\log Ag_2SO_4^\circ = -3.00 + 2\log Ag^+ \quad (18.22)$$

In soils where Ag^+ is in the range of 10^{-10} M, the ratio of $Ag_2SO_4^\circ/Ag^+$ is 10^{-23}, so the importance of this species can be fully discounted.

The sulfite complexes of silver shown in Reactions 42 and 43 of Table 18.1 can be shown to be of no great significance in soils because of the low SO_3^{2-} levels found in soils as mentioned in discussing the stability of $AgSO_3(c)$.

The AgHS° given by Reaction 44 of Table 18.1 can be combined to give the following equilibrium relationships:

$$\begin{array}{lr} & \log K° \\ Ag^+ + H^+ + S^{2-} \rightleftharpoons AgHS° & 26.94 \\ SO_4^{2-} + 8e^- + 8H^+ \rightleftharpoons S^{2-} + 4H_2O & 20.74 \\ \hline Ag^+ + 9H^+ + 8e^- + SO_4^{2-} \rightleftharpoons AgHS° + 4H_2O & 47.68 \end{array}$$

(18.23)

which gives

$$\log AgHS° = 47.68 + \log SO_4^{2-} - pH - 8(pe + pH) + \log Ag^+ \quad (18.24)$$

For a soil of pH 7 having 10^{-3} M SO_4^{2-}, this equation becomes

$$\log AgHS° = 37.68 - 8(pe + pH) + \log Ag^+ \quad (18.25)$$

From this equation the ratio of AgHS°/Ag$^+$ is unity when $pe + pH = 4.68$ and increases 10^8 for each unit decrease in $pe + pH$. Although AgHS° can be discounted as being important in soils of $pe + pH > 5$, this complex becomes extremely important below $pe + pH$ of 4.68. It should be recalled that α-Ag$_2$S(c) precipitates at $pe + pH = 4.29$. At this redox the ratio of AgHS°/Ag$^+$ is $10^{3.36}$ or 2291-fold and may delay the formation of silver sulfides unless there is sufficient Ag$^+$ to first satisfy the complexation.

Reaction 45 of Table 18.1 can be used to show that Ag(HS)$_2^-$ is not highly significant in the $pe + pH$ range above 4.29 where α-Ag$_2$S(c) precipitates. Its importance in soils can be discounted in view of the many other metal sulfide reactions that will occur in this low redox range as shown in Chapter 17.

In summary the developments in this chapter show the silver minerals and complexes that can form in chemical environments such as soils.

The oxides, sulfates, carbonates, and phosphates of silver are much too soluble to persist in soils. The minerals that can form depending upon the activities of the halide ions are, AgI(c) AgBr(c), and AgCl(c). In addition, Ag(c) can form under reduced conditions and eventually Ag$_2$S can form if reduction approaches $pe + pH$ of <4.29. The exact conditions necessary for each of these minerals to precipitate are given by equations and stability diagrams.

Silver also forms many soluble complexes with halides, ammonia, nitrate, sulfate, sulfite, and sulfide. The complexes with nitrate, sulfate, and sulfite are not sufficiently stable to be significant in soils, whereas the complexes with Br$^-$, Cl$^-$, and I$^-$ are highly significant, and the activity ranges of the latter complexes are fully described. Complexes with ammonia are only slightly significant even if NH_4^+ in fertilized soils exceeds 10^{-3} M. Under strongly reduced conditions (pe + pH < 4.7), $AgHS°$ can become a significant complex for Ag^+ and may prevent the formation of silver sulfides unless abundant silver is present.

The developments in this chapter show how the chemical equilibrium relationships of a trace element can be used to describe its solubility and possible chemical reactions in soils. Investigations are needed to test these solubility predictions.

REFERENCES

Carson, B. L. and I. C. Smith. 1975. Silver an appraisal of environmental exposure. Nat. Inst. Env. Health Sci. Tech. Rept. No. 3.

Jenne, E. A., D. C. Girvin, J. W. Ball, and J. M. Burchard. 1978. Inorganic speciation of silver in natural waters—fresh to marine. In D.A.Klein,(Ed.), Environmental Impacts of Artificial Ice as Nucleating Agents, pp. 41–61. Dowden, Hutchison, and Ross, Stroudsburg, PA.

Klein, D. A., Ed. 1978. Environmental impacts of nucleating agents used in weather modification programs. Dowden, Hutchinson, and Ross, Stroudsburg, PA.

Lindsay, W. L. and M. Sadiq. 1978. Theoretically solubility relationships of silver in soils. In D. A. Klein (Ed.) Environmental Impacts of Nucleating Agents Used in Weather Modification Programs, pp. 25–40. Dowden, Hutchinson, and Ross, Stroudsburg, PA.

PROBLEMS

18.1 Develop the equation for Fig. 18.1 showing the coexistence of $Ag_2O(c)$ and $AgO(c)$. Discuss the possible significance of these silver minerals in soils.

18.2 A soil of pH 6.5 and pe + pH of 14 has the following halide activities: Cl^- 10^{-3} M, Br^- 10^{-4} M, and I^- 10^{-6} M. From Fig. 18.2 estimate the following for this soil.
  a. The most stable silver mineral
  b. The activity of Ag^+
  c. The maximum pe + pH at which $Ag(c)$ could form.

18.3 A soil of average silver content (Table 1.1) was used to prepare a 1:100 soil to water suspension. If the halide activities in this diluted suspension were $<10^{-8}$ M and Eh was 480 mv, estimate the following for this suspension:
  a. The stable silver mineral, if any.
  b. The maximum Ag^+ concentration in solution
  c. The pe to which this soil must be lowered before $Ag(c)$ could precipitate.

18.4 Explain why Ag(c) was selected to estimate the $pe + pH$ at which α-,β-Ag$_2$S(c) forms in soils as shown in Fig. 17.4. Under what conditions, if any, might this choice not be appropriate?

18.5 Develop the equations to show how the Ag$_3$PO$_4$(c) lines in Fig. 18.3 would be modified if brushite or hydroxyapatite rather than octacalcium phosphate had been selected to control phosphate activity. Discuss the implication of Fig. 18.3 with regard to silver solubility in soils.

18.6 Develop an appropriate expression of total soluble silver in Soil A of Table 18.2 and estimate this parameter when ionic strength is 0.1 and sufficient silver is present to form AgCl(c). What is the minimum silver content needed in this soil at 10% moisture for AgCl(c) to form?

18.7 Prepare a plot of log SO$_4^{2-}$, HS$^-$, H$_2$S°, and AgHS° at pH 7.0 as a function of redox in the range where Ag(c) transforms to α-Ag$_2$S(c), and total sulfur in solution is limited at 10^{-3} M. Discuss the implications of this plot with regard to the formation of α-Ag$_2$S(c) in highly reduced soils of different total silver content.

NINETEEN

CADMIUM

The average cadmium content of the lithosphere is estimated at 0.2 ppm, and the range for soils is given as 0.01 to 0.70 ppm (Table 1.1). The selected average for soils is 0.06 ppm, which provides only $10^{-5.27}$ M cadmium at 10% moisture if all the cadmium were solubilized. Like silver, cadmium is indeed a trace element in soils.

The principal use of cadmium in industry is for metal plating. Because of its toxicity to nearly all living organisms and its lack of useful biological function, cadmium is of great potential concern as an environmental contaminant (Lagerwerff, 1972; Fulkerson and Goeller, 1973; Frebert et al., 1974).

Some solubility measurements of cadmium in soils and water have been made (Hem, 1972; Santillan-Medrano and Jurinak, 1975; Street et al., 1977). The purpose of this chapter is to develop solubility relationships for cadmium in soils based on present information. Because the solubility data on this element are limited, further investigations are necessary before a more rigorous development can be made.

19.1 OXIDATION STATES OF CADMIUM IN SOILS

Equilibrium reactions of cadmium minerals and complexes are summarized in Table 19.1. Reaction 1 of this table shows

$$Cd^{2+} + 2e^- \rightleftharpoons Cd(c) \qquad \log K° = -13.64$$

from which

$$pe = -6.82 + \tfrac{1}{2} \log Cd^{2+} \tag{19.1}$$

or

$$\log Cd^{2+} = 13.64 + 2pe \tag{19.2}$$

Equation 19.1 shows that a 1 M Cd^{2+} in solution requires a $pe = -6.82$ before Cd(c) can form. Since Cd^{2+} activity in soil is often near 10^{-7} M, a $pe < -10.3$ would be required (Eq. 19.1). Since this redox range is below the stability field of water for pH values below 10.3, Cd(c) is not expected in soils. Since no other oxidation state of cadmium is stable in the redox range of soils, the chemistry of cadmium in soils is limited to Cd(II) minerals and complexes.

19.2 CADMIUM MINERALS IN SOILS

The solubilities of various cadmium minerals are given by Reactions 2 through 12 of Table 19.1 and are plotted in Fig. 19.1. The soil-Cd line

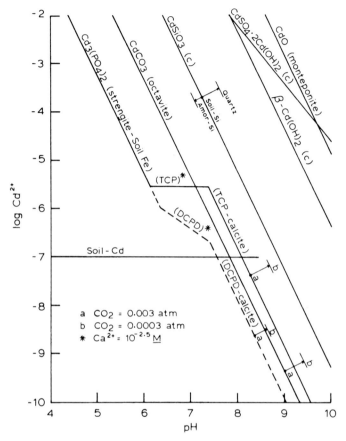

Fig. 19.1 The solubility of several cadmium minerals compared to soil-Cd at 10^{-7} M.

included here was estimated from Street et al. (1978), who found Cd^{2+} activities of approximately 10^{-7} M in the pH range of 6 to 7.5. Their findings are summarized by the reaction

$$\text{soil-Cd} \rightleftharpoons Cd^{2+} \quad \log K° = -7.00 \quad (19.3)$$

At pH values above 7.5, depending on $CO_2(g)$, Cd^{2+} activity is limited by $CdCO_3$(octavite). As shown in Fig. 19.1, octavite with a CO_2 of 0.003 atm depresses Cd^{2+} 100-fold for each unit increase in pH. At this CO_2 level, the depression begins at pH 7.84.

The mineral $CdSiO_3(c)$ is more soluble than octavite, so it is not expected to form in soils. The minerals CdO(monteponite), β-$Cd(OH)_2(c)$ and $CdSO_4 \cdot 2Cd(OH)_2(c)$ are also too soluble to form in soils.

TABLE 19.1 EQUILIBRIUM CONSTANTS OF CADMIUM MINERALS AND COMPLEXES

Reaction No.	Equilibrium Reaction	log $K°$
	Redox Reaction	
1	$Cd^{2+} + 2e^- \rightleftharpoons Cd(c)$	−13.64
	Minerals	
2	$CdO(monteponite) + 2H^+ \rightleftharpoons Cd^{2+} + H_2O$	15.14
3	$\beta\text{-}Cd(OH)_2(c) + 2H^+ \rightleftharpoons Cd^{2+} + 2H_2O$	13.65
4	$CdCO_3(octavite) + 2H^+ \rightleftharpoons Cd^{2+} + CO_2(g) + H_2O$	6.16
5	$CdSiO_3(c) + 2H^+ + H_2O \rightleftharpoons Cd^{2+} + H_4SiO_4°$	7.63
6	$CdSO_4(c) \rightleftharpoons Cd^{2+} + SO_4^{2-}$	−0.04
7	$CdSO_4 \cdot H_2O(c) \rightleftharpoons Cd^{2+} + SO_4^{2-} + H_2O$	−1.59
8	$CdSO_4 \cdot 2Cd(OH)_2(c) + 4H^+ \rightleftharpoons 3Cd^{2+} + SO_4^{2-} + 4H_2O$	22.65
9	$2CdSO_4 \cdot Cd(OH)_2(c) + 2H^+ \rightleftharpoons 3Cd^{2+} + 2SO_4^{2-} + 2H_2O$	6.73
10	$Cd_3(PO_4)_2(c) + 4H^+ \rightleftharpoons 3Cd^{2+} + 2H_2PO_4^-$	1.00
11	$CdS(greennokite) \rightleftharpoons Cd^{2+} + S^{2-}$	−27.07
12	$\text{soil-Cd} \rightleftharpoons Cd^{2+}$	−7.00*
	Hydrolysis Species	
13	$Cd^{2+} + H_2O \rightleftharpoons CdOH^+ + H^+$	−10.10
14	$Cd^{2+} + 2H_2O \rightleftharpoons Cd(OH)_2° + 2H^+$	−20.30
15	$Cd^{2+} + 3H_2O \rightleftharpoons Cd(OH)_3^- + 3H^+$	−33.01
16	$Cd^{2+} + 4H_2O \rightleftharpoons Cd(OH)_4^{2-} + 4H^+$	−47.29
17	$Cd^{2+} + 5H_2O \rightleftharpoons Cd(OH)_5^{3-} + 5H^+$	−61.93
18	$Cd^{2+} + 6H_2O \rightleftharpoons Cd(OH)_6^{4-} + 6H^+$	−76.81
19	$2Cd^{2+} + H_2O \rightleftharpoons Cd_2OH^{3+} + H^+$	−6.40
20	$4Cd^{2+} + 4H_2O \rightleftharpoons Cd_4(OH)_4^{4+} + 4H^+$	−27.92
	Halide Complexes	
21	$Cd^{2+} + Br^- \rightleftharpoons CdBr^+$	2.15
22	$Cd^{2+} + 2Br^- \rightleftharpoons CdBr_2°$	3.00
23	$Cd^{2+} + 3Br^- \rightleftharpoons CdBr_3^-$	3.00
24	$Cd^{2+} + 4Br^- \rightleftharpoons CdBr_4^{2-}$	2.90
25	$Cd^{2+} + Cl^- \rightleftharpoons CdCl^+$	1.98
26	$Cd^{2+} + 2Cl^- \rightleftharpoons CdCl_2°$	2.60
27	$Cd^{2+} + 3Cl^- \rightleftharpoons CdCl_3^-$	2.40
28	$Cd^{2+} + 4Cl^- \rightleftharpoons CdCl_4^{2-}$	2.50
29	$Cd^{2+} + I^- \rightleftharpoons CdI^+$	2.28

TABLE 19.1 (Continued)

Reaction No.	Equilibrium Reaction	log $K°$
30	$Cd^{2+} + 2I^- \rightleftharpoons CdI_2°$	3.92
31	$Cd^{2+} + 3I^- \rightleftharpoons CdI_3^-$	5.00
32	$Cd^{2+} + 4I^- \rightleftharpoons CdI_4^{2-}$	6.00
	Ammonia Complexes	
33	$Cd^{2+} + NH_4^+ \rightleftharpoons CdNH_3^{2+} + H^+$	−6.73
34	$Cd^{2+} + 2NH_4^+ \rightleftharpoons Cd(NH_3)_2^{2+} + 2H^+$	−14.00
35	$Cd^{2+} + 3NH_4^+ \rightleftharpoons Cd(NH_3)_3^{2+} + 3H^+$	−21.95
36	$Cd^{2+} + 4NH_4^+ \rightleftharpoons Cd(NH_3)_4^{2+} + 4H^+$	−30.39
	Other Complexes	
37	$Cd^{2+} + CO_2(g) + H_2O \rightleftharpoons CdHCO_3^+ + H^+$	−5.73
38	$Cd^{2+} + CO_2(g) + H_2O \rightleftharpoons CdCO_3° + 2H^+$	−14.06
39	$Cd^{2+} + NO_3^- \rightleftharpoons CdNO_3^+$	0.31
40	$Cd^{2+} + 2NO_3^- \rightleftharpoons Cd(NO_3)_2°$	0.00
41	$Cd^{2+} + H_2PO_4^- \rightleftharpoons CdHPO_4° + H^+$	−4.00
42	$Cd^{2+} + P_2O_7^{4-} \rightleftharpoons CdP_2O_7^{2-}$	8.70
43	$Cd^{2+} + SO_4^{2-} \rightleftharpoons CdSO_4°$	2.45

* Estimated from Street et al. (1978).

Also included in Fig. 19.1 is $Cd_3(PO_4)_2(c)$ which is plotted for five different conditions depending on which minerals control phosphate. The selected phosphate controls include: $FePO_4 \cdot 2H_2O$(strengite) and $Fe(OH)_3$(soil) at low pH, β-$Ca_3(PO_4)_2$(tricalcium phosphate) and soil-Ca at intermediate pH, and tricalcium phosphate (TCP), $CaCO_3$(calcite), and $CO_2(g)$ at high pH. In acid soils $Cd_3(PO_4)_2(c)$ is too soluble to account for the Cd^{2+} levels generally found. In the pH range of 6 to 7.5, $Cd_3(PO_4)_2(c)$ in equilibrium with TCP controls Cd^{2+} at approximately $10^{-5.5}$ M. In calcareous soils octavite is more stable than $Cd_3(PO_4)_2(c)$ in equilibrium with TCP or more insoluble phosphate minerals.

In recently fertilized soils $CaHPO_4 \cdot 2H_2O$(brushite) may be present and support a higher level of phosphate than does TCP (Chapter 12). Under these conditions, $Cd_3(PO_4)_2(c)$ increases in stability as shown by the dashed lines in Fig. 19.1, and Cd^{2+} would be depressed to approximately 10^{-6} to 10^{-7} M, in the pH range of 6.3 to 7.4. Above this pH, $Cd_3(PO_4)_2(c)$

in equilibrium with DCPD and calcite at 0.003 atm of $CO_2(g)$ is more stable than octavite. Since phosphate level seldomly exceeds those of DCPD, the $Cd_3(PO_4)_2(c)$-DCPD lines in Fig. 19.1 represents the lowest Cd^{2+} that can be imposed by phosphate in soils. The solubility value of $Cd_3(PO_4)_2(c)$ used in this development is that reported by Santillan-Medrano and Jurinak (1975). Investigations of other possible cadmium phosphates are needed.

The cadmium minerals. $CdSO_4(c)$, $CdSO_4 \cdot H_2O(c)$, and $2CdSO_4 \cdot$ are too soluble to appear in Fig. 19.1. These minerals are not expected to form in soils.

The solubility of CdS(greennokite) is given by Reaction 11 of Table 19.1 and can be combined to give:

		log $K°$
$Cd^{2+} + S^{2-} \rightleftharpoons$ CdS(greennokite)		27.07
$SO_4^{2-} + 8e^- + 8H^+ \rightleftharpoons S^{2-} + 4H_2O$		20.74
soil-Cd $\rightleftharpoons Cd^{2+}$		−7.00
soil-Cd $+ SO_4^{2-} + 8e^- + 8H^+ \rightleftharpoons$ CdS(greennokite) $+ 4H_2O$		40.81

(19.4)

from which

$$pe + pH = 5.10 + 0.12 \log SO_4^{2-} \tag{19.5}$$

When SO_4^{2-} is 10^{-3} M, Eq. 19.5 becomes

$$pe + pH = 4.73 \tag{19.6}$$

Thus CdS(greennokite) can form in a soil having SO_4^{2-} at 10^{-3} M when $pe + pH$ drops below 4.73.

If $CdCO_3$(octavite) rather than soil-Cd controls Cd^{2+} activity, the corresponding redox is

$$pe + pH = 6.75 + 0.12 \log SO_4^{2-} - 0.12\, CO_2(g) - 0.25\, pH \tag{19.7}$$

When SO_4^{2-} is 10^{-3} M and CO_2 is $10^{-2.52}$ atm, this equation becomes

$$pe + pH = 6.69 - 0.25\, pH \tag{19.8}$$

Equations 19.6 and 19.8 were plotted earlier in Fig. 17.4 and show the redox at which CdS(greennokite) can form in soils relative to the many other sulfides. Highly reduced conditions ($pe + pH < 4.73$) are required, yet greennokite forms before ZnS(sphalerite).

19.3 HYDROLYSIS SPECIES OF Cd(II)

The hydrolysis reactions of Cd^{2+} are given by Reactions 13 through 20 of Table 19.1. These species are plotted in Fig. 19.2 for equilibrium with soil-Cd at 10^{-7} M Cd^{2+} below pH 7.84, and with $CdCO_3$(octavite) and 0.003 atm of CO_2 above this pH. None of the hydrolysis species contribute significantly to total cadmium in solution except possibly $CdOH^+$ and $Cd(OH)_2^0$ above pH 7.5. All monomeric cadmium species change one log unit for each unit change in log Cd^{2+}. The $Cd_2(OH)^{3+}$ species changes 2 log units, whereas $Cd_4(OH)_4^{4+}$ changes 4 log units. In the presence of octavite,

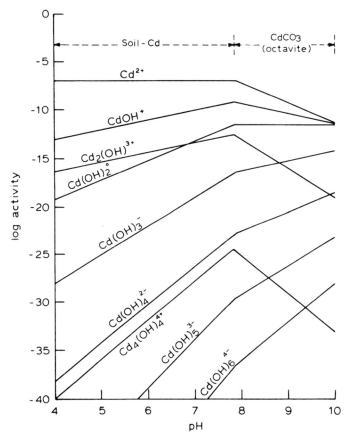

Fig. 19.2 The hydrolysis species of Cd^{2+} in equilibrium with soil-Cd or $CdCO_3$(octavite) at 0.003 atm of CO_2.

increasing log $CO_2(g)$ by one unit, decreases log Cd^{2+} and all monomeric hydrolysis species by one log unit, the dimer by 2 log units, and the tetramer by 4 log units. Thus Fig. 19.2 permits an easy diagnosis of the cadmium hydrolyses species as a function of Cd^{2+} activity, pH, and $CO_2(g)$.

19.4 HALIDE AND AMMONIA COMPLEXES OF CADMIUM

The halide and ammonia complexes of cadmium are defined by Reactions 21 through 36 and are plotted in Fig. 19.3 under the conditions that Cd^{2+} is controlled by soil-Cd or octavite at 0.003 atm of CO_2.

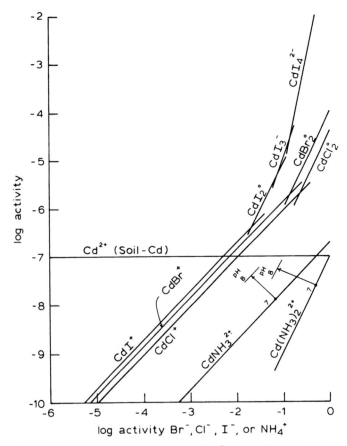

Fig. 19.3 Halide and ammonium complexes of Cd^{2+} in equilibrium with soil-Cd.

The halide complexes have similar stabilities, which decrease slightly in the order $I^- > Br^- > Cl^-$. At approximately 10^{-2} M halide activities, the complexes of CdI^+, $CdBr^+$, and $CdCl^+$ are about equal to Cd^{2+}. At lower halide activities, the complexes become less important, while at higher activities they become more important. The higher order complexes $CdCl_3^-$, $CdCl_4^{2-}$, $CdBr_3^-$, and $CdBr_4^{2-}$ are included in Table 19.1 but do not appear in Fig. 19.3 because halide activities > 1 M are necessary for them to become the dominant complexes in solution. Since such halide concentrations are seldomly encountered in soils, the graph was not extended to include them.

The ammonia complexes are less important than the halide complexes, but they increase with pH when the NH_4^+ ion is used as the reacting species. Unless the NH_4^+ is very high ($> 10^{-2}$ M), the cadmium-ammonia complexes are not significant in soils. The formation of $Cd(NH_3)_3^{2+}$ and $Cd(NH_3)_4^{2+}$ complexes require > 1 M NH_4^+ for these species to become dominant species in solution. Some temporary solubilization of cadmium from soils can be expected in the vicinity of $(NH_4)_2HPO_4$ (diammonium phosphate) fertilizer granules because NH_4^+ approaches 7 M when pH is near 8 (Lindsay et al., 1962).

19.5 OTHER CADMIUM COMPLEXES

Equilibrium reactions for other cadmium complexes are given by Reactions 37 through 43 of Table 19.1. These complexes include carbonates, nitrates, phosphates, and sulfates, which are plotted in Fig. 19.4.

An example of how the lines in Fig. 19.4 were developed is shown for $CdHCO_3^+$ by combining the following reactions:

			log $K°$
$Cd^{2+} + CO_2(g) + H_2O$	$\rightleftharpoons$	$CdHCO_3^+ + H^+$	-5.73
soil-Cd	$\rightleftharpoons$	Cd^{2+}	-7.00
soil-Cd + $CO_2(g) + H_2O$	$\rightleftharpoons$	$CdHCO_3^+ + H^+$	-12.73 (19.9)

which gives

$$\log CdHCO_3^+ = -12.73 + \log CO_2(g) + pH \quad (19.10)$$

When $\log CO_2(g) = -2.52$

$$\log CdHCO_3^+ = -15.25 + pH \quad (19.11)$$

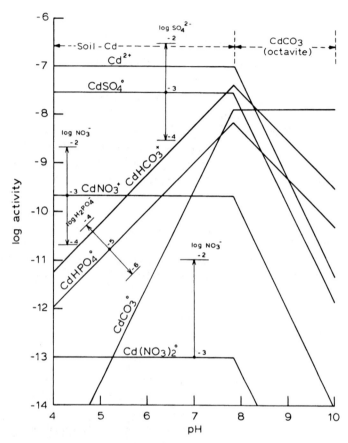

Fig. 19.4 Cadmium complexes of carbonate, nitrate, phosphate, and sulfate in equilibrium with soil-Cd and CdCO$_3$(octavite) at 0.003 atm CO$_2$.

Equation 19.11 is plotted as the CdHCO$_3^+$ line in Fig. 19.4. When Cd^{2+} is controlled by octavite

	log $K°$
CdCO$_3$(octavite) + 2H$^+$ $\rightleftharpoons$ Cd^{2+} + CO$_2$(g) + H$_2$O	6.16
Cd^{2+} + CO$_2$(g) + H$_2$O $\rightleftharpoons$ CdHCO$_3^+$ + H$^+$	−5.73
CdCO$_3$(octavite) + H$^+$ $\rightleftharpoons$ CdHCO$_3^+$	0.43

(19.12)

OTHER CADMIUM COMPLEXES

which gives

$$\log \text{CdHCO}_3^+ = 0.43 - \text{pH} \tag{19.13}$$

Equation 19.13 is plotted as the CdHCO_3^+ line above pH 7.84 where octavite is present.

The most abundant cadmium species in Fig. 19.4 is CdSO_4°, which equals that of Cd^{2+} at $10^{-2.45}$ M SO_4^{2-}. The next most significant species is CdHCO_3^+, which approaches the activity of Cd^{2+} near pH 8. Increasing $\text{CO}_2(\text{g})$ by 10-fold, raises the CdHCO_3^+ line by one log unit (Eq. 19.10) and lowers the Cd^{2+} line in equilibrium with octavite by one log unit. The CdHCO_3^+ line in equilibrium with octavite at high pH, however, is not affected by $\text{CO}_2(\text{g})$ (Eq. 19.3). Thus increasing $\text{CO}_2(\text{g})$ increases the relative importance of CdHCO_3^+. Also the CdCO_3° complex becomes important near pH 8 and increases similarly with increase in $\text{CO}_2(\text{g})$. This species maintains a constant activity of $10^{-7.9}$ M in the presence of octavite and is independent of $\text{CO}_2(\text{g})$ and pH. Thus the carbonate species in solution are important in alkaline soils, especially at elevated $\text{CO}_2(\text{g})$ levels.

The CdNO_3^+ and $\text{Cd}(\text{NO}_3)_2^\circ$ complexes can be fairly well ignored in soils as nitrate seldom reaches 10^{-2} M. The CdHPO_4° ion pair approaches significance in alkaline soils only when H_2PO_4^- is greater than 10^{-5} M, otherwise it can also be ignored.

As pointed out above, the CdHCO_3^+ and CdCO_3° lines in the presence of octavite are independent of $\text{CO}_2(\text{g})$. All other species in Fig. 19.4 decrease one log unit for each log unit increase in $\text{CO}_2(\text{g})$. Consequently this figure is useful for estimating the importance of cadmium complexes at Cd^{2+} activities different from 10^{-7} M represented by soil-Cd. For example, if the Cd^{2+} in a given soil were 10^{-5} M rather than 10^{-7} M, all lines in equilibrium with soil-Cd shift upward by 2 log units. Thus Fig. 19.4 can be used to estimate the importance of cadmium complexes at any pH, Cd^{2+}, $\text{CO}_2(\text{g})$, NO_3^-, SO_4^{2-}, or H_2PO_4^- level.

Importance of the $\text{CdP}_2\text{O}_7^{2-}$ complex defined by Reaction 42 of Table 19.1 can be seen from the reactions:

			$\log K^\circ$
$\text{Cd}^{2+} + \text{P}_2\text{O}_7^{4-}$	$\rightleftharpoons$	$\text{CdP}_2\text{O}_7^{2-}$	8.70
soil-Cd	$\rightleftharpoons$	Cd^{2+}	−7.00
$\beta\text{-Ca}_2\text{P}_2\text{O}_7(\text{c})$	$\rightleftharpoons$	$2\text{Ca}^{2+} + \text{P}_2\text{O}_7^{4-}$	−14.70
2Ca^{2+}	$\rightleftharpoons$	2 soil-Ca	2(2.50)
soil-Cd + $\text{Ca}_2\text{P}_2\text{O}_7(\text{c})$	$\rightleftharpoons$	$\text{CdP}_2\text{O}_7^{2-}$ + 2 soil-Ca	−8.00

$$\tag{19.14}$$

In soils where $P_2O_7^{4-}$ is fixed by β-$Ca_2P_2O_7(c)$, Ca^{2+} by soil-Ca, and Cd^{2+} by soil-Cd, the $CdP_2O_7^{2-}$ complex would be 10^{-8} M or about one-tenth that of Cd^{2+}. As pointed out in Chapter 12, pyrophosphate is unstable in soils and eventually disappears. Thus $CdP_2O_7^{2-}$ is not expected to be important in most soils.

19.6 NEED FOR FURTHER STUDIES

The soil-Cd line in Fig. 19.1 is only a convenient reference level for considering Cd^{2+} solubility relationships in soils. This value of 10^{-7} M Cd^{2+} is based on a limited number of soils (Street et al., 1978). In their studies, when cadmium was added at 10^{-5} to 10^{-4} M levels, it rapidly precipitated giving 10^{-7} M Cd^{2+} in solution. In contrast, Santillian-Medrano and Jurinak (1975) reported much higher levels of cadmium (10^{-4} to 10^{-3} M). Their rates of cadmium addition were, however, much greater and exceeded those found in most natural soils ($10^{-5.33}$ M) (Table 1.1).

It appears that $CdCO_3$(octavite) provides a limit on the solubility of Cd^{2+} in soils of high pH, especially at elevated $CO_2(g)$ (Street et al., 1978; Santillan-Medrano and Jurinak, 1975). Further investigations are necessary to examine the solubility relationships of Cd^{2+} in acid soils and to explore other possible cadmium minerals that may account for the rapid removal of cadmium from solution to less than 10^{-7} M found by Street et al. (1978).

REFERENCES

Frebert, L. T., M. Piscator, G. Norberg, and T. Kjellstrom. 1974. Cadmium in the environment, 2nd ed. CRC Press, Inc. ISBN-0-87819-018-X.

Fulkerson, W. and H. E. Goeller. 1973. Cadmium — The dissipated element. Oak Ridge National Laboratory ORNL NSF-EP-21.

Hem, J. D. 1972. Chemistry and occurrence of cadmium and zinc in surface water and groundwater. Water Resources Res. 8:661–679.

Lagerwerff, J. V. 1972. Lead, mercury, and cadmium as environmental contaminants. *In* J. J. Mortvedt, P. M. Giordano, and W. L. Lindsay, Eds., Micronutrients in Agriculture, pp. 593–646. Soil Science Society of America, Madison, Wi.

Lindsay, W. L., A. W. Frazier, and H. F. Stephenson. 1962. Identification of reaction products from phosphate fertilizers in soils. Soil Sci. Soc. Am. Proc. 26:446–452.

Santillan-Medrano, J. and J. J. Jurinak. 1975. The chemistry of lead and cadmium in soil: Solid phase formation. Soil Sci. Soc. Am. Proc. 39:851–856.

Street, J. J., W. L. Lindsay, and B. R. Sabey. 1977. The solubility and plant uptake of cadmium in soils amended with cadmium and sewage sludge. J. Env. Qual. 6:72–77.

Street, J. J., B. R. Sabey, and W. L. Lindsay. 1978. Influence of pH, phosphorus, cadmium, sewage sludge, and incubation time on the solubility and plant uptake of cadmium. J. Env. Qual. 7:286-290.

PROBLEMS

19.1 Develop the necessary equations and plot the $Cd_3(PO_4)_2(c)$ solubility for a soil in equilibrium with hydroxyapatite and soil-Ca and/or calcite when CO_2 is 0.003 atm. Compare this plot to Fig. 19.1 and discuss its significance.

19.2 Indicate how each of the activity lines in Fig. 19.2 changes with Cd^{2+} activity, pH, and $CO_2(g)$.

19.3 Develop an equation for the $Cd_2(OH)^{3+}$ line in Fig. 19.2 and estimate the contribution of this species to total cadmium in solution resulting from Cd^{2+} and its hydrolysis species for a soil of pH 7.5 having $10^{-4.5}$ M Cd^{2+}.

19.4 Develop an equation and estimate the total soluble cadmium in a soil of pH 7.0 containing $10^{-2.3}$ M I^-, Br^-, Cl^-, and SO_4^{2-} which maintains equilibrium with soil-Ca and β-TCP with CO_2 present at 0.003 atm.

TWENTY

LEAD

The average lead content of the lithosphere is estimated at 16 ppm, and the common range for soils is 2 to 200 ppm. The selected average lead content of soils of 10 ppm corresponds to $10^{-3.32}$ M Pb if all the lead were to dissolve in the soil solution at 10% moisture (Table 1.1).

Because lead is toxic to both man and animals, there is considerable concern regarding this element as a contaminant in the environment (Page et al., 1971; Lagerwerff, 1972; Cartwright et al. 1976). The objective of this chapter is to depict the solubility relationships of lead minerals and complexes in soil environments in a meaningful way. This development is based on limited information and will require revision as more extensive solubility relationships become available.

20.1 SOLUBILITY OF LEAD MINERALS

The equilibrium reactions of lead minerals and complexes are summarized in Table 20.1. The solubilities of lead oxides, carbonates, and sulfates are plotted in Fig. 20.1. The oxides PbO(yellow) and PbO(red) are the most soluble minerals of those depicted here. Even at pH 8 these oxides require nearly 10^{-3} M Pb^{2+} for equilibrium. The hydroxide $Pb(OH)_2$(c) is considerably more stable at pH 8 maintaining approximately 10^{-8} M Pb^{2+}. Both $PbCO_3$(cerussite) and $Pb_3(CO_3)_2(OH)_2$(c) have identical solubilities at 0.0003 atm of CO_2. Increasing CO_2 causes cerussite to be more stable whereas reducing it makes $Pb(OH)_2$(c) the stable phase. Since CO_2 in soils is generally higher than that of the atmosphere, cerussite will be the more stable mineral in soils.

Lead oxides of higher oxidation states, which include PbO_2(c) and Pb_3O_4(c), are of intermediate solubility in highly oxidized soils (pe + pH = 20.61) and increase in solubility with decreasing redox. At pe + pH = 18.81, Pb_3O_4(c) and PbO_2(c) have equal solubilities and are only slightly more stable than the PbO minerals. It is virtually impossible for the higher oxidation state lead oxides to form in soils since soils are nearly always more reduced than pe + pH of 20.61.

Lead forms many mixed minerals containing carbonate, sulfate, oxide, and chloride. These mixed minerals are of intermediate solubility as shown in Fig. 20.1. In general their solubilities at pH 7.5 decrease in the order:

$Pb_2CO_3Cl_2$(c) > $PbSO_4 \cdot 3PbO$(c) > $PbCO_3 \cdot PbO$(c)
> $PbCO_3 \cdot 2PbO$(c) > $PbSO_4 \cdot PbO$(c) > $Pb_3(CO_3)_2(OH)_2$(c).

Except for the latter mineral, these compounds are generally too soluble to be important in soils. In the presence of 10^{-3} M SO_4^{2-}, the mineral $PbSO_4$ (anglesite) limits Pb^{2+} at $10^{-4.79}$ M and its solubility changes one log unit for each log unit change in SO_4^{2-}.

TABLE 20.1 EQUILIBRIUM REACTIONS FOR LEAD MINERALS AND COMPLEXES AT 25°C

Reaction No.	Equilibrium Reaction	$\log K°$
	Oxides, Carbonates, Sulfates	
1	$PbO(\text{yellow}) + 2H^+ \rightleftharpoons Pb^{2+} + H_2O$	12.89
2	$PbO(\text{red}) + 2H^+ \rightleftharpoons Pb^{2+} + H_2O$	12.72
3	$Pb(OH)_2(c) + 2H^+ \rightleftharpoons Pb^{2+} + 2H_2O$	8.16
4	$Pb_3O_4(c) + 8H^+ + 2e^- \rightleftharpoons 3Pb^{2+} + 4H_2O$	73.79
5	$PbO_2(c) + 4H^+ + 2e^- \rightleftharpoons Pb^{2+} + 2H_2O$	49.68
6	$PbCO_3(\text{cerussite}) + 2H^+ \rightleftharpoons Pb^{2+} + CO_2(g) + H_2O$	4.65
7	$Pb_2CO_3Cl_2(\text{phosgenite}) + 2H^+ \rightleftharpoons 2Pb^{2+} + CO_2(g) + H_2O + 2Cl^-$	−1.80
8	$Pb_3(CO_3)_2(OH)_2(c) + 6H^+ \rightleftharpoons 3Pb^{2+} + 2CO_2(g) + 4H_2O$	17.51
9	$PbCO_3 \cdot PbO(c) + 4H^+ \rightleftharpoons 2Pb^{2+} + CO_2(g) + 2H_2O$	17.39
10	$PbSO_4(\text{anglesite}) \rightleftharpoons Pb^{2+} + SO_4^{2-}$	−7.79
11	$PbSO_4 \cdot PbO(c) + 2H^+ \rightleftharpoons 2Pb^{2+} + SO_4^{2-} + H_2O$	−0.19
12	$PbSO_4 \cdot 2PbO(c) + 4H^+ \rightleftharpoons 3Pb^{2+} + SO_4^{2-} + 2H_2O$	11.01
13	$PbSO_4 \cdot 3PbO + 6H^+ \rightleftharpoons 4Pb^{2+} + SO_4^{2-} + 3H_2O$	22.30
	Silicates	
14	$PbSiO_3(c) + 2H^+ + H_2O \rightleftharpoons Pb^{2+} + H_2SiO_4°$	5.94
15	$Pb_2SiO_4(c) + 4H^+ \rightleftharpoons 2Pb^{2+} + H_4SiO_4°$	18.45
	Phosphates	
16	$Pb(H_2PO_4)_2(c) \rightleftharpoons Pb^{2+} + 2H_2PO_4^-$	−9.85
17	$PbHPO_4(c) + H^+ \rightleftharpoons Pb^{2+} + H_2PO_4^-$	−4.25

18	$Pb_3(PO_4)_2(c) + 4H^+ \rightleftharpoons 3Pb^{2+} + 2H_2PO_4^-$	−5.26
19	$Pb_4O(PO_4)_2(c) + 6H^+ \rightleftharpoons 4Pb^{2+} + 2H_2PO_4^- + H_2O$	2.24
20	$Pb_5(PO_4)_3OH(c)(hydroxypyromorphite) + 7H^+ \rightleftharpoons 5Pb^{2+} + 3H_2PO_4^- + H_2O$	−4.14
21	$Pb_5(PO_4)_3Br(c)(bromopyromorphite) + 6H^+ \rightleftharpoons 5Pb^{2+} + 3H_2PO_4^- + Br^-$	−19.49
22	$Pb_5(PO_4)_3Cl(c)(chloropyromorphite) + 6H^+ \rightleftharpoons 5Pb^{2+} + 3H_2PO_4^- + Cl^-$	−25.05
23	$Pb_5(PO_4)_3F(c)(fluoropyromorphite) + 6H^+ \rightleftharpoons 5Pb^{2+} + 3H_2PO_4^- + F^-$	−12.98

Other Minerals

24	soil-Pb $\rightleftharpoons Pb^{2+}$	−8.50*
25	$PbMoO_4(wulfenite) \rightleftharpoons Pb^{2+} + MoO_4^{2-}$	−16.04
26	$PbS(galena) \rightleftharpoons Pb^{2+} + S^{2-}$	−27.51
27	$Pb^{2+} + 2e^- \rightleftharpoons Pb(c)$	−4.33

Hydrolysis Species

28	$Pb^{2+} + H_2O \rightleftharpoons PbOH^+ + H^+$	−7.70
29	$Pb^{2+} + 2H_2O \rightleftharpoons Pb(OH)_2^\circ + 2H^+$	−17.75
30	$Pb^{2+} + 3H_2O \rightleftharpoons Pb(OH)_3^- + 3H^+$	−28.09
31	$Pb^{2+} + 4H_2O \rightleftharpoons Pb(OH)_4^{2-} + 4H^+$	−39.49
32	$2Pb^{2+} + H_2O \rightleftharpoons Pb_2OH^{3+} + H^+$	−6.40
33	$3Pb^{2+} + 4H_2O \rightleftharpoons Pb_3(OH)_4^{2+} + 4H^+$	−23.89
34	$4Pb^{2+} + 4H_2O \rightleftharpoons Pb_4(OH)_4^{4+} + 4H^+$	−20.89
35	$6Pb^{2+} + 8H_2O \rightleftharpoons Pb_6(OH)_8^{4+} + 8H^+$	−43.58

Halide Complexes

36	$Pb^{2+} + Br^- \rightleftharpoons PbBr^+$	1.77
37	$Pb^{2+} + 2Br^- \rightleftharpoons PbBr_2^\circ$	2.60
38	$Pb^{2+} + 3Br^- \rightleftharpoons PbBr_3^-$	3.00

(*Continued*)

TABLE 20.1 (*Continued*)

Reaction No.	Equilibrium Reaction	log $K°$
39	$Pb^{2+} + 4Br^- \rightleftharpoons PbBr_4^{2-}$	2.30
40	$Pb^{2+} + Cl^- \rightleftharpoons PbCl^+$	1.60
41	$Pb^{2+} + 2Cl^- \rightleftharpoons PbCl_2^0$	1.78
42	$Pb^{2+} + 3Cl^- \rightleftharpoons PbCl_3^-$	1.68
43	$Pb^{2+} + 4Cl^- \rightleftharpoons PbCl_4^{2-}$	1.38
44	$Pb^{2+} + F^- \rightleftharpoons PbF^+$	1.49
45	$Pb^{2+} + 2F^- \rightleftharpoons PbF_2^0$	2.27
46	$Pb^{2+} + 3F^- \rightleftharpoons PbF_3^-$	3.42
47	$Pb^{2+} + 4F^- \rightleftharpoons PbF_4^{2-}$	3.10
48	$Pb^{2+} + I^- \rightleftharpoons PbI^+$	1.92
49	$Pb^{2+} + 2I^- \rightleftharpoons PbI_2^0$	3.15
50	$Pb^{2+} + 3I^- \rightleftharpoons PbI_3^-$	3.92
51	$Pb^{2+} + 4I^- \rightleftharpoons PbI_4^{2-}$	4.50
	Other Complexes	
52	$Pb^{2+} + NO_3^- \rightleftharpoons PbNO_3^+$	1.17
53	$Pb^{2+} + 2NO_3^- \rightleftharpoons Pb(NO_3)_2^0$	1.40
54	$Pb^{2+} + H_2PO_4^- \rightleftharpoons PbH_2PO_4^+$	1.50
55	$Pb^{2+} + H_2PO_4^- \rightleftharpoons PbHPO_4^0 + H^+$	−4.10
56	$Pb^{2+} + P_2O_7^{4-} \rightleftharpoons PbP_2O_7^{2-}$	11.30
57	$Pb^{2+} + SO_4^{2-} \rightleftharpoons PbSO_4^0$	2.62
58	$Pb^{2+} + 2SO_4^{2-} \rightleftharpoons Pb(SO_4)_2^{2-}$	3.47

* Selected reference level for Pb^{2+} in soils not containing $PbCO_3$(cerussite) based on Fig. 20.2.

SOLUBILITY OF LEAD MINERALS

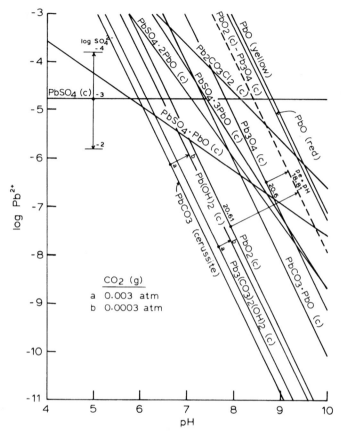

Fig. 20.1 The solubility of various lead oxides, carbonates, and sulfates when SO_4^{2-} and Cl^- are 10^{-3} M and CO_2 is 0.003 atm or as specified.

Of the minerals included in Fig. 20.1 $PbSO_4$(anglesite) is the most stable below pH 6, whereas $PbCO_3$(cerussite) is most stable at higher pH values. Changing slopes and shifts in the solubility lines for the various minerals in Fig. 20.1 are logical reflections of the ratios of Pb/SO_4, Pb/Cl, Pb/CO_3 present in these minerals.

The solubilities of several lead silicates and phosphates are given by Reactions 14 through 23 of Table 20.1 and are plotted in Fig. 20.2. Cerussite is included for comparison. The silicates Pb_2SiO_4(c) and $PbSiO_3$(c) lines are shown for equilibrium with quartz at 10^{-4} M $H_4SiO_4^\circ$ and with shifts indicated for soil-Si and kaolinite-gibbsite (Chapter 5). These lead silicates are too soluble to be expected in soils.

Lead is capable of forming numerous phosphate minerals as shown

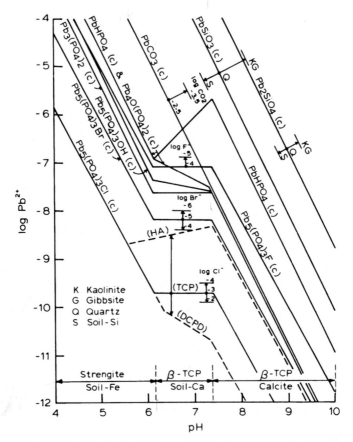

Fig. 20.2 The solubility of various lead silicates and phosphates compared to $PbCO_3$(cerussite) when phosphate is controlled by various solid phases as indicated and $CO_2(g)$ is 0.003 atm.

in Fig. 20.2. Nriagu (1974) examined the solubilities of many lead phosphates and pointed out how they may be important in controlling lead in natural environments. Since the average lead content of soils (10 ppm) is generally much less than that of phosphorus (600 ppm), it is highly possible that phosphate may control Pb^{2+} solubility.

In developing Fig. 20.2 phosphate activity was fixed by strengite and soil-Fe at low pH, by β-tricalcium phosphate (β-TCP) and soil-Ca at intermediate pH, and by β-TCP, calcite, and 0.003 atm of CO_2 at high pH (Fig. 12.8). The β-TCP reference was arbitrarily selected as intermediate between the phosphate maintained by $CaHPO_4 \cdot 2H_2O$(DCPD) as an upper limit and by hydroxyapatite(HAP) as a lower limit. The dashed lines in Fig. 20.2 show the effect of these limits on Pb^{2+} activity as phosphate shifts equilibrium from DCPD to HA.

SOLUBILITY OF LEAD MINERALS

In general lead phosphates decrease in solubility in the order

$$Pb(H_2PO_4)_2(c) > PbHPO_4(c) > Pb_4O(PO_4)_2(c)$$
$$> Pb_5(PO_4)_3F(c) > Pb_5(PO_4)_3OH(c) > Pb_3(PO_4)_2(c)$$
$$> Pb_5(PO_4)_3Br(c) > Pb_5(PO_4)_3Cl(c).$$

This comparison is based on 10^{-3} M Cl^-, 10^{-4} M F^-, and 10^{-5} M Br^-. These levels were selected because CaF_2(fluorite) in soils maintains approximately 10^{-4} M F^- (Chapter 12) and Br^- is generally less abundant than either Cl^- or F^-. Obviously $Pb_5(PO_4)_3Cl$(chloropyromorphite) is the most insoluble of the lead phosphate minerals and has the capability of controlling Pb^{2+} solubility throughout the pH range of most soils. In this text an arbitrary level of $10^{-8.5}$ M Pb^{2+} was selected as an appropriate Pb^{2+} activity expected for soils in the pH range of 5.5 to 7.5 where hydroxyapatite and chloropyromorphite control phosphate and lead activities, respectively. This level of lead is represented by the simplified equation

$$\text{soil-Pb} \rightleftharpoons Pb^{2+} \qquad \log K^\circ = \sim -8.5 \qquad (20.1)$$

Because few measurements are available to test this Pb^{2+} solubility relationship, Eq. 20.1 must be considered tentative. An exact solubility relationship would have a slight pH dependency and vary with the degree of crystallinity of both hydroxyapatite and chloropyromorphite. When phosphate in soils is more soluble than that represented by hydroxyapatite, Pb^{2+} activity is depressed further. Because of the many complexities, Eq. 20.1 is used as a convenient approximation of Pb^{2+} activity in soils in the pH range of 5.5 to 7.5.

In soils below pH 6.0, Pb^{2+} solubility is increased as phosphate is depressed by trivalent cations such as Fe^{3+} and Al^{3+}. At pH values somewhat below 4.0, $PbSO_4$(anglesite) is more stable than chloropyromorphite depending upon SO_4^{2-} activity. For calcareous soils the following equilibria give the Pb^{2+} activity in equilibrium with chloropyromorphite:

	$\log K^\circ$
$Pb_5(PO_4)_3Cl(c)$ $+ 6H^+ \rightleftharpoons 5Pb^{2+} + 3H_2PO_4^- + Cl^-$	-25.05
$5Ca^{2+} + 3H_2PO_4^- + H_2O \rightleftharpoons Ca_5(PO_4)_3OH(HA) + 7H^+$	-14.46
$5CaCO_3(\text{calcite}) + 10H^+ \rightleftharpoons 5Ca^{2+} + 5CO_2(g) + 5H_2O$	$5(9.74)$
$Pb_5(PO_4)_3Cl(c) + 5CaCO_3(\text{calcite}) + 9H^+ \rightleftharpoons 5Pb^{2+} + 5Ca_5(PO_4)_3OH(HA) + 5CO_2(g) + 5H_2O + Cl^-$	9.19

$$(20.2)$$

which reduces to

$$\log \text{Pb}^{2+} = 1.84 - \log \text{CO}_2(g) - 0.2 \log \text{Cl}^- - 1.80\,\text{pH} \quad (20.3)$$

Equation 20.3 is plotted as the dashed line farthest to the right in Fig. 20.2. It represents the Pb^{2+} activity controlled by $\text{Pb}_5(\text{PO}_4)_3\text{Cl(c)}$ and HA in the presence of calcite and 0.003 atm of CO_2. Increasing $\log \text{CO}_2(g)$ by one unit depresses both $\text{Pb}_5(\text{PO}_4)_3\text{Cl(c)}$ and PbCO_3(cerussite) solubility lines by one log unit. Thus chloropyromorphite is always slightly more stable than cerussite. Should phosphate be more soluble than that permitted by hydroxyapatite, the solubility line for chloropyromorphite would be depressed further. Decreasing Cl^- activity shifts the chloropyromorphite line upward 0.2 log unit for each unit increase in $\log \text{Cl}^-$ bringing the chloropyromorphite and cerussite lines closer together.

Although the solubility relationships of lead in soils are quite complex, cerussite and soil-Pb (Reactions 6 and 24 of Table 20.1) give fair estimates of Pb^{2+} activities expected in soils. These equations are used later in this chapter for depicting lead complexes in solution.

Two additional lead minerals are included in Table 20.1: PbMoO_4 (wulfenite) defined by Reaction 25 and PbS(galena) defined by Reaction 26. Since the average molybdenum content of soils (2 ppm) is generally less than lead (10 ppm), lead is more likely to control molybdate solubility than the reverse. For this reason the solubility relationships of wulfenite will be discussed with molybdenum in Chapter 22.

In the case of PbS(galena), the following equilibrium reactions can be combined to obtain the redox at which galena can form:

For acid soils:

		$\log K^\circ$
$\text{Pb}^{2+} + \text{S}^{2-} \rightleftharpoons \text{PbS(galena)}$		27.51
$\text{SO}_4^{2-} + 8e^- + 8\text{H}^+ \rightleftharpoons \text{S}^{2-} + 4\text{H}_2\text{O}$		20.74
soil-Pb $\rightleftharpoons \text{Pb}^{2+}$		-8.50
soil-Pb + SO_4^{2-} + $8e^-$ + 8H^+ $\rightleftharpoons$ PbS(galena) + $4\text{H}_2\text{O}$		39.75

$$(20.4)$$

from which

$$pe + pH = 4.97 + \tfrac{1}{8}\log \text{SO}_4^{2-} \quad (20.5)$$

When SO_4^{2-} is 10^{-3} M,

$$pe + pH = 4.59 \quad (20.6)$$

SOLUBILITY OF LEAD MINERALS

For alkaline soils:

	log $K°$
$Pb^{2+} + S^{2-} \rightleftharpoons$ PbS(galena)	27.51
$SO_4^{2-} + 8e^- + 8H^+ \rightleftharpoons S^{2-} + 4H_2O$	20.74
$PbCO_3$(cerussite) $+ 2H^+ \rightleftharpoons Pb^{2+} + CO_2(g) + H_2O$	4.65
$PbCO_3(c) + SO_4^{2-} + 8e^- + 10H^+ \rightleftharpoons$ PbS(galena) $+ CO_2(g) + 5H_2O$	52.90 (20.7)

from which

$$pe + pH = 6.61 + \tfrac{1}{8} \log SO_4^{2-} - \tfrac{1}{8} \log CO_2(g) + \tfrac{1}{4} pH \quad (20.8)$$

When $SO_4^{2-} = 10^{-3}$ M and $CO_2(g)$ is 0.003 atm,

$$pe + pH = 6.55 - 0.25 pH \quad (20.9)$$

Equations 20.6 and 20.9 were plotted previously as the PbS-soil-Pb and PbS-PbCO$_3$ lines in Fig. 17.3. These lines show the redox (pe + pH) at which galena can first form. Above these redox levels, depending on pH, galena cannot form, and below these redox levels, Pb^{2+} solubility is controlled by galena and whatever SO_4^{2-} or other sulfur species persists (Chapter 17).

The question arises whether or not Pb(c) may be expected in soils. The following equations are convenient for answering this question:

$$Pb^{2+} + 2e^- \rightleftharpoons Pb(c) \quad \log K° = -4.33 \quad (20.10)$$

which gives

$$pe = -2.16 + 0.5 \log Pb^{2+} \quad (20.11)$$

or

$$\log Pb^{2+} = 4.33 + 2pe \quad (20.12)$$

Since Pb^{2+} in soils is expected to approach $10^{-8.5}$ M, Eq. 20.11 becomes

$$pe = -2.16 + 0.5(-8.5) = -6.41 \quad (20.13)$$

Thus only at $pe < -6.41$ could Pb(c) form. At pH 7 this represents a pe + pH of 0.59 which is very low for soil environments. Should sufficient sulfur be present to form galena, the activity of Pb^{2+} would be sufficiently depressed that Pb(c) could not form within the stability field of water even at pH values > 6.41.

20.2 HYDROLYSIS SPECIES OF LEAD

The hydrolysis reactions of Pb(II) are given by Reactions 28 through 35 of Table 20.1 and are plotted in Fig. 20.3. The Pb^{2+} activities used in developing this plot were soil-Pb at $10^{-8.5}$ M Pb^{2+} in acid soils and $PbCO_3$(cerussite) with 0.003 atm of CO_2 at higher pH. Rationales for these choices were discussed in previous sections.

Below pH 8.0 only Pb^{2+} and $PbOH^+$ contribute significantly to total lead in solution. Since soils seldom attain pH values above 9, $Pb(OH)_2^\circ$ and $Pb(OH)_3^-$ barely approach significance even at high pH. The monomeric

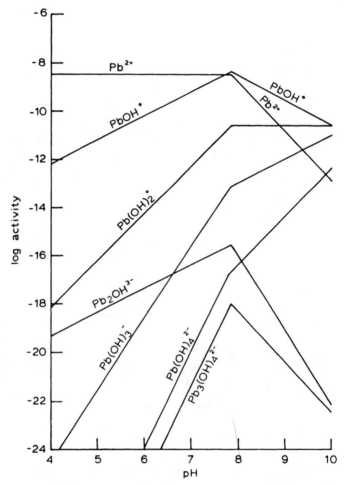

Fig. 20.3 The hydrolysis species of Pb(II) in equilibrium with soil-Pb at $10^{-8.5}$ M Pb^{2+} or $PbCO_3$(cerussite) at 0.003 atm CO_2.

lead species depicted in Fig. 20.3 increase one log unit for each log unit increase in Pb^{2+}, but their relative positions remain fixed. Although Pb_2OH^{3+} would increase 2 log units and $Pb_3(OH)_4^{2+}$ 3 log units for each log unit increase in Pb^{2+}, these species still do not contribute significantly to total lead in solution even in polluted soils where Pb^{2+} may approach $10^{-3} M$.

20.3 HALIDE COMPLEXES OF LEAD

The halide complexes of Pb(II) are given by Reactions 36 through 51 of Table 20.1 and are plotted in Fig. 20.4. The reference level of Pb^{2+} used for

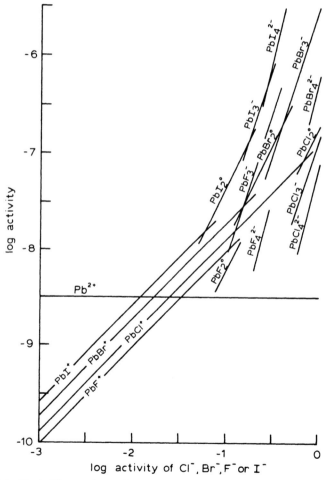

Fig. 20.4 The halide complexes of lead in equilibrium with soil-Pb at $10^{-8.5} M$ Pb^{2+}.

this development was soil-Pb at $10^{-8.5}$ M Pb^{2+}. In general the halide complexes have similar stabilities but decrease slightly in the order $PbI^+ >$ $PbBr^+ > PbCl^- > PbF^-$. At halide activities $> 10^{-4}$ M, all of these complexes begin to contribute significantly to total lead in solution. In the halide range of 10^{-2} to $10^{-1.5}$ M the halide complexes are approximately equal to free Pb^{2+}. Only in the halide activity range $> 10^{-1.3}$ M do the higher order complexes such as PbI_2°, PbI_3^-, PbI_4^{2-}, PbF_3^-, $PbBr_2^\circ$, and $PbCl_2^\circ$ exceed the simple 1:1 complexes. Generally the halide complexes of Pb^{2+} are not highly significant in soils.

Although Fig. 20.4 was developed for a Pb^{2+} activity of $10^{-8.5}$ M, it can be used to obtain halide complexes at any Pb^{2+} activity. Increasing or

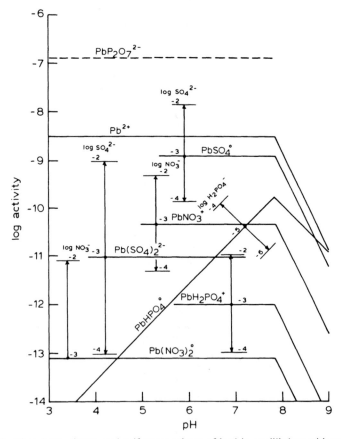

Fig. 20.5 Phosphate, nitrate, and sulfate complexes of lead in equilibrium with soil-Pb and $PbCO_3$(cerussite) at 0.003 atm of CO_2. The $PbP_2O_7^{2-}$ line is in metastable equilibrium with β-Ca_2P_2O-(c) and soil-Ca.

decreasing Pb^{2+} activity by one log unit raises or lowers all of the lines in this figure by one log unit accordingly.

20.4 OTHER COMPLEXES OF LEAD

Several other complexes of lead are given by Reactions 52 through 58 of Table 20.1 and are plotted in Fig. 20.5. The Pb^{2+} activities used to develop this plot were soil-Pb at low pH and $PbCO_3$(cerussite) with 0.003 atm of CO_2 at high pH.

The $PbSO_4^°$ complex is significant and equals that of Pb^{2+} at $10^{-2.62}$ M SO_4^{2-}. The $Pb(SO_4)_2^{2-}$ species become significant only as SO_4^{2-} increases above 10^{-3} M. The $PbNO_3^+$ complex becomes slightly significant as NO_3^- increases above 10^{-3} M while the $Pb(NO_3)_2^°$ can be ignored so long as $NO_3^- < 10^{-2}$ M. The $PbH_2PO_4^+$ complex is not significant in the phosphate range generally encountered in soils while the $PbHPO_4^°$ ion pair is significant above pH 7. Shifts in the $PbHPO_4^°$ line in Fig. 20.5 can be readily obtained by knowing the $H_2PO_4^-$ activity.

Another line included in Fig. 20.5 is that of $PbP_2O_7^{2-}$ shown by the dashed line at approximately 10^{-7} M. This line corresponds to a $P_2O_7^{4-}$ activity of $10^{-9.7}$ M depicted earlier in Fig. 12.13 as the equilibrium level with β-$Ca_2P_2O_7$(c) and soil-Ca. In the presence of this pyrophosphate mineral, $PbP_2O_7^{2-}$ would be a highly significant solution species. A broken line is used to depict this species because eventually β-$Ca_2P_2O_7$(c) will disappear as pyrophosphate dissolves and hydrolyzes to orthophosphates (Chapter 12). The $PbP_2O_7^{2-}$ species will then diminish and become insignificant in the phosphate range generally encountered in soils.

REFERENCES

Cartwright, B., R. H. Merry, and K. G. Tiller. 1976. Heavy metal contamination of soils around a lead smelter at Port Pirie, South Australia. Aust. J. Soil Res. 15:69-81.

Lagerwerff, J. N. 1972. Lead, mercury, and cadmium as environmental contaminants. In J. J. Mortvedt, P. M. Giordano, W. L. Lindsay, Eds., Micronutrients in Agriculture. pp. 593-636. Soil Science Society of America, Madison, Wi.

Nriagu, J. O. 1974. Lead orthophosphates. IV. Formation and stability in the environment. Geochim. Cosmochim. Acta, 38:887-898.

Page, A. L., T. J. Ganje, and M. S. Joshi. 1971. Lead quantities in plants, soil, and air near some major highways in southern California. Hilgardia 41(1):1-41.

Santillan-Medrano, J. and J. J. Jurinak. 1975. The chemistry of lead and cadmium in soil: Solid phase formation. Soil Sci. Soc. Am. Proc. 39:851-856.

PROBLEMS

20.1 Develop the equation and plot the line for $Pb_2CO_3Cl_2(c)$ shown in Fig. 20.1. Indicate by arrows and short lines how this line shifts for a 10-fold increase and a 10-fold decrease in Cl^- activity and $CO_2(g)$.

20.2 Prove that $PbO_2(c)$ and $Pb_3O_4(c)$ have identical solubilities at $pe + pH$ 18.81.

20.3 Develop the necessary equations and show how the $Pb_5(PO_4)_3Cl(c)$ lines in Fig. 20.2 would shift if octacalcium phosphate rather than β-tricalcium phosphate had been used to control phosphate activity.

20.4 For a calcareous soil of pH 8 what conditions would have to exist for chloropyromorphite, hydroxyapatite, and cerussite to coexist?

20.5 Discuss the pros and cons of using soil-Pb at $10^{-8.5}$ M and cerussite as solubility controls of Pb^{2+} in soils.

20.6 Develop an equation to show how $CO_2(g)$ affects the horizontal portion of the $Pb(OH)_2^\circ$ line in Fig. 20.3.

20.7 Develop the necessary equations and designate the fewest parameters that are necessary to measure in order that Pb^{2+} activity can be obtained from total lead in soil solution. Assume that the major anions are $< 10^{-2}$ M and that soil pH values lie between 4.0 and 8.5.

20.8 Develop the necessary equations and show how the $PbP_2O_7^{2-}$ line in Fig. 20.5 shifts as β-$Ca_2P_2O_7(c)$ disappears and pyrophosphate attains equilibrium with a) β-tricalcium phosphate or b) hydroxyapatite.

TWENTY-ONE

MERCURY

The mercury content of the lithosphere is estimated at 0.1 ppm, and the range commonly reported for soils is 0.01 to 0.3 ppm. The selected average for soils is 0.03 ppm which corresponds to $10^{-5.83}$ M Hg if it were all solubilized at 10% moisture (Table 1.1). Soils overlying HgS(red, cinnabar) deposits may contain as much as 40,000 ppm (Shacklette, 1965).

Mercury is generally classified as a chalcophilic element, that is, one that tends to concentrate in sulfide minerals. It is one of the most ubiquitous of all heavy metals because it vaporizes sufficiently to enter the atmosphere. It is estimated that 100,000 tons of mercury are returned to the earth annually by rainfall compared to only 10,000 tons of mercury that are mined annually for industrial purposes (Shacklette, 1965).

In recent years mercury has caused toxicity problems to man especially from fish feeding in areas where mercury contaminated wastes have been discarded. Such problems have stimulated considerable research to understand the behavior of mercury in the environment (Frieberg and Vostal, 1972; USGS, 1970; Wallace et al., 1971).

The objective of this chapter is to examine the various mineral and solution species of mercury to predict its behavior in soils and natural aqueous environments.

Mercury exists primarily in three oxidation states: 0, +1, and +2. Because of the numerous reactions of this element, the following order will be followed: (1) stability of Hg(II) minerals and complexes, (2) stability of Hg(I) minerals and complexes, (3) stability of elemental mercury, (4) stability of mercury sulfides, (5) effect of redox on stability of mercury minerals and complexes, and (6) organic mercury reactions.

21.1 STABILITY OF Hg(II) MINERALS AND COMPLEXES

The equilibrium relationships of many different forms of mercury are summarized in Table 21.1. The reactions in this table are arranged in the order in which they are used in the discussion. An excellent summary of the thermodynamic properties and reactions of mercury is given by Hepler and Olofsson (1975).

The solubilities of Hg(II) minerals are given by Reactions 1 through 8 of Table 21.1 and are plotted in Fig. 21.1. The activity ranges for the anions of these minerals were selected as typical of those expected for soils. The halide minerals $HgCl_2(c)$ and $HgBr_2(c)$ are generally more soluble than the oxides and hydroxides, which are more soluble than $HgI_2(c)$. Below pH 6 $HgCl_2(c)$ and $HgBr_2(c)$ may be more stable than the HgO minerals depending on the activities of Cl^- and Br^-. The solubility of $HgSO_4(c)$ (Reaction 8 of Table 21.1) is too high to appear in Fig. 21.1.

The oxides of Hg(II) in order of decreasing solubility include:

$Hg(OH)_2(c)$ > HgO(red, hexagonal)
> HgO(yellow, orthorhombic) > HgO(red, orthorhombic).

The solubilities of all of these minerals are very similar. In this text the designation HgO(red) refers to the orthorhombic form given by Reaction 4 of Table 21.1, which is slightly the most stable.

Of the Hg(II) minerals included in Table 21.1 $HgI_2(c)$ is the most insoluble. One might except this mineral to control Hg^{2+} activity in soils. However, if the molar ratio of Hg to I in soils is >0.5, precipitation of $HgI_2(c)$ would utilize all of the available iodide and lower the activity of I^- sufficiently that HgO(red) could form and control Hg^{2+} activity. Since most soils contain

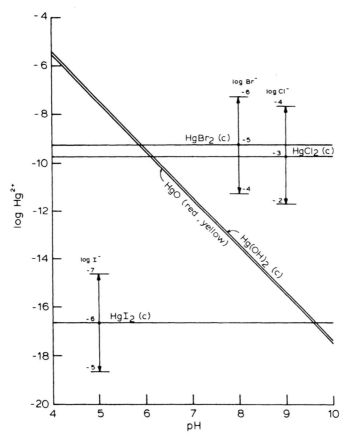

Fig. 21.1 The solubility of Hg(II) minerals.

TABLE 21.1 EQUILIBRIUM REACTIONS OF MERCURY AT 25°C

Reaction No.	Equilibrium Reaction	log $K°$
	Hg(II) Minerals	
1	$HgBr_2(c) \rightleftharpoons Hg^{2+} + 2Br^-$	−19.28
2	$HgCl_2(c) \rightleftharpoons Hg^{2+} + 2Cl^-$	−15.71
3	$HgI_2(c) \rightleftharpoons Hg^{2+} + 2I^-$	−28.62
4	$HgO(\text{red, orthorhombic}) + 2H^+ \rightleftharpoons Hg^{2+} + H_2O$	2.44
5	$HgO(\text{red, hexagonal}) + 2H^+ \rightleftharpoons Hg^{2+} + H_2O$	2.48
6	$HgO(\text{yellow, orthorhombic}) + 2H^+ \rightleftharpoons Hg^{2+} + H_2O$	2.46
7	$Hg(OH)_2(c) + 2H^+ \rightleftharpoons Hg^{2+} + 2H_2O$	2.59
8	$HgSO_4(c) \rightleftharpoons Hg^{2+} + SO_4^{2-}$	−3.34
	Hg(II) Complexes and Gases	
9	$Hg^{2+} + Br^- \rightleftharpoons HgBr^+$	9.74
10	$Hg^{2+} + 2Br^- \rightleftharpoons HgBr_2°$	17.51
11	$Hg^{2+} + 2Br^- \rightleftharpoons HgBr_2(g)$	12.37
12	$Hg^{2+} + 3Br^- \rightleftharpoons HgBr_3^-$	19.69
13	$Hg^{2+} + 4Br^- \rightleftharpoons HgBr_4^{2-}$	20.96
14	$Hg^{2+} + Cl^- \rightleftharpoons HgCl^+$	6.76
15	$Hg^{2+} + 2Cl^- \rightleftharpoons HgCl_2°$	13.16
16	$Hg^{2+} + 2Cl^- \rightleftharpoons HgCl_2(g)$	7.71
17	$Hg^{2+} + 3Cl^- \rightleftharpoons HgCl_3^-$	14.05
18	$Hg^{2+} + 4Cl^- \rightleftharpoons HgCl_4^{2-}$	14.47
19	$Hg^{2+} + I^- \rightleftharpoons HgI^+$	12.83
20	$Hg^{2+} + 2I^- \rightleftharpoons HgI_2°$	24.07
21	$Hg^{2+} + 2I^- \rightleftharpoons HgI_2(g)$	21.24
22	$Hg^{2+} + 3I^- \rightleftharpoons HgI_3^-$	27.77
23	$Hg^{2+} + 4I^- \rightleftharpoons HgI_4^{2-}$	29.79
24	$Hg^{2+} + Br^- + Cl^- \rightleftharpoons HgBrCl°$	15.93
25	$Hg^{2+} + Br^- + I^- \rightleftharpoons HgBrI°$	21.05
26	$Hg^{2+} + Br^- + 3I^- \rightleftharpoons HgBrI_3^{2-}$	28.19
27	$Hg^{2+} + 2Br^- + 2I^- \rightleftharpoons HgBr_2I_2^{2-}$	26.35
28	$Hg^{2+} + 3Br^- + I^- \rightleftharpoons HgBr_3I^{2-}$	24.08
29	$Hg^{2+} + Br^- + H_2O \rightleftharpoons HgBrOH° + H^+$	5.50
30	$Hg^{2+} + Cl^- + I^- \rightleftharpoons HgClI°$	19.29
31	$Hg^{2+} + I^- + H_2O \rightleftharpoons HgIOH° + H^+$	8.58
32	$Hg^{2+} + F^- \rightleftharpoons HgF^+$	1.56
33	$Hg^{2+} + 2F^- \rightleftharpoons HgF_2(g)$	−7.34
34	$Hg^{2+} + H_2O \rightleftharpoons HgOH^+ + H^+$	−3.40
35	$Hg^{2+} + 2H_2O \rightleftharpoons Hg(OH)_2° + 2H^+$	−6.19
36	$Hg^{2+} + 3H_2O \rightleftharpoons Hg(OH)_3^- + 3H^+$	−21.10

TABLE 21.1 (*Continued*)

Reaction No.	Equilibrium Reaction	log $K°$
37	$Hg^{2+} + Cl^- + H_2O \rightleftharpoons HgClOH° + H^+$	3.23
38	$Hg^{2+} + NO_3^- \rightleftharpoons HgNO_3^+$	0.33
39	$Hg^{2+} + 2NO_3^- \rightleftharpoons Hg(NO_3)_2°$	-1.36
40	$Hg^{2+} + SO_4^{2-} \rightleftharpoons HgSO_4°$	1.41
41	$Hg^{2+} + NH_4^+ \rightleftharpoons HgNH_3^{2+} + H^+$	-0.48
42	$Hg^{2+} + 2NH_4^+ \rightleftharpoons Hg(NH_3)_2^{2+} + 2H^+$	-1.06
43	$Hg^{2+} + 3NH_4^+ \rightleftharpoons Hg(NH_3)_3^{2+} + 3H^+$	-9.34
44	$Hg^{2+} + 4NH_4^+ \rightleftharpoons Hg(NH_3)_4^{2+} + 4H^+$	-17.83
	Hg(I) Minerals	
45	$Hg_2Br_2(c) \rightleftharpoons Hg_2^{2+} + 2Br^-$	-22.14
46	$Hg_2Cl_2(calomel) \rightleftharpoons Hg_2^{2+} + 2Cl^-$	-17.85
47	$Hg_2I_2(c) \rightleftharpoons Hg_2^{2+} + 2I^-$	-28.33
48	$Hg_2F_2(c) \rightleftharpoons Hg_2^{2+} + 2F^-$	-2.22
49	$Hg_2CO_3(c) + 2H^+ \rightleftharpoons Hg_2^{2+} + CO_2(g) + H_2O$	4.19
50	$Hg_2(OH)_2(c) + 2H^+ \rightleftharpoons Hg_2^{2+} + 2H_2O$	5.26
51	$Hg_2HPO_4(c) + H^+ \rightleftharpoons Hg_2^{2+} + H_2PO_4^-$	-5.20
52	$Hg_2SO_4(c) \rightleftharpoons Hg_2^{2+} + SO_4^{2-}$	-6.20
	Hg(I) Complexes and Gases	
53	$\frac{1}{2}Hg_2^{2+} + Br^- \rightleftharpoons HgBr(g)$	-16.80
54	$\frac{1}{2}Hg_2^{2+} + Cl^- \rightleftharpoons HgCl(g)$	-20.50
55	$\frac{1}{2}Hg_2^{2+} + F^- \rightleftharpoons HgF(g)$	-33.16
56	$\frac{1}{2}Hg_2^{2+} + I^- \rightleftharpoons HgI(g)$	-11.13
57	$Hg_2^{2+} + P_2O_7^{4-} \rightleftharpoons Hg_2P_2O_7^{2-}$	9.48
58	$Hg_2^{2+} + H_2O + P_2O_7^{4-} \rightleftharpoons Hg_2OHP_2O_7^{3-} + H^+$	1.71
59	$Hg_2^{2+} + 2P_2O_7^{4-} \rightleftharpoons Hg_2(P_2O_7)_2^{6-}$	2.00
	Hg(l), Hg°, and Redox	
60	$Hg(l) \rightleftharpoons Hg(g)$	-5.58
61	$2Hg(l) \rightleftharpoons Hg_2(g)$	-11.95
62	$Hg(l) \rightleftharpoons Hg°$	-6.52
63	$Hg(l) + H^+ + e^- \rightleftharpoons HgH(g)$	-37.39
64	$Hg(l) \rightleftharpoons \frac{1}{2}Hg_2^{2+} + e^-$	-13.46
65	$Hg(l) \rightleftharpoons Hg^{2+} + 2e^-$	-28.86
66	$Hg° \rightleftharpoons Hg(g)$	0.94
67	$Hg° \rightleftharpoons \frac{1}{2}Hg_2^{2+} + e^-$	-6.93
68	$Hg° \rightleftharpoons Hg^{2+} + 2e^-$	-22.34
69	$\frac{1}{2}Hg_2^{2+} \rightleftharpoons Hg^{2+} + e^-$	-15.40

TABLE 21.1 (*Continued*)

Reaction No.	Equilibrium Reaction	log $K°$
	Sulfide Minerals and Complexes	
70	$Hg_2S(c) \rightleftharpoons Hg_2^{2+} + S^{2-}$	-54.77
71	α-HgS(red, cinnabar) $\rightleftharpoons Hg^{2+} + S^{2-}$	-52.03
72	β-HgS(black, metacinnabar) $\rightleftharpoons Hg^{2+} + S^{2-}$	-51.66
73	$Hg^{2+} + 2H^+ + 2S^{2-} \rightleftharpoons Hg(HS)_2^\circ$	63.58
74	$Hg^{2+} + 2S^{2-} \rightleftharpoons HgS_2^{2-}$	51.00
	Organics	
75	$Hg^{2+} + 2CH_4(g) \rightleftharpoons Hg(CH_3)_2(g) + 2H^+$	-14.53
76	$Hg^{2+} + CH_4(g) + NH_4^+ \rightleftharpoons HgCH_3NH_3^{2+} + 3H^+ + 2e^-$	-17.79
77	$Hg^{2+} + 2CH_4(g) + 2NH_4^+ \rightleftharpoons Hg(CH_3NH_2)_2^{2+} + 6H^+ + 4e^-$	-35.04
78	$Hg^{2+} + CH_4(g) + NH_4^+ + Cl^- \rightleftharpoons HgClCH_3NH_4^+ + 3H^+ + 2e^-$	-10.88
79	$Hg^{2+} + CH_3COO^- \rightleftharpoons Hg(CH_3COO)^+$	3.09
80	$Hg_2^{2+} + 2C_2O_4^{2-} \rightleftharpoons Hg_2(C_2O_4)_2^{2-}$	6.98
81	$Hg_2C_2O_4(c) \rightleftharpoons Hg_2^{2+} + C_2O_4^{2-}$	-12.77

more iodine than mercury (5 ppm of iodine compared to 0.03 ppm of mercury, Table 1.1) only polluted soils whose molar ratio of mercury to iodine is > 0.5 or when extremely low activities of I^- are present can HgO(red) possibly form (Fig. 21.1).

Since the total mercury content of soils is generally very low, let us examine the complexes of Hg(II) to see if they are sufficiently stable to prevent the formation of either $HgI_2(c)$ or HgO(red). The complexes of Hg(II) are defined by Reactions 9 through 44 of Table 21.1. The halide complexes are plotted in Fig. 21.2 and 21.3 when equilibrium is maintained with $HgI_2(c)$ at 10^{-6} I^- which fixed Hg^{2+} at $10^{-16.62}$ M. These two figures show the numerous Hg(II) complexes that form with the halide anions. The most stable of these complexes is HgI_2°, which is present at $10^{-4.55}$ M. In the presence of $HgI_2(c)$, the activity of this complex is fixed for all combinations of Hg^{2+} and I^- as shown by

$$
\begin{array}{rcl}
 & & \log K° \\
HgI_2(c) \rightleftharpoons Hg^{2+} + 2I^- & & -28.62 \\
Hg^{2+} + 2I^- \rightleftharpoons HgI_2^\circ & & 24.07 \\
\hline
HgI_2(c) \rightleftharpoons HgI_2^\circ & & -4.55 \quad (21.1)
\end{array}
$$

STABILITY OF Hg(II) MINERALS AND COMPLEXES

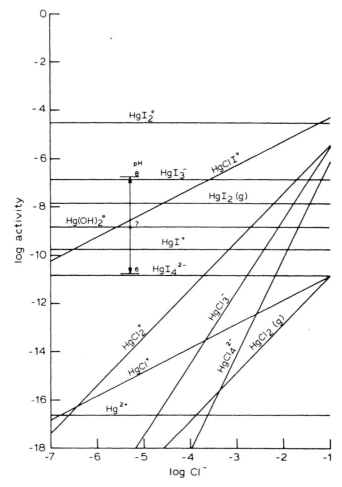

Fig. 21.2 Chloride and iodide complexes with Hg(II) in equilibrium with $HgI_2(c)$ at 10^{-6} M I^- which fixes Hg^{2+} at $10^{-16.62}$ M. The $Hg(OH)_2^0$ complex is included for comparison with Fig. 21.5.

This means that $HgI_2(c)$ can only form in soils containing sufficient mercury to first satisfy the equilibrium requirements of the HgI_2^0 complex. For a soil at 10% moisture this is equivalent to 0.58 ppm of mercury (dry weight basis), and only in soils containing more than this amount of mercury is it possible for $HgI_2(c)$ to form. Similarly, it would require > 0.71 ppm of iodine (dry weight basis). Other Hg(II) halide complexes may also contribute to total soluble mercury if the halide activities are sufficiently high (Fig. 21.2 and 21.3). The most likely species include $HgClI^0$, $HgBrI^0$, and $HgBr_2^0$.

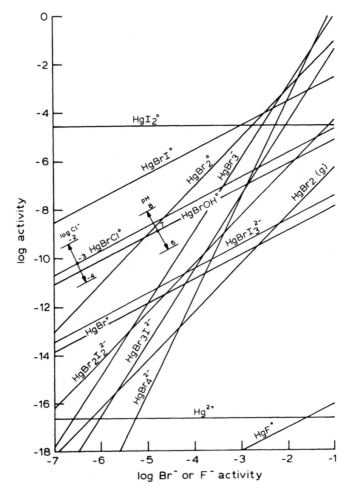

Fig. 21.3 Mercuric complexes of bromide and fluoride in equilibrium with $HgI_2(c)$ and $10^{-6}\,M\,I^-$ which fixes Hg^{2+} at $10^{-16.62}\,M$. The $HgI_2°$ complex is included for comparison to Fig. 21.2.

Even though the lines in Fig. 21.2 and 21.3 were drawn for equilibrium with $HgI_2(c)$ at $10^{-6}\,M\,I^-$, which fixes Hg^{2+} at $10^{-16.62}\,M$, they can easily be adjusted to reflect other conditions as well. For example, if I^- activity were 10^{-7} rather than $10^{-6}\,M$, the accompanying Hg^{2+} activity would increase 2 log units to $10^{-14.62}\,M$. All lines in these two figures would then shift up two log units due to change in Hg^{2+} activity and down by one log unit for each atom of I^- present in the complex. The general expression for such shifts is given by the equation:

$$\Delta \log a_i = (n - 2)\Delta \log I^- \qquad (21.2)$$

where a_i is the activity of any Hg(II) halide complex shown in Fig. 21.2 or 21.3 and n is the number of iodine atoms present in that complex. For the example cited, the HgBrI° line in Fig. 21.3 would shift by one log unit:

$$\Delta \log a_i = (1 - 2)(-1) = 1 \qquad (21.3)$$

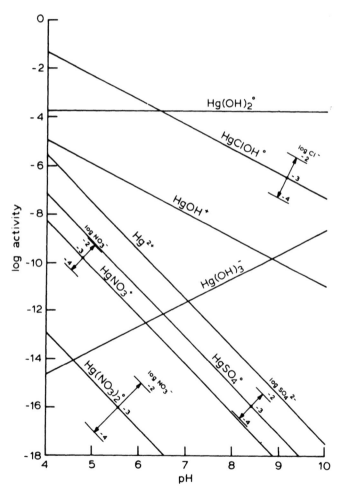

Fig. 21.4 Complexes of nitrate, sulfate, and hydrolysis species of Hg(II) in equilibrium with HgO(red).

Thus Eq. 21.2 can be used in connection with Fig. 21.2 and 21.3 to estimate the halide complexes in equilibrium with $HgI_2(c)$ at all combinations of halide or Hg^{2+} activities.

Now let us consider the situation in which I^- activity lowers sufficiently that $HgI_2(c)$ cannot precipitate, but HgO(red) can (see Fig. 21.1). The hydrolysis and other complexes of Hg(II) in equilibrium with HgO(red) are shown in Fig. 21.4. Before HgO(red) can form, $Hg(OH)_2^0$ must reach

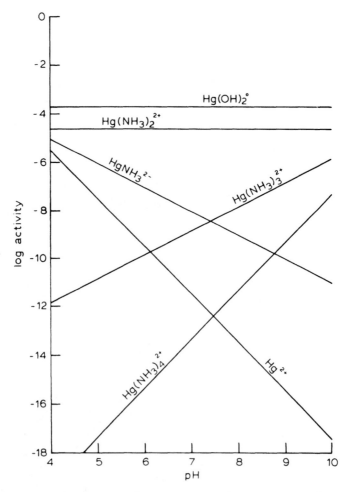

Fig. 21.5 Ammonia complexes of Hg^{2+} in equilibrium with HgO(red) and 10^{-3} M NH_4^+ compared to $Hg(OH)_2^0$.

STABILITY OF Hg(I) MINERALS AND COMPLEXES 353

$10^{-3.75}$ M as shown by the relationship:

	log $K°$
$Hg^{2+} + 2H_2O \rightleftharpoons Hg(OH)_2° + 2H^+$	-6.19
$HgO(red) + 2H^+ \rightleftharpoons Hg^{2+} + H_2O$	2.44
$HgO(red) + H_2O \rightleftharpoons Hg(OH)_2°$	-3.75 (21.4)

This means that $Hg(OH)_2°$ must reach $10^{-3.75}$ M before HgO(red) can precipitate. For a soil at 10% moisture this requires a total soil mercury content of >3.61 ppm of mercury (dry weight basis). If other halide complexes contribute significantly to soluble mercury, an even higher total mercury content would be required (Fig. 21.2 through 21.5). For these reasons HgO(red) is not expected to form in soils unless they have been highly contaminated with mercury.

The ammonia complexes of Hg(II) are plotted in Fig. 21.5 on the basis that equilibrium is attained with HgO(red) and with NH_4^+ fixed at 10^{-3} M. The $Hg(NH_3)_2^{2+}$ complex is the most stable solution species and falls at $10^{-4.62}$ M. The $Hg(OH)_2°$ species was included for comparison. The ammonia complexes in this figure shift with changes in NH_4^+. A log unit change in NH_4^+ causes a one log unit shift for each atom of NH_3 present in the complex. The activities of $Hg(NH_3)_2^{2+}$ and $Hg(OH)_2°$ become equal when NH_4^+ reaches $10^{-2.56}$ M. Since NH_4^+ activity in soils fluctuates widely due to the mineralization of organic matter and its oxidation to NO_3^- (Chapter 16), the Hg(II) ammonia complexes in Fig. 21.5 will rise and fall accordingly.

21.2 STABILITY OF Hg(I) MINERALS AND COMPLEXES

The stabilities of Hg(I) minerals are given by Reactions 45 through 52 of Table 21.1 and are plotted in Fig. 21.6. Again the anion activities selected to develop this plot are typical of those found in many soils.

Of the Hg(I) minerals, $Hg_2I_2(c)$ is the most stable. A 10^{-6} M activity of I^- depresses Hg_2^{2+} to $10^{-16.33}$ M. The minerals $Hg_2Br_2(c)$ and $Hg_2Cl_2(c)$ support $10^{-12.14}$ and $10^{-11.85}$ M Hg_2^{2+}, respectively, when $Br^- = 10^{-5}$ M and $Cl^- = 10^{-3}$ M. A one log unit change in halide activity causes a 2 log unit shift in Hg_2^{2+}. Other Hg(I) minerals in order of decreasing solubility include $Hg_2SO_4(c) > Hg_2CO_3(c) > Hg_2(OH)_2(c) > Hg_2HPO_4(c)$. The solubility of $Hg_2HPO_4(c)$ shifts with changes in phosphate activity. The minerals $Hg_2SO_4(c)$, $Hg_2(OH)_2(c)$, and $Hg_2CO_3(c)$ are too soluble to precipitate in soils. Decreasing phosphate solubility below that of β-TCP allows

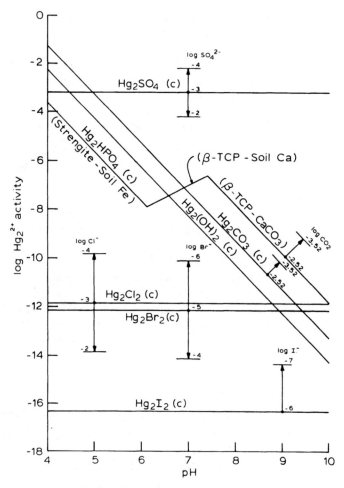

Fig. 21.6 The solubility of Hg(I) minerals.

Hg$_2$HPO$_4$(c) to become even more soluble, and less likely to precipitate in soils. Of the minerals depicted in Fig. 21.6, only the halides appear as likely precipitants of mercury.

Solution complexes of Hg(I) are sufficiently weak that Hg$_2^{2+}$ is the major Hg(I) species in solution. Gaseous halide complexes with Hg(I) given by Reactions 53 through 56 of Table 21.1 are also highly unstable in soils.

Pyrophosphate complexes of Hg(I) are given by Reactions 57 through 59 of Table 21.1. In soils, P$_2$O$_7^{4-}$ is generally near $10^{-9.7}$ M if pyrophosphate minerals are present (Chapter 12). Even at this activity of P$_2$O$_7^{4-}$, the mercury

STABILITY OF ELEMENTAL MERCURY

complexes are unimportant because Hg_2^{2+} is generally $< 10^{-16}$ M. Eventual hydrolysis of pyrophosphate to orthophosphate makes these complexes even less important.

The halide minerals shown in Fig. 21.6 must be considered possible stable phases. The mineral $Hg_2I_2(c)$ is the most stable, but appropriate activities of I^-, Br^-, and Cl^- could cause the other halides to become stable. Because of the high stability of the Hg(II) complexes (Section 21.1), these species must be considered even in the presence of Hg(I) minerals.

21.3 STABILITY OF ELEMENTAL MERCURY

Mercury exists as a liquid at ordinary temperatures and pressures. Equilibrium reactions involving elemental mercury and its different oxidation states are summarized by Reactions 60 through 69 of Table 21.1.

The vapor pressure of Hg(l) is given by

$$Hg(l) \rightleftharpoons Hg(g) \qquad \log K^\circ = -5.58 \tag{21.5}$$

which indicates that liquid mercury maintains an equilibrium vapor pressure of $10^{-5.58}$ atm. This is approximately 2×10^{-3} mm of mercury. Although small, it is still sufficient to account for a significant cycling of mercury into the atmosphere.

The solubility of Hg(l) in water is given by the reaction

$$Hg(l) \rightleftharpoons Hg^\circ \qquad \log K^\circ = -6.52 \tag{21.6}$$

which indicates that Hg(l) will dissolve in water to the extent of $10^{-6.52}$ M Hg°. Aqueous solutions that are undersaturated with Hg(l) will have correspondingly lower Hg° activities.

Liquid mercury oxidizes to give Hg_2^{2+} according to the reaction

$$Hg(l) \rightleftharpoons \tfrac{1}{2}Hg_2^{2+} + e^- \qquad \log K^\circ = -13.42 \tag{21.7}$$

for which

$$pe = 13.46 + \tfrac{1}{2} \log Hg_2^{2+} \tag{21.8}$$

or

$$\log Hg_2^{2+} = -26.92 + 2pe \tag{21.9}$$

Oxidation of Hg_2^{2+} to Hg^{2+} is expressed by the reaction

$$\tfrac{1}{2}Hg_2^{2+} \rightleftharpoons Hg^{2+} + e^- \qquad \log K^\circ = -15.40 \tag{21.10}$$

for which

$$\log Hg^{2+} = -15.40 + pe + \tfrac{1}{2} \log Hg_2^{2+} \tag{21.11}$$

Obviously redox relationships must be considered in determining the stability of elemental mercury relative to other mineral forms of this element.

The stabilities of Hg(I) halides relative to Hg(l) are plotted in Fig. 21.7 as a function of pe. When Cl^- is 10^{-3} M, Br^- is 10^{-5} M, and I^- is 10^{-6} M, $Hg_2I_2(c)$ is the most stable mineral above $pe = 5.30$ (equivalent to $pe + pH$ of 12.30 at pH 7.0). Below this pe, Hg(l) is the stable phase. The pe at which this transition occurs increases one unit for each log unit increase in I^-. The activity of Cl^- or Br^- must decrease or that of I^- decrease for Hg_2Cl_2 (calomel) or $Hg_2Br_2(c)$ to be stable rather than $Hg_2I_2(c)$.

Since the halide complexes of Hg(II) are very stable, they must be examined to see if they can prevent the precipitation of $Hg_2I_2(c)$ even though it appears from Fig. 21.7 that this mineral could very likely form.

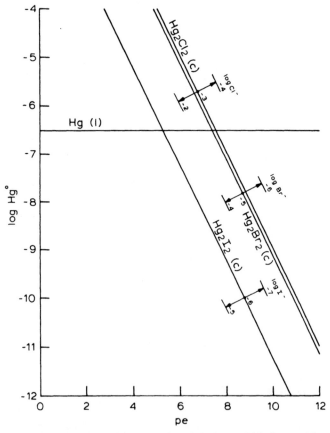

Fig. 21.7 The stability of Hg(I) halide minerals and the pe at which they transform to Hg(l).

STABILITY OF ELEMENTAL MERCURY

As shown by Fig. 21.2 and 21.3 the major solution species of mercury, where $Hg_2I_2(c)$ can be expected to form, include the following:

$$[Hg]_{Total} = [HgI_2^\circ] + [HgBrI^\circ] + [HgClI^\circ] \quad (21.12)$$

Since only uncharged species are involved, activity coefficients are unity; therefore, concentrations can be equated to activities. Expressing each term on the right side of Eq. 21.12 by its equilibrium constant expression gives

$$[Hg]_{Total} = 10^{24.07}(Hg^{2+})(I^-)^2 + 10^{21.05}(Hg^{2+})(Br^-)(I^-) \\ + 10^{19.29}(Hg^{2+})(Cl^-)(I^-) \quad (21.13)$$

Substitution of the reference levels for the halide anions into Eq. 21.13 permits the mole fraction of each contributing species to be estimated. For example,

$$\frac{[HgI_2^\circ]}{[Hg]_{Total}} = \frac{10^{12.07}}{10^{12.07} + 10^{10.05} + 10^{10.29}} = 0.975 \quad (21.14)$$

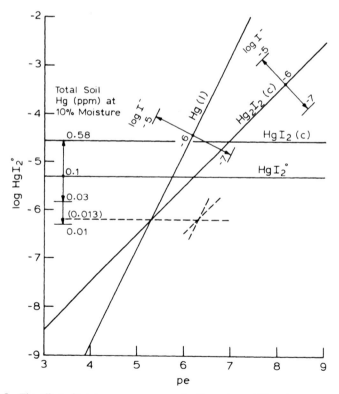

Fig. 21.8 The effect of total mercury content of soils on the stability of $Hg_2I_2(c)$ in soils.

This means that 97.5% of the mercury in solution would be present as HgI_2°. Similar expressions for the other species show them to be 1.6% for $HgClI^\circ$ and 0.9% $HgBrI^\circ$.

Knowing that total mercury in solution is 97.5% HgI_2°, we can plot the relationship shown in Fig. 21.8 where the HgI_2° activities in equilibrium with total soluble mercury, Hg(l), and Hg_2I_2(c) can be compared. From this plot it can be determined that soils containing less than 0.013 ppm of total mercury (dry weight basis) cannot precipitate Hg_2I_2(c) but will form Hg(l) as the redox drops below pe of 5.30. Soils having greater than 0.013 ppm of total mercury can precipitate Hg_2I_2(c) over a narrow redox range. For example, a soil containing 0.1 ppm of mercury could have Hg_2I_2(c) present in the pe range of 6.19 to 5.30, whereas soils with >0.58 ppm of total mercury would precipitate HgI_2(c) above pe = 6.95 and Hg_2I_2(c) between pe 6.95 and 5.30.

Although Fig. 21.8 is drawn for 10^{-6} M I^-, it can be used for any I^- activity. Decreasing I^- by one log unit shifts the Hg(l) and Hg_2I_2(c) lines one pe unit to the right, but their relative positions remain unchanged as shown by the short dashed lines intersecting along $10^{-6.19}$ M HgI_2° which equates to 0.013 ppm of soil Hg at 10% moisture. Thus Fig. 21.8 can be used to determine whether or not HgI_2(c) and Hg_2I_2(c) can form, and the range of their stabilities as well as the redox at which Hg(l) can form depending on the activity of I^- and the total soil mercury.

21.4 SOLUBILITY OF MERCURY SULFIDES IN SOILS

As soils containing sulfate are reduced, S^{2-} increases and permits the precipitation of most metals as sulfides (Chapter 17). For mercury the relationships are summarized in Fig. 21.9 when SO_4^{2-} is 10^{-3} M. In the region at the top of this plot Hg(l) is the stable phase, but as the pe + pH drops to line a, Hg(l) is transformed to Hg_2S(c) according to the reaction

		log K°
$2\,Hg(l) \rightleftharpoons Hg_2^{2+} + 2e^-$		2(−13.46)
$Hg_2^{2+} + S^{2-} \rightleftharpoons Hg_2S(c)$		54.77
$SO_4^{2-} + 8e^- + 8H^+ \rightleftharpoons S^{2-} + 4H_2O$		20.74
$2\,Hg(l) + SO_4^{2-} + 6e^- + 8H^+ \rightleftharpoons Hg_2S(c) + 4H_2O$		48.59

(21.15)

from which

$$pe + pH = 8.10 + \tfrac{1}{6}\log SO_4^{2-} - \tfrac{1}{3}pH \quad (21.16)$$

SUMMARY REDOX DIAGRAM FOR MERCURY

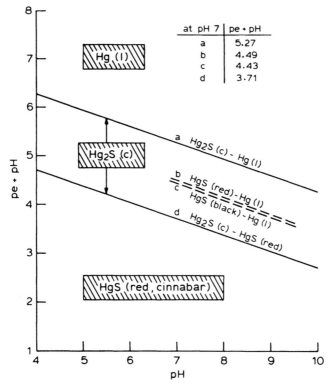

Fig. 21.9 The redox at which various mercury sulfides form and are stable when SO_4^{2-} is 10^{-3} M.

At pH 7 this transformation occurs at $pe + pH$ of 5.27, and $Hg_2S(c)$ remains the stable mineral phase from line a to line d, which at pH 7 corresponds to a $pe + pH$ of 3.71. Below line d, HgS(red, cinnabar) is the stable mercury phase. Had $Hg_2S(c)$ not formed, HgS(red) would have formed at line b. The mineral HgS(black) is always metastable to HgS(red).

Figure 21.9 is useful for showing the precise redox at which the various mercury sulfides can form and the redox range in which they are stable.

21.5 SUMMARY REDOX DIAGRAM FOR MERCURY

A summary diagram showing the stable phases and solution species of mercury as they are affected by redox for a typical soil is shown in Fig. 21.10.

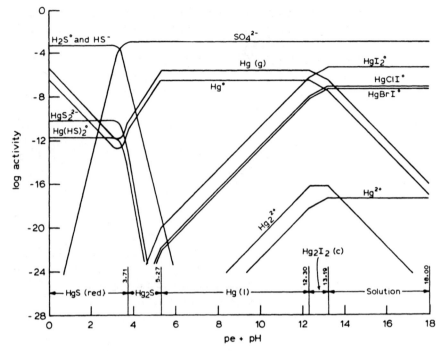

Fig. 21.10 The effect of redox on the stability of mercury in a soil of pH 7.0 containing 0.1 ppm of mercury when total sulfur and Cl-activity in solution = 10^{-3}, Br$^-$ = 10^{-5}, and I$^-$ = 10^{-6} M.

For this example, the following soil conditions were stipulated:

1. Total Hg = 0.1 ppm
2. pH   = 7.0
3. Cl$^-$ = 10^{-3} M
4. Br$^-$ = 10^{-5} M
5. I$^-$ = 10^{-6} M
6. Total S = 10^{-3} M

Let us examine the many possible solid, liquid, solution, and gaseous species of inorganic mercury over the $pe + pH$ range of 0 to 20 and express their equilibrium relationships in quantative terms.

As seen from Fig. 21.10 all of the mercury in this soil is in solution above $pe + pH$ of 13.19. The major solution species consist of $10^{-5.32}$ M HgI$_2^0$, $10^{-7.10}$ M HgClI^0, and $10^{-7.34}$ M HgBrI^0. These represent 97.45, 1.62, and 0.93% respectively, of the $10^{-5.31}$ M total mercury (0.1 ppm dry soil basis dissolved at 10% moisture). These species are stable and control the various

SUMMARY REDOX DIAGRAM FOR MERCURY

other mercury species in solution in the redox range above $pe + \mathrm{pH}$ 13.19. The distribution of mercury in these species can be calculated from Eq. 21.13 and 21.14.

As $pe + \mathrm{pH}$ lowers below 13.19, $Hg_2I_2(c)$ precipitates and remains a stable phase until $pe + \mathrm{pH}$ drops to 12.30. The corresponding pe range is 6.19 to 5.30 (Fig. 21.8). In this redox range the solution species change to reflect equilibrium with solid phase $Hg_2I_2(c)$.

At $pe + \mathrm{pH}$ of 12.30 elemental $Hg(l)$ forms as a stable phase (Fig. 21.8). Since $Hg(l)$ is accompanied by $10^{-6.52}$ M $Hg°$ (Eq. 21.6), 6.2% of the total Hg at $10^{-5.31}$ M will be present as $Hg°$ and 93.8% as $Hg(l)$. The vapor pressure of mercury expressed as $Hg(g)$ is maintained at $10^{-5.58}$ atm of mercury vapor (Eq. 21.5). As reduction continues in the presence of $Hg(l)$, both the $Hg(II)$ and the $Hg(I)$ solution species decrease.

At $pe + \mathrm{pH}$ of 5.27, $Hg(l)$ becomes unstable and transforms into $Hg_2S(c)$ according to Reaction 21.15. With further reduction $Hg(g)$ and $Hg°$ decline. At $pe + \mathrm{pH}$ of 3.71 $Hg_2S(c)$ becomes metastable with respect to HgS(red, cinnabar) which precipitates as the stable mercury phase. The latter transformation occurs at the redox represented by line d in Fig. 21.9.

In the example problem, total sulfur in solution was fixed at 10^{-3} M. The reduction of sulfate can be represented by the reaction

$$SO_4^{2-} + 8e^- + 10H^+ \rightleftharpoons H_2S° + 4H_2O \qquad \log K° = 40.67$$

(21.17)

for which

$$\log H_2S° = 40.67 + \log SO_4^{2-} - 8(pe + \mathrm{pH}) - 2\mathrm{pH} \qquad (21.18)$$

From this equation a soil of pH 7 will have equal $H_2S°$ and SO_4^{2-} at $pe + \mathrm{pH}$ 3.334. Above this redox SO_4^{2-} is the predominant sulfur species, whereas below it, SO_4^{2-} is sharply reduced to $H_2S°$ (Eq. 21.18) and to HS^-, both of which are equal at pH 7.02 (Reaction 26 of Table 17.2). These stability relationships are shown by the SO_4^{2-}, $H_2S°$, and HS^- lines in Fig. 21.10. Their affect on the mercury species in solution below $pe + \mathrm{pH}$ of 3.334 can be seen from Fig. 21.10. The complexes HgS_2^{2-} and $Hg(HS)_2°$ are fixed at $10^{-10.25}$ M and $10^{-11.67}$ M, respectively, by $H_2S°$ and HS^- which are each present at $10^{-3.30}$ M. Both $Hg°$ and $Hg(g)$ increase with decrease in redox. As shown previously (Fig. 17.3) S(rhombic) cannot form as long as total sulfur is $< 10^{-1.55}$ M.

The above example depicted by Fig. 21.10 demonstrates the usefulness of equilibrium relationships in determining the solubility and stability relationships of mercury in soils. Similar procedures can be used to examine the mercury status of soils having different mercury contents and other soil characteristics.

21.6 ORGANIC MERCURY REACTIONS

Mercury is known to form a wide range of organometallic complexes in soils and sediments. These reactions result from the tendency of mercury to form covalent bonds with various organic groups. Since the mechanism by which mercury is complexed by organic groups and concentrated in biological systems is now known (Saha and McKinley, 1973; Rogers, 1976, 1977; Summers and Silver, 1978). No attempt will be made here to develop the stability relationships of these complexes even though they are undoubtedly very important in soils. Instead the reader is referred to the detailed review of Hepler and Olofsson (1975) who documented much of what is presently known about many of the known organic mercury complexes.

REFERENCES

Frieberg, L. T. and J. J. Vostal. 1972. Mercury in the Environment. CRC Press, Inc., Cleveland, Oh.

Hepler, L. G. and G. Olofsson. 1975. Mercury: thermodynamic properties, chemical equilibria, and standard potentials. Chem. Rev. 75:585–602.

Rogers, R. D. 1976. Methlation of mercury in agricultural soils. J. Environ. Qual. 5:454–458.

Rogers, R. D. 1977. A biological methylation of mercury in soil. J. Environ. Qual. 6:463–467.

Saha, J. G. and K. S. McKinlay. 1973. Use of mercury in agriculture and its relationship to environmental pollution. Toxicol. Environ. Chem. Rev. 1:271–290.

Shacklette, H. T. 1965. Bryophytes associated with mineral deposits and solutions in Alaska: U.S. Geol. Survey Bull. 1198-C.

Summers, A. O. and S. Silver. 1978. Microbial transformations of metals. Ann. Rev. Microbiol. 32:637–672.

United States Geological Survey. 1970. Mercury in the environment. U.S. Geol. Survey Prof. Paper 713.

Wallace, R. A., William Fulkerson, W. D. Shults, and W. S. Lyons. 1971. Mercury in the environment: The human element. ORNL EP-1.

Wood, J. M. 1975. Biological cycles for elements in the environment. Die Naturwissensch. 62:357–364.

PROBLEMS

21.1 Estimate the limiting halide activities and the minimum mercury content required in a well-oxidized soil of pH 7 in order that $HgCl_2(c)$ can be a stable mineral.

21.2 If $CaHPO_4 \cdot 2H_2O$(brushite) were present in a soil of pH 6.5, what activity of Hg_2^{2+} could be maintained by $Hg_2HPO_4(c)$?

21.3 Under what conditions can $HgI_2(c)$ and $Hg_2I_2(c)$ coexist in a soil of pH 7.0?

21.4 A soil attains equilibrium with Hg(l) and Hg_2Cl_2(calomel) in 0.01 M KCl solution. What can you conclude about the following?
   a. Activities of the halide anions.
   b. The redox of this suspension.
   c. The solubility of mercury in this suspension.

21.5 A soil containing 2 ppm of total mercury is wetted to 10% moisture and shows an I^- activity of $10^{-7} M$. Under what conditions, if any, is Hg_2I_2(c) stable in this soil?

21.6 Develop the equation for line d in Fig. 21.9 and explain why HgS(black) is never a stable phase in soils at 25°C.

21.7 Explain how the stability diagram in Fig. 21.10 would change for each of the following modifications of the example soil:
   a. 1.0 ppm of total mercury.
   b. 0.01 ppm of total mercury.
   c. $10^{-7} M\ I^-$.
   d. $10^{-4} M$ total sulfur in solution.

TWENTY-TWO

MOLYBDENUM

The molybdenum content of the lithosphere is estimated at 2.3 ppm, whereas the common range for soils is 0.2 to 5 ppm. An accepted average for soils is 2 ppm, which provides $10^{-3.68}$ M Mo if it were to dissolve completely in the soil solution at 10% moisture.

Molybdenum is an essential element for both plants and animals and has been widely studied. Although this element is not highly toxic, it has been the subject of considerable investigation from an environmental standpoint (Chappell and Petersen, 1976, 1977). Molybdenum reactions have been described as some of the most complex of any chemical element (Cotton and Wilkinson, 1972). It forms many complexes and polymerized species at high concentrations, but in solutions of $< 10^{-4}$ M the polymeric forms are not of great importance (Jenkins and Wain, 1963).

Thermodynamic constants for many of the recognized chemical species of molybdenum have not been critically measured and only estimated values are available in the literature (Titley, 1963). The development in this chapter is limited to those species whose thermodynamic properties have been fairly well established. As future measurements become available, they can be included to expand the development given here.

The objective of this chapter is to show how solubility data for molybdenum that is presently available can be used to predict the chemical reactions of molybdenum in soils. The discussion is arranged as follows: (1) molybdenum species in solution, (2) the stability of molybdenum minerals, and (3) the effect of redox on molybdenum solubility.

22.1 MOLYBDENUM SPECIES IN SOLUTION

The dissociation constants of molybdic acid and the hydrolysis reactions of MoO_4^{2-} are given by Reactions 1 through 6 of Table 22.1. The aqueous solution species are plotted in Fig. 22.1 in equilibrium with $H_2MoO_4(c)$. Above pH 4.24 MoO_4^{2-} is the major solution species. The solution species generally decrease in the order $MoO_4^{2-} > HMoO_4^- > H_2MoO_4^0 > MoO_2(OH)^+ > MoO_2^{2+}$. In the pH range of 3 to 5 the first three species contribute significantly to total molybdenum in solution. The latter two ions can generally be ignored in soils.

The molybdenum levels in soil solution are generally much lower than those depicted here, yet this graph can still be used to show the relative abundance of the various ions regardless of total molybdenum or pH. For example, if the MoO_4^{2-} activity in a soil at pH 5 drops from $10^{-3.37}$ M to $10^{-6.37}$ M, all of the solution species decrease by 3 log units, however their ratios remain fixed. Thus Fig. 22.1 can be used to obtain the activities of all species when pH and the activity of any species is known.

TABLE 22.1 EQUILIBRIUM REACTIONS OF MOLYBDENUM AT 25°C

Reaction No.	Equilibrium Reaction	log $K°$
	Molybdic Acid Dissociation	
1	$H_2MoO_4^\circ \rightleftharpoons H^+ + HMoO_4^-$	−4.00
2	$HMoO_4^- \rightleftharpoons H^+ + MoO_4^{2-}$	−4.24
3	$H_2MoO_4^\circ \rightleftharpoons 2H^+ + MoO_4^{2-}$	−8.24
4	$H_2MoO_4(c) \rightleftharpoons 2H^+ + MoO_4^{2-}$	−13.37
	Hydrolysis Reactions	
5	$MoO_4^{2-} + 3H^+ \rightleftharpoons MoO_2(OH)^+ + H_2O$	8.20
6	$MoO_4^{2-} + 4H^+ \rightleftharpoons MoO_2^{2+} + 2H_2O$	8.65
	Oxides and Mo(c)	
7	$MoO_3(molybdite) + H_2O \rightleftharpoons MoO_4^{2-} + 2H^+$	−12.10
8	$MoO_2(c) + 2H_2O \rightleftharpoons MoO_4^{2-} + 2e^- + 4H^+$	−30.02
9	$Mo(c) + 4H_2O \rightleftharpoons MoO_4^{2-} + 6e^- + 8H^+$	−19.73
	Metal Molybdates	
10	$Ag_2MoO_4(c) \rightleftharpoons 2Ag^+ + MoO_4^{2-}$	−11.55
11	$CaMoO_4(c) \rightleftharpoons Ca^{2+} + MoO_4^{2-}$	−7.94
12	$CuMoO_4(c) \rightleftharpoons Cu^{2+} + MoO_4^{2-}$	−6.48
13	$FeMoO_4(c) \rightleftharpoons Fe^{2+} + MoO_4^{2-}$	−7.70
14	$MgMoO_4(c) \rightleftharpoons Mg^{2+} + MoO_4^{2-}$	−0.62
15	$MnMoO_4(c) \rightleftharpoons Mn^{2+} + MoO_4^{2-}$	−4.13
16	$PbMoO_4(wulfenite) \rightleftharpoons Pb^{2+} + MoO_4^{2-}$	−16.04
17	$ZnMoO_4(c) \rightleftharpoons Zn^{2+} + MoO_4^{2-}$	−4.94
	Soil-Mo and Sulfides	
18	soil-Mo $\rightleftharpoons MoO_4^{2-} + 0.8H^+$	−12.40*
19	$MoS_2(molybdenite) + 4H_2O \rightleftharpoons MoO_4^{2-} + 2S^{2-} + 2e^- + 8H^+$	−96.49

* Estimated from Vlek and Lindsay (1977).

STABILITY OF MOLYBDENUM MINERALS IN SOILS

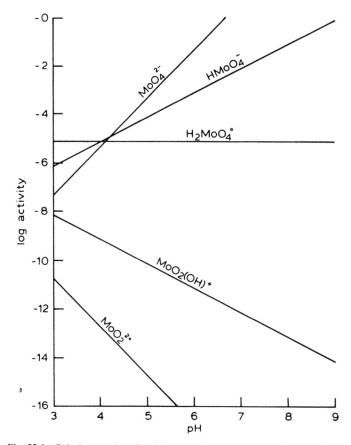

Fig. 22.1 Solution species of molybdenum in equilibrium with $H_2MoO_4(c)$.

Thermodynamic constants for Mo^{3+} and MoO_2^+ have only been estimated (Titley, 1963), so they are not included in this development. Both species can be expected to occur under quite reduced conditions and deserve further examination.

22.2 STABILITY OF MOLYBDENUM MINERALS IN SOILS

The solubility relationships of several molybdenum minerals are given by Reactions 7 through 19 of Table 22.1. Minerals whose solubility is not affected by redox are plotted in Fig. 22.2. The selection of solid phases used to fix the activities of the various cations has been documented in earlier chapters.

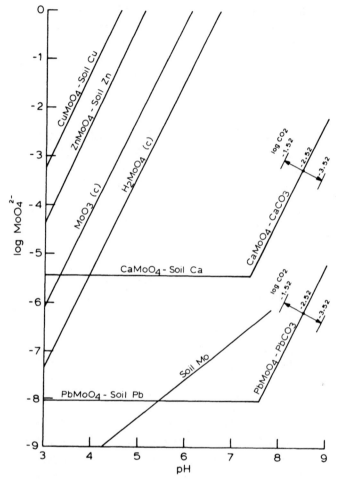

Fig. 22.2 The solubility of several molybdenum minerals which are not affected by redox when equilibrium is established with various soil cations (see text).

The equilibrium reactions of these controlling phases were also given earlier (Reactions 53 through 74 of Table 17.2).

Solubility of the minerals in Fig. 22.2 generally decrease in the order:

$CuMoO_4(c) > ZnMoO_4(c) > MoO_3$(molybdite)
$> H_2MoO_4$(molybdic acid) $> CaMoO_4(c) > PbMoO_4$(wulfenite).

At high pH the latter two minerals are affected by CO_2 because $CaCO_3$ (calcite) and $PbCO_3$(cerussite) are used to control Ca^{2+} and Pb^{2+} activities, respectively. Of the minerals shown here, wulfenite is the most stable.

The soil-Mo line in Fig. 22.2 was developed from the experimental findings of Vlek and Lindsay (1977). The specific data used for this purpose are the dark data points reported in their Fig. 2 showing MoO_4^{2-} activities of Colorado soils that did not receive molybdenum amendments. The equation representing these data can be expressed by the reaction

$$\text{soil-Mo} \rightleftharpoons MoO_4^{2-} + 0.8\,H^+ \quad \log K° = -12.40 \quad (22.1)$$

This expression of MoO_4^{2-} solubility in soils covers the pH range of 5.3 to 8.0. The relationship represented by Eq. 22.1 is an empirical expression of MoO_4^{2-} solubility based on a limited number of soils. Further studies are necessary to explore the relationship in greater detail.

The soil-Mo line is very close to that of $PbMoO_4$-soil-Pb. Since the soil-Pb line can be expected to shift as phosphate and other parameters shift (Chapter 20), these in turn will affect the solubility of wulfenite.

Vlek and Lindsay (1977) equilibrated soil suspensions in the presence and absence of wulfenite and found that wulfenite generally increased the solubility of MoO_4^{2-} in most of the soils studied. They concluded that their soils were initially undersaturated with respect to this molybdenum mineral. Further careful investigation of these solubility relationships is needed.

Iron oxides are known to react with molybdate, and $Fe_2(MoO_4)_3$ (ferrimolybdite) had been suggested as the likely reaction product (Jones, 1957; Sarafian and Furbish, 1965). Although Titley (1963) estimated thermodynamic constants for this compound, no reliable measurements are yet available. It would appear that both ferrimolybdite and Mo_3O_8(islemannite) could be important molybdenum minerals in some soils. Determination of the thermodynamic solubilities of these minerals is greatly needed.

22.3 THE EFFECT OF REDOX ON MOLYBDENUM SOLUBILITY

For the mineral $MoO_2(c)$, molybdenum is present in +4 oxidation state (Reaction 8, Table 22.1) and in Mo(c) the oxidation state is zero. In addition to these minerals molybdenum reacts with cations and anions that are also affected by redox. The effect of redox on the stability of various molybdenum minerals is shown in Fig. 22.3. This graph was prepared for pH 7.0.

Both $MnMoO_4(c)$ and $FeMoO_4(c)$ are too soluble to precipitate in soils. The controlling manganese mineral at the low redox represented here is $MnCO_3$(rhodocrocite) (Chapter 11) and the controlling iron mineral is Fe_3O_4(magnetite). Also included is soil-Fe which is metastable to magnetite (Chapter 10) and is, therefore, represented by dashed lines.

Elemental Mo(c) can only form at very low redox ($pe + pH < 1$). Since such low redox values are extremely rare for soils, Mo(c) can be discounted as a likely soil mineral.

The soil-Mo line in Fig. 22.3 limits MoO_4^{2-} activity at $10^{-6.80}$ M for a soil of pH 7.0. The pH-dependency of this line shows it to decrease 0.8 log unit for each unit increase in pH. All other lines in this figure shift upward 2 log units for each unit decrease in pH. Below $pe + pH$ of 4.61, MoS_2 (molybdenite) rapidly becomes more stable than soil-Mo. Precipitation of molybdenite lowers MoO_4^{2-} activity to extremely low levels as $pe + pH$ drops below 4.61. The equations relating to these equilibria include:

$\log K°$

$MoS_2(\text{molybdenite}) + 4H_2O \rightleftharpoons MoO_4^{2-} + 2S^{2-} + 2e^- + 8H^+$ -96.49

$2S^{2-} + 8H_2O \rightleftharpoons 2SO_4^{2-} + 16e^- + 16H^+$ $2(-20.74)$

$MoS_2(c) + 12H_2O \rightleftharpoons MoO_4^{2-} + 2SO_4^{2-} + 18e^- + 24H^+$ -137.97

(22.2)

from which

$$\log MoO_4^{2-} = -137.97 - 2 \log SO_4^{2-} + 6pH + 18(pe + pH) \quad (22.3)$$

Equation 22.3 shows how SO_4^{2-}, pH, and $pe + pH$ affect the activity of MoO_4^{2-} in the presence of molybdenite.

The redox at which soil-Mo converts to molybdenite is obtained from the reactions:

$\log K°$

$MoO_4^{2-} + 2SO_4^{2-} + 18e^- + 24H^+ \rightleftharpoons MoS_2(\text{molybdenite}) + 12H_2O$ 137.97

$\text{soil-Mo} \rightleftharpoons MoO_4^{2-} + 0.8H^+$ -12.40

$\text{soil-Mo} + 2SO_4^{2-} + 18e^- + 23.2H^+ \rightleftharpoons MoS_2(c) + 12H_2O$ 125.57

(22.4)

from which

$$pe + pH = 6.98 + 0.11 \log SO_4^{2-} - 0.29 pH \quad (22.5)$$

Equation 22.5 was plotted earlier in Fig. 17.4 to show the redox at which MoS_2(molybdenite) precipitates relative to other metal sulfides. At pH 7 with 10^{-3} M SO_4^{2-} this occurs at $pe + pH = 4.61$. Changes in SO_4^{2-} and

THE EFFECT OF REDOX ON MOLYBDENUM SOLUBILITY

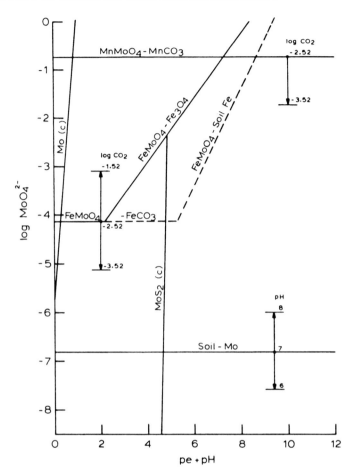

Fig. 22.3 The effect of redox on the stability of molybdenum minerals at pH 7 and 10^{-3} M SO_4^{2-}.

pH cause only slight changes in this $pe + pH$ value, but $MoS_2(c)$ is strongly affected by redox because both S and Mo are reduced.

It would be virtually impossible for plants to obtain sufficient molybdenum from $MoS_2(c)$ at redox levels below 4.61 if it were not for two important principles. First, soils seldom become much more reduced without first being depleted of SO_4^{2-} due to the precipitation of metal sulfides including pyrite (Chapter 17). Secondly, plants that are capable of growing in reduced environments generally have the capability of transporting O_2 through the stem to the roots and thereby maintain a slightly more oxidized environment in the immediate vicinity of roots than in the surrounding soil. Under these localized conditions, the highly insoluble reduced minerals become much more soluble. Therefore, nutrients become more available in the micro-environments of plant roots that are slightly more oxidized.

REFERENCES

Chappell, W. R. and K. K. Peterson. 1976. Molybdenum in the Environment. Vol. 1 The Biology of Molybdenum. Marcel Dekker, New York.

Chappell, W. R. and K. K. Petersen. 1977. Molybdenum in the Environment. Vol. 2. The Geochemistry, Cycling, and Industrial Uses of Molybdenum. Marcel Dekker, New York.

Cotton, F. A. and G. Wilkinson. 1972. Advanced Inorganic Chemistry. Third Edition. Interscience Publishers, New York.

Jenkins, I. L. and A. G. Wain. 1963. Molybdenum species in acid solution. J. Appl. Chem. (London) 13:561–564.

Jones, L. H. P. 1957. The solubility of molybdenum in simplified systems and aqueous soil suspensions. J. Soil Sci. 8:313–327.

Sarafian, P. G. and W. J. Furbish. 1965. Solubilities of natural and synthetic ferrimolybdite. Am. Miner. 50:223–226.

Titley, S. R. 1963. Some behavioral aspects of molybdenum in the supergene environment. Soc. Min. Eng. Trans. 226:199–204.

Vlek, P. L. G. and W. L. Lindsay. 1977. Thermodynamic stability and solubility of molybdenum minerals in soils. Soil Sci. Soc. Am. J. 41:42–46.

PROBLEMS

22.1 Develop the equations and plot the $PbMoO_4$(wulfenite) line in Fig. 22.2 when Pb^{2+} is controlled by $Pb_5(PO_4)_3Cl(c)$ when either strengite or hydroxyapatite, depending upon which is most stable, controls phosphate and soil-Ca and soil-Fe control Ca^{2+} and Fe^{3+}, respectively. Discuss these mineral combinations with respect to the proposed soil-Mo line in this same figure.

22.2 Develop the equations and calculate the activity of MoO_4^{2-} in a highly reduced soil in which pyrite is just beginning to precipitate and SO_4^{2-} is 10^{-4} M. Consider magnetite to be controlling Fe^{2+} activity. Explain how plants are able to obtain molybdenum from such a reduced environment.

TWENTY-THREE

ORGANIC TRANSFORMATIONS

Organic matter is an important constituent of soils and is largely composed of carbon (approximately 45% by weight). The carbon content of the lithosphere is estimated at 950 ppm, and the average content of soils is estimated at 20,000 ppm or 2% (Table 1.1). This is equivalent to approximately 4.5% organic matter.

The organic matter content of soils is highly variable. It includes living and dead organisms, their decomposition products, and humic substances. Humus is quite resistant to breakdown and generally persists in soils for many years. The organic matter content of soils is mainly affected by temperature and rainfall, ranging generally from 1 to 8% in most mineral soils.

Organic matter is thermodynamically unstable in soils and sooner or later oxidizes to $CO_2(g)$ and H_2O, leaving the minerals to leach away or to precipitate as soil minerals. During the decomposition of organic matter, electrons and protons are released. They are normally accepted by oxygen, but if oxygen is depleted, the electrons and protons that are released react with the surrounding soil, causing dissolution and reduction of the most readily reducible substances present. As soils are reduced, organic compounds are formed that are not present in well-oxidized soils.

In this chapter stability diagrams have been developed to show the effect of redox on the stability of several simple organic compounds that are important to living organisms and organic matter transformations. Organic matter persists in soils because it is continually renewed through photosynthesis. Thermodynamic diagrams are developed to show the overall energy relationships of organic matter transformations, but they do not include the detailed biochemical pathways by which the reactions are carried out.

23.1 OXIDATION STATES OF CARBON

The oxidation state of carbon in organic compound varies from $+4$ to -4. A summary of the oxidation states of several organic species is given in Table 23.1. Carbon dioxide and carbonates represent the highest oxidation state of carbon while CH_4(methane) represents the most reduced state. Other positive oxidation states includes oxalate ($+3$), formates and carbon monoxide ($+2$), while the zero oxidation state includes acetate, glucose, and graphite. The more highly reduced products having negative oxidation states include acetalydehyde and acetylene (-1), ethylene, methane, and ethanol (-2), ethane (-3), and methane (-4). The reduced species become important only in highly reduced environments.

TABLE 23.1 THE OXIDATION STATE OF CARBON IN SEVERAL ORGANIC SPECIES

Compound	Oxidation State
$CO_2(g)$, $H_2CO_3^\circ$, HCO_3^-, CO_3^{2-} (carbonate)	+4
$H_2C_2O_4^\circ$, $HC_2O_4^-$, $C_2O_4^{2-}$ (oxalate)	+3
$HCOOH(g)$, $HCCOH^\circ$, $HCOO^-$ (formate)	+2
$CO(g)$, CO° (carbon monoxide)	+2
CH_3COOH°, CH_3COO^- (acetate)	0
$C_6H_{12}O_6$ (glucose)	0
C (graphite)	0
CH_3CHO (g, acetaldehyde)	−1
C_2H_2 (g, acetylene)	−1
C_2H_4 (g, ethylene)	−2
CH_3OH° (methanol)	−2
$CH_3CH_2OH(g)$, $CH_3CH_2OH^\circ$ (ethanol)	−2
C_2H_6 (g, ethane)	−3
CH_4 (g, methane), CH_4° (methane)	−4

23.2 PRODUCTS OF GLUCOSE METABOLISM

Glucose is the primary product of photosynthesis and is a component of many polysaccharides. It is a high-energy compound from which most organisms are able to synthesize complex organic compounds essential for normal growth and function. The overall equilibrium reactions between aqueous glucose and other organic species are given by Reactions 1 through 13 of Table 23.2.

Glucose can undergo both oxidative and reductive reactions. It can also be transformed into acetic acid, which involves no net transfer of electrons. Oxidation products include carbon dioxide, oxalate, formate, and carbon monoxide, whereas reduction products include acetaldehyde, acetylene, ethylene, methanol, ethanol, ethane, acetylene, and methane. Although equations can be written for all combinations of organic species, all such reactions are not catalyzed under natural conditions. For example, the formation of methane directly from glucose and the organisms that bring about this reaction in a direct way are not known (Lehninger, 1972). Organisms have specific pathways for carrying out oxidative and reductive reactions so as to capture much of the energy released along these pathways and utilize it to carry out vital cell functions and to synthesize essential metabolites and cell structures.

TABLE 23.2 EQUILIBRIUM REACTIONS OF CARBON SPECIES AT 25°C

Reaction No.	Equilibrium Reaction	log $K°$
	Reactions of α-D-Glucose	
1	$C_6H_{12}O_6°(glucose) + 6H_2O \rightleftharpoons 6CO_2(g) + 24e^- + 24H^+$	5.03
2	$C_6H_{12}O_6°(glucose) + 6H_2O \rightleftharpoons 3C_2O_4^- + 18e^- + 24H^+$	−55.28
3	$C_6H_{12}O_6°(glucose) + 6H_2O \rightleftharpoons 6HCOO^- + 12e^- + 12H^+$	−40.54
4	$C_6H_{12}O_6°(glucose) \rightleftharpoons 6CO(g) + 12e^- + 12H^+$	−16.06
5	$C_6H_{12}O_6°(glucose) \rightleftharpoons 3CH_3COO^- + 3H^+$	33.94
6	$C_6H_{12}O_6°(glucose) \rightleftharpoons 6C(graphite) + 6H_2O$	89.11
7	$C_6H_{12}O_6°(glucose) + 6e^- + 6H^+ \rightleftharpoons 3CH_3CHO(g) + 3H_2O$	35.38
8	$C_6H_{12}O_6°(glucose) + 6e^- + 6H^+ \rightleftharpoons 3C_2H_2(g) + 6H_2O$	−20.86
9	$C_6H_{12}O_6°(glucose) + 12e^- + 12H^+ \rightleftharpoons 3C_2H_4(g) + 6H_2O$	53.30
10	$C_6H_{12}O_6°(glucose) + 12e^- + 12H^+ \rightleftharpoons 6CH_3OH°$	24.19
11	$C_6H_{12}O_6°(glucose) + 12e^- + 12H^+ \rightleftharpoons 3CH_3CH_2OH° + 3H_2O$	59.97
12	$C_6H_{12}O_6°(glucose) + 18e^- + 18H^+ \rightleftharpoons 3C_2H_6(g) + 6H_2O$	106.42
13	$C_6H_{12}O_6°(glucose) + 24e^- + 24H^+ \rightleftharpoons 6CH_4(g) + 6H_2O$	142.51
	Reactions of Acetate	
14	$CH_3COO^- + 2H_2O \rightleftharpoons 2CO_2(g) + 8e^- + 7H^+$	−9.64
15	$CH_3COO^- + 2H_2O \rightleftharpoons C_2O_4^{2-} + 6e^- + 7H^+$	−29.74
16	$CH_3COO^- + 2H_2O \rightleftharpoons 2HCOO^- + 4e^- + 5H^+$	−24.83
17	$CH_3COO^- \rightleftharpoons 2CO(g) + 4e^- + 3H^+$	−16.66
18	$CH_3COO^- + H^+ \rightleftharpoons \frac{1}{3}C_6H_{12}O_6°(glucose)$	−11.31
19	$CH_3COO^- + H^+ \rightleftharpoons C(graphite) + 2H_2O$	18.39
20	$CH_3COO^- + 2e^- + 3H^+ \rightleftharpoons CH_3CHO(g) + H_2O$	0.48
21	$CH_3COO^- + 2e^- + 3H^+ \rightleftharpoons C_2H_2(g) + 2H_2O$	−18.27
22	$CH_3COO^- + 4e^- + 5H^+ \rightleftharpoons C_2H_4(g) + 4H_2O$	6.45
23	$CH_3COO^- + 4e^- + 5H^+ \rightleftharpoons 2CH_3OH°$	−3.25
24	$CH_3COO^- + 4e^- + 5H^+ \rightleftharpoons CH_3CH_2OH° + H_2O$	8.68
25	$CH_3COO^- + 6e^- + 7H^+ \rightleftharpoons C_2H_6(g) + 2H_2O$	24.16
26	$CH_3COO^- + 8e^- + 9H^+ \rightleftharpoons 2CH_4(g) + 2H_2O$	36.19
	Oxidation to Carbon Dioxide	
27	$C_2O_4^{2-} \rightleftharpoons 2CO_2(g) + 2e^-$	20.10
28	$HCOO^- \rightleftharpoons CO_2(g) + 2e^- + H^+$	7.60
29	$CO(g) + H_2O \rightleftharpoons CO_2(g) + 2e^- + 2H^+$	3.51
30	$CH_3COO^- + 2H_2O \rightleftharpoons 2CO_2(g) + 8e^- + 7H^+$	−9.64
31	$C_6H_{12}O_6°(glucose) + 6H_2O \rightleftharpoons 6CO_2(g) + 24e^- + 24H^+$	5.03

TABLE 23.2 (*Continued*)

Reaction No.	Equilibrium Reaction	log $K°$
32	$C(graphite) + 2H_2O \rightleftharpoons CO_2(g) + 4e^- + 4H^+$	−14.01
33	$CH_3CHO(g) + 3H_2O \rightleftharpoons 2CO_2(g) + 10e^- + 10H^+$	−10.12
34	$C_2H_2(g) + 4H_2O \rightleftharpoons 2CO_2(g) + 10e^- + 10H^+$	8.63
35	$C_2H_4(g) + 4H_2O \rightleftharpoons 2CO_2(g) + 12e^- + 12H^+$	−16.09
36	$CH_3OH° + H_2O \rightleftharpoons CO_2(g) + 6e^- + 6H^+$	−3.19
37	$CH_3CH_2OH° + 3H_2O \rightleftharpoons 2CO_2(g) + 12e^- + 12H^+$	−15.95
38	$C_2H_6(g) + 4H_2O \rightleftharpoons 2CO_2(g) + 14e^- + 14H^+$	−33.80
39	$CH_4(g) + 2H_2O \rightleftharpoons CO_2(g) + 8e^- + 8H^+$	−22.91
	Reduction to Methane	
40	$CO_2(g) + 8e^- + 8H^+ \rightleftharpoons CH_4(g) + 2H_2O$	22.91
41	$C_2O_4^{2-} + 14e^- + 16H^+ \rightleftharpoons 2CH_4(g) + 4H_2O$	65.93
42	$HCOO^- + 6e^- + 7H^+ \rightleftharpoons CH_4(g) + 2H_2O$	30.51
43	$CO(g) + 6e^- + 6H^+ \rightleftharpoons CH_4(g) + H_2O$	26.43
44	$CH_3COO^- + 8e^- + 9H^+ \rightleftharpoons 2CH_4(g) + 2H_2O$	36.19
45	$C_6H_{12}O_6°(glucose) + 24e^- + 24H^+ \rightleftharpoons 6CH_4(g) + 6H_2O$	142.51
46	$C(graphite) + 4e^- + 4H^+ \rightleftharpoons CH_4(g)$	8.90
47	$CH_3CHO(g) + 6e^- + 6H^+ \rightleftharpoons 2CH_4(g) + H_2O$	35.71
48	$C_2H_2(g) + 6e^- + 6H^+ \rightleftharpoons 2CH_4(g)$	54.46
49	$C_2H_4(g) + 4e^- + 4H^+ \rightleftharpoons 2CH_4(g)$	29.74
50	$CH_3OH° + 2e^- + 2H^+ \rightleftharpoons CH_4(g) + H_2O$	19.72
51	$CH_3CH_2OH° + 4e^- + 4H^+ \rightleftharpoons 2CH_4(g) + H_2O$	27.51
52	$C_2H_6(g) + 2e^- + 2H^+ \rightleftharpoons 2CH_4(g)$	12.03
	Gases and Acid Dissociations	
53	$H_2C_2O_4° \rightleftharpoons H^+ + HC_2O_4^-$	0.10
54	$HC_2O_4^- \rightleftharpoons H^+ + C_2O_4^{2-}$	−4.27
55	$H_2C_2O_4° \rightleftharpoons 2H^+ + C_2O_4^{2-}$	−4.18
56	$HCOOH° \rightleftharpoons H^+ + HCOO^-$	−3.74
57	$HCOOH° \rightleftharpoons HCOOH(g)$	−3.75
58	$CO(g) \rightleftharpoons CO°$	−3.02
59	$CO_2(g) \rightleftharpoons CO(g) + \frac{1}{2}O_2(g)$	−45.07
60	$CH_3COOH° \rightleftharpoons H^+ + CH_3COO^-$	−4.76
61	$CH_3CH_2OH° \rightleftharpoons CH_3CH_2OH(g)$	−2.36
62	$CH_4(g) \rightleftharpoons CH_4°$	−2.87

In Fig. 23.1 are summarized the overall equilibrium reactions of glucose with many of its possible metabolic products. The reference level of glucose used to construct this diagram was 10^{-4} M. The equilibrium level of various organic species are shown as a function of redox (pe + pH).

It is apparent from Fig. 23.1 that a great potential exists for the oxidation of $C_6H_{12}O_6^\circ$(glucose) to CO_2(g) and H_2O. Throughout the redox range of pe + pH > 0, extremely high pressures of CO_2(g) would be required to prevent the oxidation of glucose. The higher the redox, the greater is the energy gradient favoring the oxidation of glucose to CO_2.

There is also a tremendous potential for CH_3COO^-(acetate) to form from glucose, and impossibly high concentration of 10^{17} M CH_3COO^- would be required to reverse this equilibrium reaction. This emphasizes the tremendous energy gradients that are involved. There is no transfer of electrons in going from glucose to acetate, so redox does not affect this transformation (Reaction 5, Table 23.2). Since $C_2O_4^{2-}$(oxalate) and $CHOO^-$ (formate) are oxidative products of glucose, the potential for them to form increases with increase in redox.

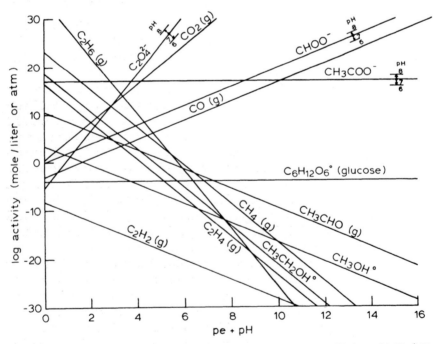

Fig. 23.1 Effect of redox on the activity of various organic species in equilibrium with 10^{-4} M α-D-glucose.

Potential reduction products of glucose metabolism include CH_3CHO (acetaldehyde), C_2H_2(acetylene), C_2H_4(ethane), C_2H_4(ethylene), $CH_3OH°$ (methanol), $CH_3CH_2OH°$(ethanol) and CH_4(methane). Under equilibrium conditions represented in Fig. 23.1, $CH_4(g)$ can exceed 1 atm at $pe + pH$ values < 5.8. Methane is known to escape from marsh areas where organic matter is abundant and highly anaerobic conditions prevail. On the other hand, $CH_4(g)$ cannot form to any great extent in environments having $pe + pH$ values > 7. Other reduced compounds such as CH_3CHO(acetaldehyde), $CH_3CH_2OH°$(ethanol), C_2H_6(ethane) or C_2H_4(ethylene) and possibly $CH_3OH°$(methanol) may accumulate under certain reduced conditions.

Although Fig. 23.1 was constructed to represent 10^{-4} M glucose, the graph can be used to predict the equilibrium levels of organic constituents at any glucose level. For example, shifts in the $CH_4(g)$ line when glucose is 10^{-5} rather than 10^{-4} M can be obtained from Reaction 13 of Table 23.2 which gives the relationship

$$6 \log CH_4(g) = 142.51 + \log C_6H_{12}O_6° - 24(pe + pH) \quad (23.1)$$

or

$$\log CH_4(g) = 23.68 + 0.17 \log C_6H_{12}O_6° - 4(pe + pH) \quad (23.2)$$

Thus a change in glucose from 10^{-4} to 10^{-5} M would shift the $CH_4(g)$ line by $(-1) \times \frac{1}{6} = -0.17$ log unit. Similar shifts for other lines can readily be determined from the ratio of carbon atoms in the organic species relative to glucose.

Equilibrium relationships between solution and gaseous species and the dissociation of organic acids are given by Reactions 53 through 62 of Table 23.2. These equations can be combined with those used to develop Fig. 23.1 to locate lines of various species not included in this figure. For example, combining Reactions 11 and 61 gives a line for $CH_3CH_2OH(g)$ that is parallel to that for $CH_3CH_2OH°$(ethanol) line but displaced downward by 2.36 log units. Various organic acid species can also be estimated. For example, $CH_3COOH°$ dissociates to give CH_3COO^- (Reaction 60) and the two species have equal activities at pH 4.76. Thus the $CH_3CHOOH°$ line would lie parallel to that of CH_3COO^- and would rise or fall one log unit for each unit decrease or increase in pH from pH 4.76.

It must be emphasized that not all reactions that can be written actually occur because catalysts or enzymes are often needed to lower the activation energy of the substrates. On the other hand, a reaction that is thermodynamically impossible cannot occur even if the catalyst is present. Equilibrium relationships aid in distinguishing between possible and impossible reactions and can be used to show how various factors shift equilibrium conditions. For example, CH_4, $H_3CH_2OH°$, and $CH_3OH°$ cannot form to any great extent in oxidizing environments above $pe + pH = 10$, but in more reduced environments they can form provided the necessary metabolic pathways and catalysts are available.

23.3 REACTIONS OF ACETIC ACID

Acetic acid is one of the most common intermediates in anaerobic respiration. All forms of organic matter, including carbohydrates, fats, amino acids, etc., etc., are converted to acetic acid during their decomposition (Hobson et al., 1966; Takai and Kamura, 1966). Although other acids such as succinic, lactic, butyric, and propionic acids are also produced, their concentrations are generally much less than acetic acid (Hobson et al., 1966). Studies with radioisotopes have shown that methane is produced largely via acetic acid and reflects the importance of acetic acid in anaerobic systems.

Figure 23.2 shows the equilibrium relationships between acetate at 10^{-4} M and various other organic metabolites of glucose at pH 7. Reaction 14

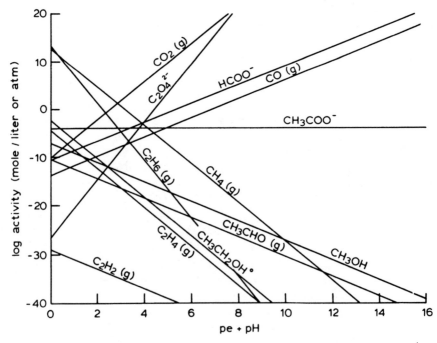

Fig. 23.2 Effect of redox on the activity of various organic species in equilibrium with 10^{-4} M CH_3COO^- (acetate) at pH 7.0.

through 26 of Table 23.2 were used to develop this diagram. Under well-oxidized conditions acetate can go to $CO(g)$, $HCOO^-$, $C_2O_4^{2-}$, and eventually to $CO_2(g)$. As $pe + pH$ drops below 6, various reduced species are stable; the most important of these is $CH_4(g)$. Only as $pe + pH$ drops below 4, can $CH_4(g)$ exceed 10^{-3} atm. As $pe + pH$ drops below 2.9, $CH_4(g)$ becomes more stable than $CO_2(g)$. Certainly redox is expected to affect the ratio of carbon dioxide to methane released from anaerobic digesters.

Ethanol is sometimes formed from acetic acid. This reaction can substitute for methane production when methane-producing organisms are absent (Hobson et al., 1966). In this case acetic acid acts as the proton acceptor.

23.4 OXIDATION TO $CO_2(g)$ AND REDUCTION TO $CH_4(g)$

Figure 23.3 shows the levels of the various carbon species in equilibrium with 0.003 atm of CO_2. These relationships were obtained from Reactions 27 through 39 of Table 23.2. The low concentrations of substrates in equilibrium

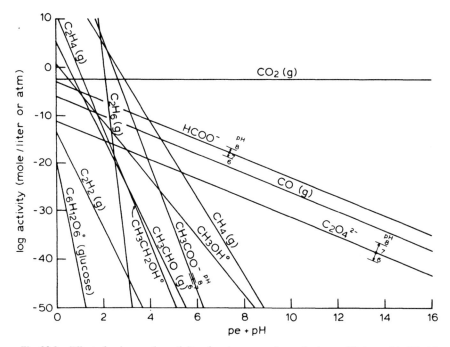

Fig. 23.3 Effect of redox on the activity of various organic species in equilibrium with $CO_2(g)$ at 0.003 atm.

with $CO_2(g)$ indicate that there is a net movement of carbon out of the soil during the time that organic substances are being oxidized. Appropriate organisms must also be present to catalyze these reactions.

Only under highly reducing conditions (pe + pH < 3) do acetate and methane reach significant levels. In open systems such as soils, methane can easily escape once it has formed.

Equilibrium reactions showing the reduction of various organic species to methane are given by Reactions 40 through 52 of Table 23.2. According to Hobson et al. (1966) methane can be produced from carbon dioxide, formic acid, acetic acid, or ethanol, as well as other low-molecular weight organic acids. Some have concluded that elemental hydrogen must be present for methane production. Thermodynamically this is not necessary as carbon can act both as electron and proton donors and acceptors as illustrated by the following reactions:

$$\begin{array}{rcl} & & \log K° \\ \frac{1}{2}CH_3COOH° + H_2O & \rightleftharpoons CO_2(g) + 4H^+ + 4e^- & -7.20 \\ \frac{1}{2}CH_3COOH° + 4H^+ + 4e^- & \rightleftharpoons CH_4(g) + H_2O & 15.72 \\ \hline CH_3COOH° & \rightleftharpoons CH_4(g) + CO_2(g) & 8.52 \end{array}$$

(23.3)

The positive log $K°$ for this reaction indicates that the reaction can proceed to the right when reactants and products are in their standard states. When $CO_2(g)$ is 0.003 atm and $CH_3COOH°$ is $10^{-6.24}$ M (corresponding to 10^{-4} M CH_3COO^- at pH 7), the equilibrium $CH_4(g)$ would be $10^{4.80}$ atm, indicating that this reaction proceeds spontaneously to the right provided appropriate catalysts are present to mediate the reaction.

When methane is produced during anaerobic digestion of organic wastes, there is a net loss of carbon from the system. This is advantageous in waste treatment because it reduces the bulk of organic material that must be disposed. In addition, some of the energy can be recovered as $CH_4(g)$ and utilized as an energy source.

23.5 STABILITY OF GRAPHITE

So far only aqueous and gaseous species of carbon have been considered. Now let us examine the equilibrium relationships of C(graphite) as a solid-phase carbon compound. Under what conditions is it stable in soils? From Eq. 32 of Table 23.2 comes the equilibrium relationship

$$CO_2(g) + 4e^- + 4H^+ \rightleftharpoons C(graphite) + 2H_2O \qquad \log K° = 14.01$$
(23.4)

which gives

$$pe + pH = 3.50 + \tfrac{1}{4} \log CO_2(g) \qquad (23.5)$$

When $CO_2(g) = 0.003$ atm

$$pe + pH = 2.87 \qquad (23.6)$$

This means that C(graphite) and 0.003 atm of CO_2 are in equilibrium when $pe + pH = 2.87$. Below this redox $CO_2(g)$ is unstable and would eventually transform to C(graphite) whereas above this redox C(graphite) is unstable and tends to oxidize to $CO_2(g)$ and water. Increasing $CO_2(g)$ raises the $pe + pH$ at which C(graphite) can form. In highly reduced natural environments amorphous carbon and coal generally accumulate as solid-phase carbon products. Coal, oil, and gas deposits in the earth are manifestations of the transformation of organic matter into reduced carbon compounds when oxygen is excluded.

In coal-mining operations, the process is reversed. The overburden materials are exposed to the atmosphere where oxygen raises the $pe + pH$ and oxidative processes proceed. It is only a matter of time until the reduced species are oxidized.

REFERENCES

Hobson, P. N., S. Bousfield, and R. Summers. 1966. Anaerobic digestion of organic matter. CRC Critical Rev. Environ. Control 4:131–191.

Lehninger, A. L. 1972. Biochemistry. Worth Publishers, New York.

McCarty, P. L. 1972. Energetics of organic matter degradation. *In* R. Mitchell, Ed., Water Pollution Microbiology, pp. 92–118. John Wiley, New York.

Takai, Y. and T. Kamura. 1966. The mechanism of reduction in waterlogged paddy soil. Fol. Microbiol. 11:304–313.

PROBLEMS

23.1 By means of equations demonstrate that the lines for methane and ethanol in Fig. 23.1 and 23.2 are parallel, but in Fig. 23.3 they are not.

23.2 How would the line for $CO_2(g)$ in Fig. 23.1 change if glucose were 10^{-8} rather than 10^{-4} M?

23.3 Determine where the line for $H_2C_2O_4^\circ$ would appear in Fig. 23.1. Explain why this line is independent of pH.

23.4 Develop a general equation to show how the lines in Fig. 23.2 shift for changes in acetate.

23.5 Explain why oxidative metabolic pathways generally capture more energy than anaerobic pathways.

23.6 What conditions are necessary for the synthesis of ethanol?

23.7 Even though $CO(g)$ can occur as a reaction product in the oxidation of glucose why is it seldom found as such?

23.8 What is the equilibrium level of $CO(g)$ in air?

23.9 Calculate the equilibrium $CH_4(g)$ in an anaerobic digester where CH_3COO^- is 10^{-2} M and $CO_2(g)$ is 0.4 atm.

23.10 What is the minimum equilibrium $pe + pH$ at which C(graphite) can form when $CO_2(g)$ is present at 10 atm?

APPENDIX

SELECTED STANDARD FREE ENERGIES OF FORMATION FOR USE IN SOIL SCIENCE

M. SADIQ AND W. L. LINDSAY

Selected standard free energies of formation (ΔG_f°, kcal mole^{-1}) of important minerals and chemical species of interest in soils are summarized in the following tables. Details of the literature review and the basis for selecting each of the ΔG_f° values are documented in a more detailed publication (Sadiq and Lindsay, 1979).

The tables that follow include the chemical species, the literature reference from which the selected value is based, and a remarks column indicating whether solubility data (°K) or thermodynamic data (ΔH_f° and S°) were used. When other free energy values were used in arriving at the reported value, the species are indicated. In a few cases the original sources could not be traced, and they are indicated by a question mark. In all cases the ΔG_f° values were recalculated to maintain internal consistency among our selections.

The values reported here represent our best efforts in reviewing the literature as of April 1, 1978 and reflect the most recent values for Al^{3+}, Fe^{3+}, Fe^{2+}, gibbsite, silicates, and many other chemical species that have changed since most previous compilations have been made. Only data for 298.16°K and zero ionic strength are included.

Even though we have endeavored to select only the most reliable data, we have no doubt erred in many cases. Hopefully these errors can be recognized and corrected in the future. We welcome feedback and comments from our readers and hope that you will assist us in expanding and improving this compilation.

Ag (Silver)

Species	ΔG_f (kcal mol^{-1})	Source	Remarks
Ag(c)	0.00	Standard state	ΔH_f
		Furukawa et al. (1968)	S
Ag$^+$	18.42	Anderson et al. (1971)	K : AgCl(c), Cl$^-$
		Wagman and Kilday (1973)	S : AgCl(c)
Ag^{2+}	64.30	Wagman et al. (1969)	?
AgBr	−12.42	Berne and Leden (1953)	K : Ag$^+$, Br$^-$
AgBr$_2^-$	−41.33	Berne and Leden (1953)	K : AgBr
AgBr$_3^{2-}$	−68.06	Smith and Martell (1976)	K : Ag$^+$, Br$^-$
AgBr$_4^{3-}$	−93.34	Smith and Martell (1976)	K : Ag$^+$, Br$^-$
AgBr(c)	−23.24	Ramette (1972)	K : Ag$^+$, Br$^-$
Ag$_2$CO$_3$(c)	−104.47	Kelley and Anderson (1935)	K : Ag$^+$, CO$_3^{2-}$
AgCl	−17.47	Jonte and Martin (1952)	K : Ag$^+$, Cl$^-$
AgCl$_2^-$	−51.49	Jonte and Martin (1952)	K : AgCl
AgCl$_3^{2-}$	−84.42	Smith and Martell (1976)	K : Ag$^+$, Cl$^-$

Ag (Silver) (Continued)

Species	ΔG_f° (kcal mol^{-1})	Source	Remarks
$AgCl_4^{3-}$	−115.38	Smith and Martell (1976)	K : Ag^+, Cl^-
AgCl(c)	−26.25	Pankratz (1970)	ΔH_f, S
AgF	−48.99	Connick and Paul (1961)	K : Ag^+, F^-
AgF(c)	−50.03	Peacock (1959)	K : Ag^+, F^-
AgI	−2.94	Smith and Martell (1976)	K : Ag^+, I^-
AgI_2^-	−22.30	Smith and Martell (1976)	K : Ag^+, I^-
AgI_3^{2-}	−37.27	Smith and Martell (1976)	K : Ag^+, I^-
AgI_4^{3-}	−50.74	Smith and Martell (1976)	K : Ag^+, I^-
AgI(c)	−15.89	McIntyre and Amis (1968)	E : Ag(c) AgI(c)
Ag_2MoO_4(c)	−178.75	Smith and Martell (1976)	K : Ag^+, MoO_4^{2-}
$AgNH_3^+$	7.49	Vosburgh and McClure (1943)	K : Ag^+, NH_3
$Ag(NH_3)_2^+$	−4.07	Vosburgh and McClure (1943)	K : Ag^+, NH_3
$AgNO_3$	−7.95	Smith and Martell (1976)	K : Ag^+, NO_3^-
$AgNO_3$(c)	−7.96	Parker (1965) Latimer et al. (1938)	ΔH_f : Ag^+, NO_3^- S
AgOH	−21.84	Biedermann and Sillen (1960)	K : Ag_2O(c)
$Ag(OH)_2^-$	−62.20	Biedermann and Sillen (1960)	K : Ag_2O(c)
$Ag(OH)_3^{2-}$	−85.91	Dirkse and Graven (1959)	K : AgO(c)
$Ag_3(OH)_4$	−120.78	Pleskov and Kabanov (1957)	K : Ag_2O(c)
AgO(c)	3.23	Dirkse (1962)	E : Ag_2O(c) AgO(c)
AgOH(c)	−29.69	Antikainen and Dryssen (1960)	K : Ag^+
$Ag(OH)_2$(c)	−45.86	Dirkse and Graven (1959)	K : AgO(c)
Ag_2O(c)	−2.67	Gregor and Pitzer (1962b)	E : Ag(c) Ag_2O(c)
Ag_2O_3(c)	28.82	Stehlik (1963)	K : Ag^+
Ag_3PO_4(c)	−211.71	Wagman et al. (1969)	K : Ag^+, PO_4^{3-}
AgHS	2.19	Cloke (1963)	K : Ag^+, H_2S(g)
$Ag(HS)_2^-$	−3.08	Cloke (1963)	K : Ag^+, H_2S(g)
$AgSO_3^-$	−105.40	Smith and Martell (1976)	K : Ag^+, SO_3^{2-}
$AgSO_4^-$	−161.30	Hopkins and Wulff (1965)	K : Ag^+, SO_4^{2-}
Ag_2SO_3	−91.18	Chateau et al. (1956a)	K : Ag^+, SO_3^{2-}
Ag_2SO_4	−141.11	Wagman et al. (1969)	K : Ag^+, SO_4^{2-}
α-Ag_2S(c)	−9.51	**Mustafaev and Abbasov (1970)**	ΔG_r : Ag(c), $S_{(rh)}$
β-Ag_2S(c)	−9.45	Rosenquist (1949)	ΔG_r : H_2S(g), Ag(c)
Ag_2SO_3(c)	−98.18	Chateau et al. (1956b)	K : Ag^+, SO_3^{2-}
Ag_2SO_4(c)	−147.67	Lietzke and Hall (1967)	K : Ag^+, SO_4^{2-}

Al (Aluminum)

Species	ΔG_f°	Source	Remarks
Al(c)	0.0	Standard state	ΔH_f°
		CODATA (1976)	S°
Al^{3+}	−117.33	Singh (1974)	K°: $Al(OH)_3$(gib)
AlF^{2+}	−194.78	Baumann (1969)	K°: Al^{3+}, F^-
AlF_2^+	−270.38	Baumann (1969)	K°: Al^{2+}, F^-
AlF_3°	−343.83	Baumann (1969)	K°: AlF_2^+, F^-
AlF_4^-	−415.01	Baumann (1969)	K°: AlF_3°, F^-
$Al(NO_3)_3^\circ$	−197.42	Wagman et al. (1968)	K°: Al^{3+}, NO_3^-
$AlOH^{2+}$	−167.17	Frink and Peech (1963)	K°: Al^{3+}
$Al(OH)_2^+$	−218.02	Smith and Martell (1976)	K°: Al^{3+}
$Al(OH)_3^\circ$	−266.94	Smith and Martell (1976)	K°: Al^{3+}
$Al(OH)_4^-$	−312.25	Kittrick (1966)	K°: $Al(OH)_3$(gib)
$Al(OH)_5^{2-}$	−354.06	Marino et al. (1976)	K°: Al^{3+}
$Al_2(OH)_2^{4+}$	−337.54	Smith and Martell (1976)	K°: Al^{3+}
$Al(OH)_3$(amorp)	−274.21	Kittrick (1966)	K°: Al^{3+}
α-$Al(OH)_3$(bayerite)	−275.78	Russell et al. (1955)	K°: $Al(OH)_4^-$
γ-$Al(OH)_3$(gibbsite)	−276.43	Hemingway and Robie (1977)	ΔH_f°
		Hemingway et al. (1977a)	S°
$Al(OH)_3$(nordstrandite)	−276.3	Hem et al. (1973)	K°: Al^{3+}
α-AlOOH(diaspore)	−219.9	Haas and Holdaway (1970)	ΔH_f°, S°
γ-AlOOH(boehmite)	−219.61	Russell et al. (1955)	K°: $Al(OH)_4^-$
α-Al_2O_3(corundum)	−378.17	CODATA (1976)	ΔH_f°
		Furukawa et al. (1956)	S°
γ-Al_2O_3(c)	−373.38	Yokokawa and Kleppa (1964)	ΔH_f°
		Chase et al. (1974b)	S°
$AlPO_4$(berlinite)	−388.50	Jaza et al. (1958)	ΔH_f°
		Egan and Wakefield (1960)	S°
$AlPO_4 \cdot 2H_2O$(variscite)	−505.97	Bache (1963)	K°: Al^{3+}, $H_2PO_4^-$
		Lindsay et al. (1959)	
$H_6K_3Al_5(PO_4)_8 \cdot 18H_2O$ (K-Taranakite)	−4014.76	Taylor and Gurney (1961)	K°: K^+, Al^{3+}, $H_2PO_4^-$
$H_6(NH_4)_3Al_5(PO_4)_8 \cdot 18H_2O$ (NH_4-Taranakite)	−3864.84	Taylor and Gurney (1961)	K°: NH_4^+, Al^{3+}, $H_2PO_4^-$
$AlSO_4^+$	−299.64	Behr and Wendt (1962)	K°: Al^{3+}, SO_4^{2-}
		Richburg and Adams (1970)	
$Al(SO_4)_2^-$	−475.82	Behr and Wendt (1962)	K°: Al^{3+}, SO_4^{2-}
		Izatt et al. (1969)	
$Al_2(SO_4)_3^\circ$	−765.94	Wagman et al. (1968)	K°: Al^{3+}, SO_4^{2-}
$Al_2(SO_4)_3$(c)	−740.08	Young (1945)	ΔH_f°: $Al_2(SO_4)_3 \cdot 6H_2O$(c)
		Shomate (1945)	S°
$Al_2(SO_4)_3 \cdot 6H_2O$(c)	−1103.93	Young (1945)	ΔH_r°: H_2SO_4, Al(c)
		Shomate (1945)	S°
$KAl_3(SO_4)_2(OH)_6$ (alunite)	−1111.38	Kelley et al. (1946)	ΔH_f°, S°

AlSi (Aluminosilicate)

Species	ΔG_f°	Source	Remarks
Al_2SiO_5(andalusite)	−584.25	Holm and Kleppa (1966a,b,c)	ΔH_r°: Al_2O_3(cor), SiO_2(q)
		Holm and Kleppa (1966a,b,c)	S°
Al_2SiO_5(kyanite)	−583.38	Shearer and Kleppa (1973)	ΔH_r°: Al_2O_3(cor), SiO_2(q)
		Holm and Kleppa (1966b)	S°
Al_2SiO_5(sillimanite)	−582.81	Charlu et al. (1975)	ΔH_r°: Al_2O_3(cor), SiO_2(q)
		Holm and Kleppa (1966b)	S°
$Al_2Si_2O_5(OH)_4$(dickite)	−908.55	Robie and Waldbaum (1968)	ΔG_r°: kaolinite
		King and Weller (1961a)	S°
$Al_2Si_2O_5(OH)_4$ (halloysite)	−904.77	Robie and Waldbaum (1968)	ΔG_r°: kaolinite, $Al(OH)_3$(gib)
		King and Weller (1961a)	S°
$Al_2Si_2O_5(OH)_4$ (kaolinite)	−909.23	Hemingway and Robie (1977)	ΔH_r°: SiO_2(q), $Al(OH)_3$(gib)
		King and Weller (1961a)	S°
$Al_2Si_4O_{10}(OH)_2$ (pyrophyllite)	−1261.17	Hemingway and Robie (1977)	ΔG_r°: kaolinite SiO_2(q)
		Robie et al. (1976)	S°
$CaAl_2SiO_6$(pyroxene)	−745.13	Robie and Waldbaum (1968)	ΔH_f°, S°
$CaAl_2Si_2O_8$ (anorthite)	−960.68	Hemingway and Robie (1977)	ΔH_f°
		King (1957b)	S°
$CaAl_2Si_2O_8$ (hexagonal anorthite)	−956.90	Barany (1962a)	ΔH_r°: anorthite
		Hemingway and Robie (1977)	S°
$CaAl_2Si_2O_8$(Ca-glass)	−946.24	Hemingway and Robie (1977)	ΔH_f°, S°
$CaAl_2Si_4O_{12} \cdot 2H_2O$ (wairakite)	−1482.55	Zen (1972)	ΔG_r°: quartz, anorthite
$CaAl_2Si_2O_8 \cdot 2H_2O$ (lawsonite)	−1081.95	Hemingway and Robie (1977)	ΔH_r°: SiO_2(q), HCl, CaO(lime), $AlCl_3 \cdot 6H_2O$(c)
		King and Weller (1961b)	S°
$Ca_2Al_4Si_8O_{24} \cdot 7H_2O$ (leonhardite)	−3155.37	Hemingway and Robie (1977)	ΔH_f°
		Kelley and King (1961)	S°
$KAlSiO_4$(kaliophilite)	−479.70	Hemingway and Robie (1977)	ΔH_f°
		Kelley and King (1961)	S°
$KAlSi_2O_6$(leucite)	−687.62	Hemingway and Robie (1977)	ΔH_f°, S°
$KAlSi_3O_8$(K-glass)	−885.34	Hemingway and Robie (1977)	H_f°, S°

(*Continued*)

AlSi (Aluminosilicate) (Continued)

Species	ΔG_f°	Source	Remarks
$KAlSi_3O_8$(microcline)	-894.71	Hemingway and Robie (1977)	ΔH_f°
		Openshaw et al. (1976)	S°
$KAlSi_3O_8$(high sanidine)	-894.16	Waldbaum (1966)	ΔH_f°: microcline
		Hemingway and Robie (1977)	S°
$KAl_2(AlSi_3O_{10})(OH)_2$ (muscovite)	-1339.15	Hemingway and Robie (1977)	ΔH_f°
		Robie et al. (1976)	S°
$K_{0.6}Mg_{0.25}Al_{2.3}Si_{3.5}O_{10}(OH)_2$(illite)	-1304.44	Helgeson (1969)	$\Delta H_f^\circ, S^\circ$
$KMg_3AlSi_3O_{10}F_2$ (fluorphlogopite)	-1461.63	Hemingway and Robie (1977)	$\Delta H_f^\circ, S^\circ$
$Mg_2Al_4Si_5O_{18}$ (Mg-cordierite)	-2075.26	Cheronosky (1974)	ΔG_f°: clinochlore, forsterite, spinal
$Mg_{0.2}(Si_{3.81}Al_{1.71}Fe(III)_{0.22}Mg_{0.29})O_{10}(OH)_2$(Mg-montmorillonite)	-1258.84	Weaver et al. (1971)	K°: Fe^{3+}, Mg^{2+}, Al^{3+}, $H_4SiO_4^\circ$
$(Mg_{2.71}Fe(II)_{0.02}Fe(III)_{0.46}Ca_{0.06}K_{0.1})(Si_{2.91}Al_{1.14})O_{10}(OH)_2$(vermiculite)	-1324.38	Nriagu (1975)	Estimated: $Mg(OH)_2(c)$ $Fe(OH)_2(c)$, $Fe(OH)_3$(amorp), $Al(OH)_3$(gib) $H_4SiO_4(c)$, $Ca(OH)_2(c)$, $Mg(OH)_2(c)$
$Mg_5Al_2Si_3O_{10}(OH)_8$ (chlorite)	-1975.56	Zen (1972) Faweett and Yoder (1966)	$\Delta H_f^\circ, S^\circ$
$NaAlSiO_4$(nepheline)	-477.24	Hemingway and Robie (1977)	ΔH_f°
		Kelley et al. (1953)	S°
$NaAlSi_3O_8$(Na-glass)	-876.32	Hemingway and Robie (1977)	$\Delta H_f^\circ, S^\circ$
$NaAlSi_3O_8$(high albite)	-886.15	Hemingway and Robie (1977)	$\Delta H_f^\circ, S^\circ$
$NaAlSi_3O_8$(low albite)	-887.41	Hemingway and Robie (1977)	$\Delta H_f^\circ, S^\circ$
$NaAlSi_2O_6$(jadeite)	-682.17	Robie and Waldbaum (1968)	ΔH_f°
		Kelley et al. (1953)	S°
$NaAlSi_2O_6 \cdot H_2O$ (analcime)	-737.44	Thompson (1973)	$\Delta H_f^\circ, S^\circ$
$NaAl_3Si_3O_{10}(OH)_2$ (paragonite)	-1328.83	Zen (1972)	$\Delta H_f^\circ, S^\circ$
$Na_{0.33}Al_{2.33}Si_{3.67}O_{10}(OH)_2$(beidellite)	-1281.21	Helgeson (1969)	$\Delta H_f^\circ, S^\circ$: Na^+, Al^{3+}, $H_4SiO_4^\circ$

Br (Bromine)

Species	ΔG_f°	Source	Remarks
Br^-(bromide ion)	-24.87	CODATA (1971)	$\Delta H_f^\circ, S^\circ$
HBr(g)	-12.77	CODATA (1976)	$\Delta H_f^\circ, S^\circ$
Br_2(g, bromine)	0.751	CODATA (1976)	$\Delta H_f^\circ, S^\circ$
Br_2(l, bromine)	0.0	Standard state	ΔH_f°
		Hildenbrand et al. (1958)	S°

C (Carbon)

Species	ΔG_f°	Source	Remarks
C(graphite)	0.00	Standard state	ΔH_f°
		DeSorbo and Tyler (1953)	S°
CO(g, carbon monoxide)	-32.78	CODATA (1976)	$\Delta H_f^\circ, S^\circ$
CO°	-28.66	Wagman et al. (1968)	$K^\circ: CO(g)$
CO_2(g, carbon dioxide)	-94.26	CODATA (1976)	$\Delta H_f^\circ, S^\circ$
CO_3^{2-}(carbonate ion)	-126.18	Adams (1971)	$K^\circ: HCO_3^-$
		Garrels et al. (1961)	
HCO_3^-(bicarbonic ion)	-140.27	Read (1975)	$K^\circ: H_2CO_3^\circ$
$H_2CO_3^\circ$(carbonic acid)	-148.95	Harned and Davis (1943)	$K^\circ: CO_2(g)$
CH_4(g, methane)	-12.14	Prosen and Rossini (1945)	ΔH_f°
		Kelley and King (1961)	S°
CH_4°	-8.23	Wagman et al. (1968)	$\Delta H_f^\circ, S^\circ$
C_2H_2(g, acetylene)	50.00	Rossini et al. (1953)	ΔH_f°
		Gordon (1963)	S°
C_2H_4(g, ethylene)	16.28	Rossini et al. (1953)	$\Delta H_f^\circ, S^\circ$
C_2H_6(g, ethane)	-7.87	Prosen and Rossini (1945)	ΔH_f°
		Rossini et al. (1953)	S°
CH_3CHO(g, acetaldehyde)	-32.26	Dolliver et al. (1938)	ΔH_f°
		Pitzer and Weltner (1949)	S°
CH_3OH°(methanol)	-41.93	Wagman et al. (1968)	$K^\circ: CH_4(g)$
C_2H_5OH(g, ethanol)	-40.22	Stull et al. (1969)	ΔH_f°
		Green (1961b)	S°
$C_2H_5OH^\circ$(ethanol)	-43.44	Wagman et al. (1968)	?
$HCOO^-$(formate ion)	-83.9	Wagman et al. (1968)	?
HCOOH(g, formic acid)	-83.89	Green (1961a, b)	$\Delta H_f^\circ, S^\circ$
$HCOOH^\circ$(formic acid)	-89.0	Wagman et al. (1968)	?
CH_3COO^-(acetate ion)	-88.29	Wagman et al. (1968)	?
CH_3COOH°(acetic acid)	-94.78	Harned and Ehlers (1933)	$K^\circ: CH_3COO^-$
$C_2O_4^{2-}$(oxalate ion)	-161.1	Wagman et al. (1968)	?
$HC_2O_4^-$(bioxalate ion)	-166.93	Wagman et al. (1968)	?
$H_2C_2O_4^\circ$(oxalic acid)	-166.8	Wagman et al. (1968)	?
$C_6H_{12}O_6^\circ$(α-D-glucose)	-218.58	Stecher (1968)	?
$C_6H_{12}O_6$(α-D-glucose)	-217.62	Stull et al. (1969)	$\Delta H_f^\circ, S^\circ$

Ca (Calcium)

Species	ΔG_f°	Source	Remarks
Ca(c)	0.0	Standard state CODATA (1977)	ΔH_f° S°
Ca^{2+}	−132.52	Smith and Martell (1976)	K°: $Ca(OH)_2$(c)
$CaHCO_3^+$	−274.33	Reardon et al. (1973)	K°: Ca^{2+}, HCO_3^-
$CaCO_3^\circ$	−263.00	Reardon and Langmuir (1974)	K°: Ca^{2+}, CO_3^{2-}
$CaCO_3$(aragonite)	−269.87	O'Connor (1975)	K°: Ca^{2+}, HCO_3^-
$CaCO_3$(calcite)	−270.18	Jacobson and Langmuir (1974)	K°: Ca^{2+}, CO_3^{2-}
$CaCO_3 \cdot 6H_2O$(ikaite)	−607.52	Marland (1975)	K°: $CaCO_3$(calcite)
$CaCl^+$	−162.53	Harned and Owen (1958)	K°: Ca^{2+}, Cl^-
$CaCl_2^\circ$	−195.26	Parker et al. (1971)	K°: Ca^{2+}, Cl^-
CaF_2(fluorite)	−282.58	Smith and Martell (1976)	K°: Ca^{2+}, F^-
$CaFe_2O_4$(c)	−338.09	Koehler et al. (1961) King (1954b)	ΔH_r°: CaO(c), Fe_2O_3(c) S°
$CaMoO_4$(powellite)	−343.19	Barany (1962b) Weller and King (1963)	ΔH_f° S°
$CaNO_3^+$	−152.61	Fedorov et al. (1974)	K°: Ca^{2+}, NO_3^-
$Ca(NO_3)_2^\circ$	−179.66	Fedorov et al. (1974)	K°: Ca^{2+}, NO_3^-
$CaOH^+$	−171.89	Smith and Martell (1976)	K°: Ca^+
$Ca(OH_2)^\circ$	−207.71	Greenberg and Copeland (1960)	K°: Ca^{2+}
CaO(lime)	−144.26	CODATA (1977b) Chase et al. (1974a)	ΔH_f° S°
$Ca(OH)_2$(portlandite)	−214.79	Hatton et al. (1959)	ΔH_f°, S°
$CaPO_4^-$	−386.51	Chughtai et al. (1968)	K°: Ca^{2+}, PO_4^{3-}
$CaHPO_4^\circ$	−398.29	Smith and Martell (1976)	K°: Ca^{2+}, HPO_4^{2-}
$CaH_2PO_4^+$	−406.28	Chughtai et al. (1968)	K°: Ca^{2+}, $H_2PO_4^-$
$CaP_2O_7^{2-}$	−599.04	Wolhoff and Overbeek (1959)	K°: Ca^{2+}, $P_2O_7^{4-}$
$CaHP_2O_7^-$	−607.51	Irani and Callis (1960)	K°: Ca^{2+}, $HP_2O_7^{3-}$
$CaOHP_2O_7^{3-}$	−639.49	Wolhoff and Overbeek (1959)	K°: Ca^{2+}, $P_2O_7^{4-}$
$CaHPO_4$(monetite)	−403.96	McDowell et al. (1971)	K°: Ca^{2+}, HPO_4^{2-}
$CaHPO_4 \cdot 2H_2O$ (brushite)	−516.89	Bennett and Adams (1976)	K°: Ca^{2+}, HPO_4^{2-}
$Ca(H_2PO_4)_2 \cdot H_2O$(c)	−734.48	Barton and McConnel (1974)	K°: Ca^{2+}, $H_2PO_4^-$
α-$Ca_3(PO_4)_2$(c)	−922.70	Naumov et al. (1971) Kelley and King (1961)	ΔH_r°: β-$Ca_3(PO_4)_2$(c) S°
β-$Ca_3(PO_4)_2$ (whitelockite, TCP)	−927.37	Gregory et al. (1970)	K°: Ca^{2+}, PO_4^{3-}
$Ca_8H_2(PO_4)_6 \cdot 5H_2O$ (octacalcium phosphate)	−2942.62	Moreno et al. (1960b)	K°: Ca^{2+}, PO_4^{3-}
$Ca_{10}F_2(PO_4)_6$ (fluorapatite)	−3094.73	Chien and Black (1976)	K°: Ca^{2+}, F^-, PO_4^{3-}

Ca (Calcium) (Continued)

Species	ΔG_f°	Source	Remarks
$Ca_{10}(OH)_2(PO_4)_6$ (hydroxyapatite)	−3030.24	Avnimelech et al. (1973)	K°: Ca^{2+}, PO_4^{3-}
β-$Ca_2P_2O_7$(c)	−742.33	Smith and Martell (1976)	K°: Ca^{2+}, $P_2O_7^{4-}$
CaS(oldhamite)	−113.07	Richardson and Jeffes (1952)	ΔH_f°
		Anderson (1931)	S°
$CaSO_4^\circ$	−313.62	Ainsworth (1973)	K°: Ca^{2+}, SO_4^{2-}
$CaSO_4$(anhydrous, insoluble)	−316.48	Dewing and Richardson (1959)	ΔH_f°
		Kelley and King (1961)	S°
α-$CaSO_4$(soluble)	−313.81	Parker et al. (1971)	ΔH_f°
		Kelley (1941)	S°
β-$CaSO_4$(soluble)	−312.86	Parker et al. (1971)	ΔH_f°
		Kelley (1941)	S°
$CaSO_4 \cdot 2H_2O$(gypsum)	−430.17	Gardner and Glueckauf (1970)	K°: Ca^{2+}, SO_4^{2-}
β-$CaSiO_3$(wollastonite)	−370.39	Todd (1951)	ΔH_r°: CaO(c), SiO_2(q)
		Kelley and King (1961)	S°
$CaSiO_3$ (pseudowollastonite)	−369.09	Kelley (1962)	ΔH_f°
		Kelly and King (1961)	
β-Ca_2SiO_4(larnite)	−523.66	King (1951)	ΔH_r°: CaO(c), SiO_2(q)
		Todd (1951)	S°
γ-Ca_2SiO_4(calcium olivine)	−526.12	King (1957a)	ΔH_r°: β-Ca_2SiO_3(c)
		King (1957b)	S°
$CaMg(SiO_3)_2$(diopside)	−724.62	Neuvonen (1952)	H_r°: SiO_2(q), CaO(c), MgO(c)
		King (1957b)	S°

Cd (Cadmium)

Species	ΔG_f°	Source	Remarks
Cd(c)	0.0	Standard state	ΔH_f°
		Hultgren et al. (1963, 1970)	S°
Cd^{2+}	−18.61	Burnett and Zirin (1966)	E°: Cd(c)/Cd^{2+}
$CdBr^+$	−46.41	Hertz (1960)	K°: Cd^{2+}, Br^-
$CdBr_2^\circ$	−72.44	Smith and Martell (1976)	K°: Cd^{2+}, Br^-
$CdBr_3^-$	−97.31	Smith and Martell (1976)	K°: Cd^{2+}, Br^-
$CdBr_4^{2-}$	−122.05	Smith and Martell (1976)	K°: Cd^{2+}, Br^-
$CdCO_3^\circ$	−150.38	Gardner (1975)	K°: Cd^{2+}, CO_3^{2-}
$CdHCO_3^+$	−161.74	Zirino and Yamamoto (1972)	K°: Cd^{2+}, HCO_3^-
$CdCO_3$(otavite)	−161.16	Gamsjäger et al. (1965)	K°: Cd^{2+}, CO_3^{2-}
$CdCl^+$	−52.68	Bonner et al. (1962)	K°: Cd^{2+}, Cl^-
$CdCl_2^\circ$	−84.90	Smith and Martell (1976)	K°: Cd^{2+}, Cl^-
$CdCl_3^-$	−116.00	Smith and Martell (1976)	K°: Cd^{2+}, Cl^-

(Continued)

Cd (Cadmium) (Continued)

Species	ΔG_f°	Source	Remarks
$CdCl_4^{2-}$	−147.50	Smith and Martell (1976)	K°: Cd^{2+}, Cl^-
CdI^+	−34.1	Bates and Vosburgh (1938)	K°: Cd^{2+}, I^-
CdI_2°	−48.72	Bates and Vosburgh (1938)	K°: CdI^+, I^-
CdI_3^-	−62.57	Bates and Vosburgh (1938)	K°: CdI_2°, I^-
CdI_4^{2-}	−76.31	Smith and Martell (1976)	K°: Cd^{2+}, I^-
$CdNH_3^{2+}$	−28.42	Smith and Martell (1976)	K°: Cd^{2+}, NH_3°
$Cd(NH_3)_2^{2+}$	−37.49	Smith and Martell (1976)	K°: Cd^{2+}, NH_3°
$Cd(NH_3)_3^{2+}$	−45.65	Smith and Martell (1976)	K°: Cd^{2+}, NH_3°
$Cd(NH_3)_4^{2+}$	−53.12	Smith and Martell (1976)	K°: Cd^{2+}, NH_3°
$CdNO_3^+$	−45.67	Vasil'ev (1961, 1962)	K°: Cd^{2+}, NO_3^-
$Cd(NO_3)_2^\circ$	−71.89	Smith and Martell (1976)	K°: Cd^{2+}, NO_3^-
$CdOH^+$	−61.52	Smith and Martell (1976)	K°: Cd^{2+}
$Cd(OH)_2^\circ$	−104.3	Smith and Martell (1976)	K°: Cd^{2+}
$Cd(OH)_3^-$	−143.64	Wagman et al. (1968)	K°: Cd^{2+}
$Cd(OH)_4^{2-}$	−180.85	Smith and Martell (1976)	K°: Cd^{2+}
$Cd(OH)_5^{3-}$	−217.57	Spivakovskii and Moisa (1964)	K°: $Cd(OH)_4^{2-}$
$Cd(OH)_6^{4-}$	−253.96	Spivakovskii and Moisa (1964)	K°: $Cd(OH)_5^{3-}$
Cd_2OH^{3+}	−85.18	Anotonetti et al. (1976)	K°: Cd^{2+}
$Cd_4(OH)_4^{4+}$	−263.10	Anotonetti et al. (1976)	K°: Cd^{2+}
CdO(monteponite)	−54.65	Adami and King (1965)	ΔH_r°: $CdSO_4(c)$, $H_2SO_4(7.06H_2O)$
		Millar (1928)	S°
β-$Cd(OH)_2(c)$	−113.37	Smith and Martell (1976)	K°: Cd^{2+}
		Provost and Wulff (1970)	ΔH_r°: Cd^{2+}
$CdHPO_4^\circ$	−285.00	Santillan-Medrano and Jurinak (1975)	K°: Cd^{2+}, HPO_4^{2-}
$CdP_2O_7^{2-}$	−487.72	Wolhoff and Overbeek (1959)	K°: Cd^{2+}, $P_2O_7^{4-}$
$Cd_3(PO_4)_2(c)$	−598.16	Santillan-Medrano and Jurinak (1975)	K°: Cd^{2+}, PO_4^{3-}
$CdSO_4^\circ$	−199.90	Smith and Martell (1976)	K°: Cd^{2+}, SO_4^{2-}
CdS(greennokite)	−35.03	Adami and King (1964)	ΔH_r°: Cd^{2+}, S^{2-}
		Kelley and King (1961)	S°
$CdSO_4(c)$	−196.62	Popadopoulos and Giauque (1955)	ΔH_r°: $CdO(c)$, $H_2SO_4(7.06H_2O)$
		Popadopoulos and Giauque (1955)	S°
$CdSO_4 \cdot H_2O(c)$	−255.42	Papadopoulos and Giauque (1955)	ΔH_r°, S°
$CdSO_4 \cdot 2Cd(OH)_2(c)$	−429.63	Wagman et al. (1968)	K°: $CdSO_4(c)$, $Cd(OH)_2(c)$
$2CdSO_4 \cdot Cd(OH)_2(c)$	−515.93	Wagman et al. (1968)	K°: $CdSO_4(c)$, $Cd(OH)_2(c)$
$CdSiO_3(c)$	−264.18	Barany (1959)	ΔH_r°: $CdO(c)$, $SiO_2(q)$
		King (1959)	S°

Cl (Chlorine)

Species	ΔG_f°	Source	Remarks
Cl⁻ (chloride ion)	−31.371	CODATA (1971)	$\Delta H_f^\circ, S^\circ$
HCl(g)	−22.77	CODATA (1971)	$\Delta H_f^\circ, S^\circ$
Cl₂(g, chlorine)	0.0	Standard state	ΔH_f°
		CODATA (1971)	S°

Cu (Copper)

Species	ΔG_f°	Source	Remarks
Cu(c)	0	Standard state	ΔH_f°
		Furukawa et al. (1968)	S°
Cu⁺ (cuprous ion)	12.10	Malyszko and Duda (1975)	$K^\circ: Cu^{2+}$
Cu²⁺ (cupric ion)	15.67	CODATA (1977a)	$\Delta H_f^\circ, S^\circ$
CuCO₃	−119.69	Schindler et al. (1968)	$K^\circ: Cu^{2+}, CO_3^{2-}$
Cu(CO₃)₂²⁻	−250.10	Schindler et al. (1968)	$K^\circ: Cu^{2+}, CO_3^{2-}$
CuHCO₃⁺	−127.46	Mattigod and Sposito (1977)	Est: Cu^{2+}, HCO_3^-
CuCO₃(c)	−123.65	Smith and Martell (1976)	$K^\circ: Cu^{2+}, CO_3^{2-}$
Cu₂(OH)₂(CO₃) (malachite)	−215.26	Schindler et al. (1968)	$K^\circ: Cu^{2+}, CO_3^{2-}$
Cu₃(OH)₂(CO₃)₂ (azurite)	−341.57	Schindler et al. (1968)	$K^\circ: Cu^{2+}, CO_3^{2-}$
CuCl°	−22.95	Smith and Martell (1976)	$K^\circ: Cu^+, Cl^-$
CuCl⁺	−16.25	Smith and Martell (1976)	$K^\circ: Cu^{2+}, Cl^-$
CuCl₂	−46.91	Bjerrum (1946)	$K^\circ: CuCl^+, Cl^-$
CuCl₂⁻	−58.16	Smith and Martell (1976)	$K^\circ: Cu^+, Cl^-$
CuCl₃	−76.30	Bjerrum (1946)	$K^\circ: CuCl_2, Cl^-$
CuCl₃²⁻	−89.79	Smith and Martell (1976)	$K^\circ: Cu^+, Cl^-$
Cu₂Cl₄²⁻	−119.15	Smith and Martell (1976)	$K^\circ: Cu^+, Cl^-$
α-CuFe₂O₄ (cupric ferrite)	−205.3	Barany et al. (1964)	$\Delta H_f^\circ: CuO(c), Fe_2O_3(c)$
		Barany et al. (1964)	S°
α-Cu₂Fe₂O₄ (cuprous ferrite)	−229.04	Barany et al. (1964)	$\Delta H_f^\circ: CuO(c)$
		Barany et al. (1964)	S°
CuMoO₄(c)	−193.01	Naumov et al. (1971)	$K^\circ: Cu^{2+}, MoO_4^{2-}$
CuNO₃⁺	−11.65	Smith and Martell (1976)	$K^\circ: Cu^{2+}, NO_3^-$
Cu(NO₃)₂	−37.06	Smith and Martell (1976)	$K^\circ: Cu^{2+}, NO_3^-$
CuOH⁺	−30.52	Smith and Martell (1976)	$K^\circ: Cu^{2+}$
Cu(OH)₂	−78.91	Dodig and Pavlovic (1969)	$K^\circ: Cu^{2+}$
Cu(OH)₃⁻	−117.90	Dodig and Pavlovic (1969)	$K^\circ: Cu^{2+}$
Cu(OH)₄²⁻	−157.08	Smith and Martell (1976)	$K^\circ: Cu^{2+}$
Cu₂(OH)₂²⁺	−67.47	Arnek and Patel (1968)	$K^\circ: Cu^{2+}$
CuO (tenorite)	−30.57	Mah et al. (1967)	ΔH_f°
		Kelley and King (1961)	S°
CuOH(c)	−45.54	Shchukarev et al. (1953)	$K^\circ: Cu^+$
Cu(OH)₂(c)	−85.87	Schorsch (1965)	$K^\circ: Cu^{2+}$

(Continued)

Cu (Copper) (*Continued*)

Species	ΔG_f°	Source	Remarks
Cu_2O(cuperite)	−35.45	Mah et al. (1967)	ΔH_f°
		Kelley and King (1961)	S°
$CuHPO_4^\circ$	−250.72	Sigel et al. (1967)	$K^\circ: Cu^{2+}, HPO_4^{2-}$
$CuH_2PO_4^+$	−258.35	Mattigod and Sposito (1977)	Est: $Cu^{2+}, H_2PO_4^-$
$CuP_2O_7^{2-}$	−450.63	Wagman et al. (1969)	$\Delta G_f^\circ: Cu^{2+}, P_2O_7^{4-}$
$CuHP_2O_7^-$	−461.73	Schupp et al. (1963)	$K^\circ: Cu^{2+}, HP_2O_7^{3-}$
$CuH_2P_2O_7^\circ$	−467.03	Schupp et al. (1963)	$K^\circ: Cu^{2+}, H_2P_2O_7^{2-}$
$Cu_2P_2O_7^\circ$	−425.86	Wagman et al. (1969)	$\Delta G_f^\circ: Cu^{2+}, P_2O_7^{4-}$
$Cu_3(PO_4)_2$(c)	−493.63	Wagman et al. (1969)	$\Delta G_f^\circ: Cu^+, PO_4^{3-}$
$Cu_3(PO_4)_2 \cdot 2H_2O$(c)	−609.60	Volkov et al. (1976)	$\Delta H_f^\circ, S_f^\circ: Cu_3(PO_4)_2$
$Cu_2P_2O_7$(c)	−446.66	Wagman et al. (1969)	$\Delta G_f^\circ: Cu^{2+}, SO_4^{2-}$
$CuSO_4^\circ$	−165.50	Wagman et al. (1969)	$\Delta G_f^\circ: Cu^{2+}, SO_4^{2-}$
CuS(covellite)	−13.06	Smith and Martell (1976)	$K^\circ: Cu^{2+}, S^{2-}$
Cu_2S(chalcocite)	−21.50	Smith and Martell (1976)	$K^\circ: Cu^{2+}, S^{2-}$
$CuSO_4$(chalcocyanite)	−157.21	Adami and King (1965)	$\Delta H_f^\circ: CuO(c), H_2SO_4(7H_2O)$
$CuSO_4 \cdot 5H_2O$(c)	−449.27	Larson et al. (1968)	$\Delta H_f^\circ, S^\circ$
Cu_2SO_4(c)	−156.41	Garrels and Christ (1965)	$\Delta G_f^\circ: Cu^{2+}, SO_4^{2-}$
$CuO \cdot CuSO_4$(c)	−187.61	Stuve et al. (1975)	$\Delta H_f^\circ, S^\circ$
$Cu_4(OH)_6SO_4$ (bronchantite)	−434.46	Smith and Martell (1976)	$K^\circ: Cu^{2+}, SO_4^{2-}$
$Cu_4(OH)_6SO_4 \cdot 1.3H_2O$(c)	−505.53	Sillen and Martell (1964)	$K^\circ: Cu^{2+}, SO_4^{2-}$

e⁻ (electron)

Species	ΔG_f°	Source
e^-	0.00	Standard state ($S^\circ = 15.603$ cal deg^{-1} mol^{-1})

F (Fluorine)

Species	ΔG_f°	Source	Remarks
F^- (fluoride ion)	67.93	Johnson et al. (1973)	$\Delta H_f^\circ: HF(l)$
		CODATA (1976)	S°
HF(g)	−65.82	CODATA (1976)	$\Delta H_f^\circ, S^\circ$
F_2(g, fluorine)	0.0	Standard state	ΔH_f°
		CODATA (1976)	S°

Fe (Iron)

Species	ΔG_f°	Source	Remarks
Fe(α, c)	0.00	Standard state Stull et al. (1971)	ΔH_f° S°
Fe^{2+}	-21.80	Larson et al. (1968)	$\Delta H_r^\circ, S_r^\circ: \Delta G_r^\circ$ $FeSO_4 \cdot 7H_2O(cl)$, SO_4^{2-}
Fe^{3+}	-4.02	Larson et al. (1968)	$E^\circ: Fe^{2+}/Fe^{3+}$
$FeBr^{2+}$	-28.07	Smith and Martell (1976)	$K^\circ: Fe^{3+}, Br^-$
$FeBr_2^\circ$	-71.54	Wagman et al. (1969)	$K^\circ: Fe^{3+}, Br^-$
$FeBr_3^\circ$	-78.68	Wagman et al. (1969)	$K^\circ: Fe^{3+}, Br^-$
$FeCO_3$(siderite)	-161.95	Singer and Stumm (1970)	$K^\circ: Fe^{2+}, CO_3^{2-}$
$FeCl^{2+}$	-37.41	Smith and Martell (1976)	$K^\circ: Fe^{3+}, Cl^-$
$FeCl_2^\circ$	-84.44	Rabinowitch and Stockmayer (1942)	$K^\circ: Fe^{2+}, Cl^-$
$FeCl_2^+$	-69.67	Smith and Martell (1976)	$K^\circ: Fe^{3+}, Cl^-$
$FeCl_3^\circ$	-99.19	Rabinowitch and Stockmayer (1942)	$K^\circ: FeCl_2^+, Cl^-$
$FeCl_2$(lawrencite)	-72.27	Koehler and Coughlin (1959) Wilson and Gregory (1958)	ΔH_f° S°
$FeCl_3$(molysite)	-80.06	Stull et al. (1966) Wilson and Gregory (1958)	ΔH_f° S°
FeF^{2+}	-80.14	Smith and Martell (1976)	$K^\circ: Fe^{3+}, F^-$
FeF_2°	-157.7	Wagman et al. (1969)	$K^\circ: Fe^{2+}, F^-$
FeF_2^+	-152.43	Smith and Martell (1976)	$K^\circ: Fe^{3+}, F^-$
FeF_3°	-223.77	Smith and Martell (1976)	$K^\circ: Fe^{3+}, F^-$
$FeMoO_4(c)$	-232.14	Naumov et al. (1971)	$K^\circ: Fe^{2+}, MoO_4^{2-}$
$FeNO_3^{2+}$	-32.02	Mattoo (1959)	$K^\circ: Fe^{3+}, NO_3^-$
$FeOH^+$	-69.29	Bolzan and Arvia (1963)	$K^\circ: Fe^{2+}$
$FeOH^{2+}$	-57.72	Milburn (1957)	$K^\circ: Fe^{3+}$
$Fe(OH)_2^\circ$	-113.29	Hem and Cropper (1959)	$K^\circ: FeOH^+$
$Fe(OH)_2^+$	-109.63	Smith and Martell (1976)	$K^\circ: Fe^{2+}$
$Fe(OH)_3^-$	-148.22	Smith and Martell (1976)	$K^\circ: Fe^{2+}$
$Fe(OH)_3^\circ$	-156.22	Hem and Cropper (1959)	$K^\circ: Fe(OH)_2^+$
$Fe(OH)_4^{2-}$	-185.27	Smith and Martell (1976)	$K^\circ: Fe^{2+}$
$Fe(OH)_4^-$	-201.32	Smith and Martell (1976)	$K^\circ: Fe^{3+}$
$Fe_2(OH)_2^{4+}$	-117.46	Smith and Martell (1976)	$K^\circ: Fe^{3+}$
$Fe_3(OH)_4^{2-}$	-230.23	Smith and Martell (1976)	$K^\circ: Fe^{3+}$
$Fe_{0.95}O$(wustite)	-58.68	Humphrey et al. (1952)	$\Delta H_f, S^\circ$
$FeO(c)$	-60.10	Stull et al. (1971)	$\Delta H_f^\circ, S^\circ$
α-Fe_2O_3(hematite)	-177.85	Stull et al. (1971) Gronvold and Westrum (1959)	ΔH_f° S°
γ-Fe_2O_3(maghemite)	-173.75	Sadiq et al. (1979)	$K^\circ: Fe^{3+}$
Fe_3O_4(magnetite)	-243.47	Salmon (1961) Gronvold and Westrum (1959)	ΔH_f° S°

(Continued)

Fe (Iron) (*Continued*)

Species	ΔG_f°	Source	Remarks
α-FeOOH(goethite)	−117.42	Schmalz (1959)	ΔG_r°: $Fe_2O_3(c)$
γ-FeOOH(lepidocrocite)	−115.50	Schuylenborgh (1973)	K°: Fe^{3+}
$Fe(OH)_2(c)$	−117.58	Smith and Martell (1976)	K°: Fe^{2+}
$Fe(OH)_3$(amorphous)	−169.25	Schindler et al. (1963)	K°: Fe^{3+}
$Fe(OH)_3$(soil)	−170.40	Norvell and Lindsay (1981)	K°: Fe^{3+}
$Fe_3(OH)_8$(fresh precipitate)	−459.22	Arden (1950)	K°: Fe^{2+}, $Fe(OH)_3$(amorp)
$FeHPO_4^\circ$	−288.74	Nriagu (1972b)	K°: Fe^{2+}, HPO_4^{2-}
$FeHPO_4^+$	−280.93	Lahiri (1965)	K°: Fe^{3+}, HPO_4^{2-}
$FeH_2PO_4^+$	−297.33	Nriagu (1972b)	K°: Fe^{2+}, $H_2PO_4^-$
$FeH_2PO_4^{2+}$	−283.28	Nriagu (1972a)	K°: Fe^{3+}, $H_2PO_4^-$
$FePO_4(c)$	−283.19	Wagman et al. (1969)	ΔH_f°
		Naumov et al. (1971)	S°
$FePO_4 \cdot 2H_2O$(strengite)	−398.59	Nriagu (1972a)	K : Fe^{3+}, PO_4^{3-}
$Fe_3(PO_4)_2 \cdot 8H_2O$ (vivianite)	−1058.36	Nriagu (1972b)	K°: Fe^{2+}, PO_4^{3-}
$Fe_2P_2O_7(c)$	−524.84	Teterevkov and Pechkovskii (1974)	ΔG_r°: $FePO_4(c)$
$FeSO_4^\circ$	−202.75	Izatt et al. (1969)	K°: Fe^{2+}, SO_4^{2-}
$FeSO_4^+$	−187.63	Willix (1963)	K°: Fe^{3+}, SO_4^{2-}
$Fe(SO_4)_2^-$	−367.26	Smith and Martell (1976)	K°: Fe^{3+}, SO_4^{2-}
α-$Fe_{0.95}S$(pyrrhotite, Fe-rich)	−23.98	Adami and King (1964)	ΔH_f°
		Kelley and King (1961)	S°
α-FeS(troilite)	−23.40	Smith and Martell (1976)	K°: Fe^{2+}, S^{2-}
FeS_2(pyrite)	−38.78	Gronvold and Westrum (1976)	ΔH_f°
		Gronvold and Westrum (1962)	S°
FeS_2(markasite)	−37.83	Gronvold and Westrum (1976)	ΔG_r°: Fe_2S(pyrite)
$Fe_2S_3(c)$	−66.54	Jellinek and Gorden (1924)	K°: Fe^{3+}, S^{2-}
$FeSO_4(c)$	−196.13	Wagman et al. (1969)	ΔH_f°
		Moore and Kelley (1942)	S°
$FeSO_4 \cdot 7H_2O(c)$	−599.92	Adami and Kelley (1963)	ΔH_f°
		Kelley and King (1961)	S°
$Fe_2(SO_4)_3(c)$	−537.95	Barany and Adami (1965)	ΔH_f°
		Naumov et al. (1971)	S°
$KFe_3(SO_4)_2(OH)_6$ (jarosite)	−792.66	Brown (1970)	K°: K^+, Fe^{3+}, SO_4^{2-}
$FeSiO_3(c)$	−257.60	Latimer (1952)	ΔG_f°: Fe^{2+} $SiO_2(q)$
Fe_2SiO_4(fayalite)	−329.31	Kelley (1962)	ΔG_f°: $Fe_{0.95}O$, $SiO_2(q)$

H (Hydrogen)

Species	ΔG_f°	Source	Remarks
H^+	0.0	Reference state CODATA (1976)	ΔH_f° S°
OH^-	−37.594	CODATA (1976)	$\Delta H_f^\circ, S^\circ$
$H_2(g)$	0.0	Reference state CODATA (1976)	$\Delta H_f^\circ, S^\circ$ S°
$H_2O(l)$	−56.687	CODATA (1976)	$\Delta H_f^\circ, S^\circ$
$H_2O(g)$	−54.634	CODATA (1976)	$\Delta H_f^\circ, S^\circ$

Hg (Mercury)

Species	ΔG_f°	Source	Remarks
$Hg(g)$	7.613	Hepler and Olofsson (1975)	$\Delta H_f^\circ, S^\circ$
$Hg(l)$	0.0	Standard state Kelley and King (1961)	ΔH_f° S°
Hg°	8.90	Glew and Hames (1971)	$\Delta H_f^\circ, S^\circ$
Hg^{2+}	39.365	Vanderzee and Swanson (1974)	$E^\circ: Hg_2^{2+}$
$Hg_2(g)$	16.3	Wagman et al. (1969)	
Hg_2^{2+}	36.713	Vanderzee and Swanson (1974)	$E^\circ: Hg(l)/Hg_2^{2+}$
$HgBr(g)$	16.4	Stull et al. (1971)	$\Delta H_f^\circ, S^\circ$
$HgBr_2(g)$	−27.25	Stull et al. (1971)	$\Delta H_f^\circ, S^\circ$
$HgBr^+$	1.21	Malcolm et al. (1961) Malcolm et al. (1961)	$\Delta H_r^\circ: Br^-, Hg^{2+}$ $\Delta S_r^\circ: Br^-, Hg^{2+}$
$HgBr_2^\circ$	−34.26	Williams (1954)	$K^\circ: Br^-, Hg^{2+}$
$HgBr_3^-$	−62.1	Wagman et al. (1969)	$K^\circ: Br^-, Hg^{2+}$
$HgBr_4^{2-}$	−88.71	Wagman et al. (1969)	$K^\circ: Br^-, Hg^{2+}$
$HgBrCl^\circ$	−38.6	Hepler and Olofsson (1975)	
$HgBrI^\circ$	−26.6	Hepler and Olofsson (1975)	
$HgBrI_3^{2-}$	−61.1	Hepler and Olofsson (1975)	
$HgBr_2I_2^{2-}$	−71.08	Hepler and Olofsson (1975)	
$HgBr_3I^{2-}$	−80.47	Hepler and Olofsson (1975)	
$HgBr(OH)^\circ$	−38.6	Hepler and Olofsson (1975)	
$HgBr_2(c)$	−36.67	Wagman et al. (1969) Stull et al. (1971)	ΔH_f° S°
$Hg_2Br_2(c)$	−43.22	Hepler and Olofsson (1975)	$E^\circ: Hg_2Br_2(c)/Hg(l)$
$Hg(CH_3)_2(g)$	34.9	Wagman et al. (1969)	?
$HgCH_3NH_2^{2+}$	32.5	Wagman et al. (1969)	
$Hg(CH_3NH_2)_2^{2+}$	24.9	Wagman et al. (1969)	
$HgClCH_3NH_2^+$	−8.3	Hepler and Olofsson (1975)	
$Hg(CH_3COO)^+$	−53.14	Wagman et al. (1969)	$K^\circ: Hg^{2+}, CH_3COO^-$

(*Continued*)

Hg (Mercury) (Continued)

Species	ΔG_f°	Source	Remarks
$Hg_2(CH_3COO)_2(c)$	−152.99	Hepler and Olofsson (1975)	
$Hg_2(C_2O_4)_2^{2-}$	−295.01	Hepler and Olofsson (1975)	$K^\circ: Hg_2^{2+}, C_2O_4^{2-}$
$Hg_2C_2O_4(c)$	−141.81	Wagman et al. (1969)	$K^\circ: Hg_2^{2+}, C_2O_4^{2-}$
$HgCO_3(c)$	−117.62	Saegusta (1950)	$\Delta H_f^\circ, S^\circ$
$Hg_2CO_3(c)$	−108.52	Heitman and Hogfeldt (1976)	$K^\circ: Hg_2^{2+}, CO_3^{2-}$
$HgCl(g)$	14.95	Wagman et al. (1969)	$\Delta H_f^\circ, S^\circ$
$HgCl^+$	−1.23	Hepler and Olofsson (1975)	$K^\circ: Cl^-, Hg^{2+}$
$HgCl_2^\circ$	−41.33	Wagman et al. (1969)	$K^\circ: Cl^-, HgCl^+$
$HgCl_3^-$	−73.91	Williams (1954)	$K^\circ: HgCl_2^\circ, Cl^-$
$HgCl_4^{2-}$	−105.85	Dubinskii and Shul'man (1970)	$K^\circ: HgCl_3^-, Cl^-$
$HgClI^\circ$	−30.7	Hepler and Olofsson (1975)	
$HgClOH^\circ$	−53.1	Partridge et al. (1965)	
$HgCl_2(g)$	−33.9	Johnson et al. (1966)	
$HgCl_2(c)$	−44.80	Abraham et al. (1970)	ΔH_f°
		Hepler and Olofsson (1975)	S°
$Hg_2Cl_2(calomel)$	−50.374	Hepler and Olofsson (1975)	$E^\circ: Hg_2Cl_2(c)/Hg(l), Cl^-$
		Vanderzee and Swanson (1974)	S
$HgF(g)$	−4.35	Gaydon (1953)	ΔH_f°
		Stull et al. (1971)	S°
$HgF_2(g)$	−86.48	Stull et al. (1971)	$\Delta H_f^\circ, S^\circ$
HgF^+	−30.69	Paul (1955)	$K^\circ: F^-, Hg^{2+}$
$Hg_2F_2(g)$	−102.17	Stull et al. (1971)	$\Delta H_f^\circ, S^\circ$
$HgH(g)$	51.0	Stull et al. (1971)	ΔH_f°
		Feber and Ferrick (1967)	S°
$HgI(g)$	21.16	Schumm et al. (1973)	ΔH_f°
		Stull et al. (1971)	S°
$HgI_2(g)$	−14.36	Wagman et al. (1969)	ΔH_f°
		Stull et al. (1971)	S°
HgI^+	9.49	Malcolm et al. (1961)	$\Delta H_r^\circ, \Delta S_r^\circ: Hg^{2+}, I^-$
HgI_2°	−18.22	Wagman et al. (1969)	$K^\circ: Hg^{2+}, I^-$
HgI_3^-	−35.66	Panthaleon (1958)	$K^\circ: Hg^{2+}, I^-$
HgI_4^{2-}	−50.79	Panthaleon (1958)	$K^\circ: Hg^{2+}, I^-$
$HgI(OH)^\circ$	−41.4	Hepler and Olofsson (1975)	
$HgI_2(c)$	−24.43	Wagman et al. (1969)	ΔH_f°
		Stull et al. (1971)	S°
$Hg_2I_2(c)$	−26.69	Smith and Martell (1976)	$K^\circ: Hg_2^{2+}, I^-$
$HgNH_3^{2+}$	21.03	Bjerrum (1972)	$K^\circ: Hg^{2+}, NH_3^\circ$
$Hg(NH_3)_2^{2+}$	2.83	Bjerrum (1972)	$K^\circ: NH_3^\circ, Hg(NH_3)^{2+}$
$Hg(NH_3)_3^{2+}$	−4.86	Bjerrum (1972)	$K^\circ: NH_3^\circ, Hg(NH_3)^{2+}$
$Hg(NH_3)_4^{2+}$	−12.28	Bjerrum (1972)	$K^\circ: Hg^{2+}, NH_3^\circ$
$Hg(NO_3)^+$	12.72	Hepler and Olofsson (1975)	$K^\circ: Hg^{2+}, NO_3^-$
$Hg(NO_3)_2^\circ$	−12.06	Hepler and Olofsson (1975)	$K^\circ: Hg^{2+}, NO_3^-$

Hg (Mercury) (*Continued*)

Species	ΔG_f°	Source	Remarks
$HgOH^+$	−12.69	Smith and Martell (1976)	K°: Hg^{2+}
$Hg(OH)_2^\circ$	−65.56	Smith and Martell (1976)	K°: Hg^{2+}
$Hg(OH)_3^-$	−101.92	Smith and Martell (1976)	K°: Hg^{2+}
HgO(red orthorhombic)	−13.996	Vanderzee et al. (1974) Kelley and King (1961)	ΔH_f° S°
HgO(red, hexagonal)	−13.94	Vanderzee et al. (1974)	ΔG_r° = HgO(red, orth)
Hg(yellow, orthorhombic)	−13.97	Vanderzee et al. (1974)	ΔG_r°: HgO(red, orth)
$Hg(OH)_2(c)$	−70.47	Humphrey (1951)	K°: Hg^{2+}
$Hg_2(OH)_2(c)$	−69.49	Zhuk (1954)	K°: Hg_2^{2+}
$Hg_2P_2O_7^{2-}$	−433.46	Watters and Simonaitis (1964)	K°: Hg_2^{2+}, $P_2O_7^{4-}$
$Hg_2(OH)P_2O_7^{3-}$	−479.54	Watters and Simonaitis (1964)	K°: Hg_2^{2+}, $P_2O_7^{4-}$
$Hg_2(OH)_2P_2O_7^{4-}$	−525.00	Hepler and Olofsson (1975)	K°: Hg_2^{2+}, $P_2O_7^{4-}$
$Hg_2(P_2O_7)_2^{6-}$	−880.49	Watters and Simonaitis (1964)	K°: Hg_2^{2+}, $P_2O_7^{4-}$
$Hg_2HPO_4(c)$	−242.23	Smith and Martell (1976)	K°: Hg_2^{2+}, HPO_4^{2-}
HgS_2^{2-}	10.82	Wagman et al. (1969)	K°: Hg^{2+}, S_2^{2-}
$Hg(HS)_2^\circ$	−6.34	Wagman et al. (1969)	K°: Hg^{2+}, HS°
$HgSO_4^{2-}$	−140.51	Wagman et al. (1969)	K°: Hg^{2+}, SO_4^{2-}
α-HgS(red, cinnabar)	−11.09	Hepler and Olofsson (1975) King and Weller (1962)	ΔH_f° S°
β-HgS(black, metacinnabar)	−10.59	Hepler and Olofsson (1975)	ΔH_f°, S°
$Hg_2S(c)$	−17.49	Wilcox and Bromley (1963), Kireev (1947)	ΔH_f° S°
$HgSO_4(c)$	−143.14	Wagman et al. (1969) Kostryakov (1961)	ΔH_f° S°
$Hg_2SO_4(c)$	−149.70	Gardner et al. (1969)	E°: $Hg_2SO_4(c)/Hg(l)$

I (Iodine)

Species	ΔG_f°	Source	Remarks
I^- (iodide ion)	−12.38	Johnson (1977) CODATA (1971)	ΔH_f° S°
HI(g)	0.38	Taylor and Christ (1941) CODATA (1971)	ΔH_f° S°
I_2(g, iodine)	4.63	CODATA (1971) Evans et al. (1955)	ΔH_f° S°
I_2(c, iodine)	0.0	Standard state CODATA (1971)	ΔH_f° S°
IO_3^- (iodate)	−30.6	Wagman et al. (1968)	?

K (Potassium)

Species	ΔG_f°	Source	Remarks
K(c)	0.0	Standard state CODATA (1976)	ΔH_f° S°
K^+	−67.51	CODATA (1976)	$\Delta H_f^\circ, S^\circ$
KCl°	−97.93	Paterson et al. (1971)	$K^\circ: K^+, Cl^-$
$K_2CO_3^\circ$	−261.18	Karapet'yants and Karapet'yants (1970)	$K^\circ: K^+, CO_3^{2-}$
KOH°	−104.42	Smith and Martell (1976)	$K^\circ: K^+$
$KH_2PO_4(c)$	−339.65	Luff and Reed (1978b) Kelley and King (1961)	ΔH_f° S°
$K_2HPO_4(c)$	−392.14	Luff and Reed (1978b) Luff and Reed (1978a)	ΔH_f° S°
KSO_4^-	−246.62	Smith and Martell (1976)	$K^\circ: K^+, SO_4^{2-}$

Mg (Magnesium)

Species	ΔG_f°	Source	Remarks
Mg(c)	0.0	Standard state CODATA (1977a)	ΔH_f° S°
Mg^{2+}	−109.01	Hostetler (1963)	$K^\circ: Mg(OH)_2(c)$
$MgCl_2^\circ$	−171.71	Parker et al. (1971)	$K^c: Mg^{2+}, Cl^-$
$MgCO_3^\circ$	−239.61	Nakayama (1971)	$K^\circ: Mg^{2+}, CO_3^{2-}$
$MgHCO_3^+$	−250.74	Siebert and Hostetler (1977)	$K^\circ: Mg^{2+}, HCO_3^-$
$MgCO_3$(magnesite)	−245.37	Smith and Martell (1976)	$K^\circ: Mg^{2+}, CO_3^{2-}$
$MgCO_3 \cdot 3H_2O$ (nesquehonite)	−411.62	Smith and Martell (1976)	$K^\circ: Mg^{2+}, CO_3^{2-}$
$MgCO_3 \cdot 5H_2O$ (lansfordite)	−524.82	Smith and Martell (1976)	$K^\circ: Mg^{2+}, CO_3^{2-}$
$MgCa(CO_3)_2$ (dolomite)	−518.25	Stout and Robie (1963)	$\Delta G_r^\circ: CaCO_3(c)$ $MgCO_3(c)$
$MgMoO_4(c)$	−309.70	Barany (1962b) Weller and King (1963)	ΔH_f° S°
$Mg(NO_3)_2^\circ$	−162.27	Parker et al. (1971)	$K^\circ: Mg^{2+}, NO_3^-$
$MgOH^+$	−150.08	Liu and Nancollas (1973)	$K^\circ: Mg^{2+}$
$Mg(OH)_2^\circ$	−184.21	Parker et al. (1971)	$K^\circ: Mg^{2+}$
MgO(periclase)	−136.04	CODATA (1976)	$\Delta H_f^\circ, S^\circ$
$Mg(OH)_2$(brucite)	−199.41	King et al. (1975)	$\Delta G_r^\circ: MgO(c)$
$MgHPO_4$	−375.01	Taylor et al. (1963b)	$K^\circ: Mg^{2+}, HPO_4^{2-}$
$MgHPO_4 \cdot 3H_2O$ (newberryite)	−549.04	Taylor et al. (1963b)	$K^\circ: Mg^{2+}, HPO_4^{2-}$
$MgKPO_4 \cdot 6H_2O(c)$	−776.31	Taylor et al. (1963b)	$K^c: Mg^{2+}, K^+, PO_4^{3-}$
$MgNH_4PO_4 \cdot 6H_2O$ (struvite)	−731.24	Taylor et al. (1963a)	$K^\circ: Mg^{2+}, NH_4^+,$ PO_4^{3-}
$Mg_3(PO_4)_2(c)$	−837.30	Oetting and McDonald (1963)	$\Delta H_f^\circ, S^\circ$

Mg (Magnesium) (*Continued*)

Species	ΔG_f°	Source	Remarks
$Mg_3(PO_4)_2 \cdot 8H_2O$ (boberrite)	−1304.99	Taylor et al. (1963b)	$K^\circ: Mg^{2+}, PO_4^{3-}$
$Mg_3(PO_4)_2 \cdot 22H_2O(c)$	−2096.01	Taylor et al. (1963b)	$K^\circ: Mg^{2+}, PO_4^{3-}$
$MgSO_4^\circ$	−290.00	Smith and Martell (1976)	$K^\circ: Mg^{2+}, SO_4^{2-}$
$MgS(c)$	−80.88	Wartenberg (1943)	ΔH_f°
		Chase et al. (1974b)	S°
$MgSO_4(c)$	−275.80	Knopf and Staude (1955)	ΔH_f°
		Kelley and King (1961)	S°
$MgSiO_3$(clinoenstatite)	−349.41	Torgeson and Sahama (1948)	ΔH_f°
		Kelley (1943)	S°
$Mg_{1.6}Fe(II)_{0.4}SiO_4$ (olivine)	−460.08	Huang and Keller (1972)	$K^\circ: Mg^{2+}, Mg(OH)^+, H_4SiO_4^\circ, Fe^{2+}, Fe(OH)^+$
Mg_2SiO_4(forsterite)	−491.30	Hemley et al. (1977)	$\Delta G_f^\circ: S^\circ$
$Mg_2Si_3O_6(OH)_4$ (sepiolite)	−1020.95	Christ et al. (1973)	$K^\circ: Mg^{2+}, H_4SiO_4^\circ$
$Mg_3Si_2O_5(OH)_4$ (chrysotolite)	−964.20	Hemingway and Robie (1977)	$\Delta H_f^\circ, S^\circ$
$Mg_3Si_4O_{10}(OH)_2$(talc)	−1320.56	Hemley et al. (1976a, 1976b)	$K: Mg^{2+}, H_4SiO_4^\circ$
$Mg_3Si_4O_{10}(OH)_2 \cdot 2H_2O$(vermiculite)	−1422.85	Nriagu (1975)	Est.
$Mg_6Si_4O_{10}(OH)_8$ (serpentine)	−1933.85	King et al. (1967)	$\Delta H_r^\circ, S^\circ: MgO(c), SiO_2(q)$
		King and Armstrong (1968)	

Mn (Manganese)

Species	ΔG_f°	Source	Remarks
α-Mn(c)	0.00	Standard state	ΔH_f°
		Hultgren et al. (1963)	S°
Mn^{2+}	−55.11	Covington et al. (1962)	$E^\circ: \beta\text{-}MnO_2/Mn^{2+}$
Mn^{3+}	−20.26	Grube and Huberich (1923)	$E^\circ: Mn^{2+}/Mn^{3+}$
Mn^{4+}	14.53	Lisov (1967)	$E^\circ: Mn^{2+}/Mn^{4+}$
$MnCl^+$	−87.31	Wagman et al. (1969)	$\Delta G_f^\circ: Mn^{2+}, Cl^-$
$MnCl_2^\circ$	−117.91	Wagman et al. (1969)	$\Delta G_f^\circ: Mn^{2+}, Cl^-$
$MnCO_3^\circ$	−180.32	Rossini et al. (1952)	$\Delta G_f^\circ: Mn^{2+}, CO_3^{2-}$
$MnHCO_3^+$	−197.84	Smith and Martell (1976)	$K^\circ: HCO_3^-, Mn^{2+}$
$MnCO_3$(rhodochrosite)	−195.03	Robie (1965)	$\Delta H_f^\circ, S^\circ$
$MnMoO_4(c)$	−260.59	Barany (1965)	$\Delta H_r^\circ: MnO(c), MoO_3(c)$
		Naumov et al. (1971)	S

(*Continued*)

Mn (Manganese) (Continued)

Species	ΔG_f°	Source	Remarks
$MnOH^+$	−97.35	Perrin (1962)	$K^\circ: Mn^{2+}$
$MnOH^{2+}$	−77.49	Wells and Davies (1967)	$\Delta G_f^\circ: Mn^{3+}$
$Mn(OH)_3^-$	−178.80	Fox et al. (1941)	$K^\circ: Mn(OH)_2(c)$
$Mn(OH)_4^{2-}$	−215.99	Smith and Martell (1976)	$K^\circ: Mn^{2+}$
Mn_2OH^{3+}	−152.45	Smith and Martell (1976)	$K^\circ: Mn^{2+}$
$Mn_2(OH)_3^+$	−247.69	Smith and Martell (1976)	$K^\circ: Mn^{2+}$
MnO_4^{2-}	−120.48	Carrington and Symons (1956)	$E^\circ: MnO_4^-$
MnO_4^-	−107.66	Andrews and Brown (1935)	$E^\circ: MnO_2(c)$
MnO(manganosite)	−86.71	Southard and Shomate (1942)	ΔH_f°
		Todd and Bonnickson (1951)	S°
β-MnO_2(pyrolusite)	−111.34	Coughlin (1954)	ΔH_f°
		Kelley and Moore (1943)	S°
δ-$MnO_{1.8}$(birnessite)	−108.89	Bricker (1965)	$E^\circ: Mn^{2+}$
γ-$MnO_{1.9}$(nsutite)	−109.77	Bricker (1965)	$E^\circ: Mn^{2+}$
Mn_2O_3(bixbyite)	−210.09	Otto (1964)	ΔH_f°
		King (1954a)	S°
Mn_3O_4(hausmannite)	−306.11	Brewer (1953)	ΔH_f°
		Otto (1964)	S°
γ-MnOOH(manganite)	−134.01	Bricker (1965)	$E^\circ: Mn^{2+}$
$Mn(OH)_2$(pyrochroite)	−147.76	Smith and Martell (1976)	$K^\circ: Mn^{2+}$
$MnHPO_4(c)$	−334.79	Wagman et al. (1969)	$\Delta G_f^\circ: Mn^{2+}, HPO_4^{2-}$
$Mn_3(PO_4)_2(c)$	−692.96	Stevens and Turkdogan (1954)	$\Delta H_f^\circ: MnO(c), P_2O_5(c)$
		Mah (1960)	$S^\circ: MnO(c), P_2O_5(c)$
$MnSO_4^\circ$	−236.14	Nair and Nancollis (1959)	$K^\circ: Mn^{2+}, SO_4^{2-}$
MnS(green)	−50.52	Juza et al. (1958)	ΔH_f°
		Anderson (1931)	S°
MnS(alabanite)	−52.12	Adami and King (1964)	ΔH_f°
		Kelley and King (1961)	S°
MnS_2(hauerite)	−55.53	Biltz and Weichmann (1936)	$K^\circ: MnS(alab), S_8(g)$
$MnSO_4(c)$	−228.38	Southard and Shomate (1942)	ΔH_f°
		Moore and Kelley (1942)	S°
$MnSO_4 \cdot H_2O(c)$	−289.11	Zordan and Hepler (1968)	$\Delta G_r^\circ: MnSO_4, H_2O(g)$
$Mn_2(SO_4)_3(c)$	−590.14	Karapet'yants (1955)	ΔG_f°
$MnSiO_3$(rhodonite)	−297.10	Hemingway and Robie (1977)	$\Delta H_f^\circ, S^\circ$
Mn_2SiO_4(tephroite)	−389.53	Hemingway and Robie (1977)	$\Delta H_f^\circ, S^\circ$

Mo (Molybdenum)

Species	ΔG_f°	Source	Remarks
Mo(c)	0.0	Standard state	ΔH_f°
		Stull et al. (1971)	S°
MoO_2^{2+}	−98.27	Naumov et al. (1971)	K°: MoO_3(c)
$MoO_2(OH)^+$	−154.34	Naumov et al. (1971)	K°: MoO_2^{2+}
MoO_4^{2-}	−199.84	Smith and Martell (1976)	K°: MoO_3(c)
$HMoO_4^-$	−205.62	Rohwer and Cruywagen (1963)	K°: MoO_4^{2-}
$H_2MoO_4^\circ$	−211.08	Rohwer and Cruywagen (1963)	K°: $HMoO_4^-$
MoO_2(c)	−127.41	Dellien et al. (1976a)	ΔH_f°
		King (1958a)	S°
MoO_3(c)(molybdite)	−159.66	Mah (1957)	ΔH_f°
		Staskiewitcz et al. (1955)	
		Kelley and King (1961)	S°
H_2MoO_4(c)	−218.08	Graham and Hepler (1956)	ΔH_f°
(molybdic acid)		Dellien et al. (1976a)	S°
MoS_2(c)(molybdenite)	−63.69	O'Hare et al. (1970)	ΔH_f°
		Westerum and McBride (1955)	S°

N (Nitrogen)

Species	ΔG_f°	Source	Remarks
N_2(g)	0.0	Standard state	ΔH_f°
		CODATA (1976)	S°
NH(g)	79.57	Jordan and Lonquet-Higgins (1962)	ΔH_f°
		Stull et al. (1971)	S°
NH_2(g)	42.46	Kerr et al. (1963)	ΔH_f°
		Stull et al. (1971)	S°
NH_3(g)	−3.93	CODATA (1976)	ΔH_f°
		Haar (1968)	S°
NH_3°	−6.33	Vanderzee and King (1972)	ΔH_f°: NH_3(g)
		Vanderzee and King (1972)	S°
NH_4^+	−18.99	Vanderzee et al. (1972)	ΔH_f°
		CODATA (1973)	S°
NH_4OH°	−63.06	Bates and Pinching (1950)	K°: NH_4^+
NH_2OH°	−5.60	Wagman et al. (1968)	ΔH_f°
		Karapet'yants and Karapet'yants (1970)	S°
NO(g)	20.70	Frisch (1961)	ΔH_f°
		Kelley and King (1961)	S°
NO°	24.41	Hodgman (1949)	K°: NO(g)

(*Continued*)

N (Nitrogen) (Continued)

Species	ΔG_f°	Source	Remarks
$NO_2(g)$	12.27	Wagman et al. (1968)	ΔH_f°
		Stull et al. (1971)	S°
NO_2^-	−9.02	Smith and Martell (1976)	$K^\circ: HNO_2^\circ$
HNO_2°	−13.32	Schmid and Neumann (1967)	$E^\circ: NO(g)$
$NO_3(g)$	27.75	Stull et al. (1971)	$\Delta H_f^\circ, S^\circ$
NO_3^-	−26.64	Wagman et al. (1968)	ΔH_f°
		CODATA (1976)	S°
$N_2H_4(g)$	38.04	Scott et al. (1949)	ΔH_f°
		Kelley and King (1961)	S°
$N_2H_4^\circ$	30.57	Wagman et al. (1968)	ΔH_f°
			$S^\circ: N_2(g), H_2(g)$
$N_2H_5^+$	19.70	Wagman et al. (1968)	$\Delta H_f^\circ, S^\circ$
$N_2O(g)$	24.89	Stull et al. (1971)	$\Delta H_f^\circ, S^\circ$
N_2O°	24.15	Markham and Kobe (1941)	$K^\circ: N_2O(g)$
$N_2O_3(g)$	33.24	Beattie and Bell (1957)	$\Delta H_f^\circ: NO(g), NO_2(g)$
		Devlin and Hisatsune (1961)	S°
$N_2O_4(g)$	23.39	Wourtzel (1919)	$\Delta H_f^\circ: NO_2(g)$
		Kelley and King (1961)	S°
$N_2O_5(g)$	28.18	Ray and Ogg (1957)	$\Delta H_f^\circ: NO_2(g), NO(g), N_2H_4(g)$
		Daniels and Johnston (1921)	$S^\circ: NO(g), NO_2(g), O_2(g)$
N_3^-	83.25	Wagman et al. (1968)	?

Na (Sodium)

Species	ΔG_f°	Source	Remarks
Na(c)	0.0	Standard state	ΔH_f°
		CODATA (1976)	S°
Na^+	−62.59	CODATA (1976)	$\Delta H_f^\circ, S^\circ$
$NaCl^\circ$	−93.96	Rossini et al. (1952)	$K^\circ: Na^+, Cl^-$
$NaCO_3^-$	−190.50	Garrels et al. (1961)	$K^\circ: Na^+, CO_3^{2-}$
$Na_2CO_3^\circ$	−251.38	Garrels and Christ (1965)	$K^\circ: Na^+, CO_3^{2-}$
$NaHCO_3^\circ$	−203.20	Garrels and Thompson (1962)	$K^\circ: Na^+, HCO_3^-$
$NaOH^\circ$	−99.91	Smith and Martell (1976)	$K^\circ: Na^+$
$NaSO_4^-$	−241.49	Smith and Martell (1976)	$K^\circ: Na^+, SO_4^{2-}$

O (Oxygen)

Species	ΔG_f°	Source	Remarks
O(g)	55.389	CODATA (1976)	$\Delta H_f^\circ, S^\circ$
O_2(g)	0.00	Reference state	ΔH_f°
O_2(g)	3.9	Wagman et al. (1968)	$\Delta H_f^\circ, S^\circ$
O_3(g)	39.0	Wagman et al. (1968)	$\Delta H_f^\circ, S^\circ$

P (Phosphorus)

Species	ΔG_f°	Source	Remarks
P(α, white)	0.00	Standard state	ΔH_f°
		Wagman et al. (1968)	S°
PH_3(g, phosphine)	5.00	Devyatykh and Yushin (1964)	ΔH_f°
		Wagman et al. (1968)	S°
PH_3°	2.15	Wagman et al. (1968)	$K^\circ: PH_3(g)$
HPO_3^{2-}	−191.07	Makitie and Savolainen (1968)	$K^\circ: H_2PO_3^-$
$H_2PO_3^-$	−200.33	Smith and Martell (1976)	$K^\circ: H_3PO_3^\circ$
$H_3PO_3^\circ$	−202.38	Finch et al. (1968)	$\Delta H_r^\circ: H_3PO_4(c)$
		Latimer (1952)	S°
PO_4^{3-}	−245.18	Smith and Martell (1976)	$K^\circ: HPO_4^{2-}$
HPO_4^{2-}	−262.03	Mesmer and Base (1974)	$K^\circ: H_2PO_4^-$
$H_2PO_4^-$	−271.85	Pitzer and Silvester (1976)	$K^\circ: H_3PO_4^\circ$
$H_3PO_4^\circ$	−274.78	Schumm et al. (1974)	$\Delta H_f^\circ: PCl_5(c)$
		Stephenson (1944)	S°
$(H_2PO_4)^{2-}$	−543.22	Ivakin and Voronova (1973)	$K^\circ: H_2PO_4^-$
$P_2O_7^{4-}$	−457.24	Irani and Taulli (1966)	$K^\circ: HP_2O_7^{3-}$
$HP_2O_7^{3-}$	−470.08	Näsänen (1960)	$K^\circ: H_2P_2O_7^{2-}$
$H_2P_2O_7^{2-}$	−479.22	Näsänen (1960)	$K^\circ: H_3P_2O_7^-$
$H_3P_2O_7^-$	−482.33	Smith and Martell (1976)	$K^\circ: H_4P_2O_7^\circ$
$H_4P_2O_7^\circ$	−483.42	Wu et al. (1967)	$\Delta H_r^\circ: H_2PO_4^{2-}$
		Irani and Taulli (1966)	$\Delta H_r^\circ: H_3P_2O_7^-, S^\circ$

Pb (Lead)

Species	ΔG_f°	Source	Remarks
Pb(c)	0.0	Standard state	ΔH_f°
		Kelley and King (1961)	S°
Pb^{2+}	−5.90	Smith and Martell (1976)	K°: PbO(c, yellow)
$PbBr^+$	−33.18	Biggs et al. (1955)	K°: Pb^{2+}, Br^-
$PbBr_2^\circ$	−59.19	Smith and Martell (1976)	K°: Pb^{2+}, Br^-
$PbBr_3^-$	−84.60	Smith and Martell (1976)	K°: Pb^{2+}, Br^-
$PbBr_4^{2-}$	−108.52	Smith and Martell (1976)	K°: Pb^{2+}, Br^-
$PbCO_3$(c, cerussite)	−150.51	Adami and Conway (1966)	ΔH_f°
		Anderson (1934)	S°
$Pb_2CO_3Cl_2$(c, phosgenite)	−227.94	Näsänen et al. (1963)	K°: Pb^{2+}, Cl^-, CO_2(g)
$Pb_3(CO_3)_2(OH)_2$(c)	−409.08	Randall and Spencer (1928)	K°: $Pb(OH)_3^-$, CO_3^{2-}
$PbCl^+$	−39.45	Nelson and Kraus (1954)	K°: Pb^{2+}, Cl^-
$PbCl_2^\circ$	−71.07	Nelson and Kraus (1954)	K°: $PbCl^+$, Cl^-
$PbCl_3^-$	−102.31	Nelson and Kraus (1954)	K°: $PbCl_2^\circ$, Cl^-
$PbCl_4^{2-}$	−133.27	Nelson and Kraus (1954)	K°: $PbCl_3^-$, Cl^-
PbF^+	−75.86	Wagman et al. (1968)	K°: Pb^{2+}, F^-
PbF_2°	−144.86	Talipov and Kutumova (1956)	K°: Pb^{2+}, F^-
PbF_3^-	−214.35	Talipov and Kutumova (1956)	K°: Pb^{2+}, F^-
PbF_4^{2-}	−281.85	Talipov and Kutumova (1956)	K°: Pb^{2+}, F^-
PbI^+	−20.90	Biggs et al. (1955)	K°: Pb^{2+}, I^-
PbI_2°	−34.96	Tur'yan (1962)	K°: Pb^{2+}, I^-
PbI_3^-	−48.39	Tur'yan (1962)	K°: Pb^{2+}, I^-
PbI_4^{2-}	−61.56	Tur'yan (1962)	K°: Pb^{2+}, I^-
$PbMoO_4$(wulfenite)	−227.62	Dellien et al. (1976a)	ΔH_r°: Pb^{2+}, MoO_4^{2-}
		Weller and Kelley (1964)	S°
$PbNO_3^+$	−34.14	Smith and Martell (1976)	K°: Pb^{2+}, NO_3^-
$Pb(NO_3)_2^\circ$	−61.09	Smith and Martell (1976)	K°: Pb^{2+}, NO_3^-
$PbOH^+$	−52.09	Smith and Martell (1976)	K°: Pb^{2+}
$Pb(OH)_2^\circ$	−95.06	Smith and Martell (1976)	K°: Pb^{2+}
$Pb(OH)_3^-$	−137.64	Smith and Martell (1976)	K°: Pb^{2+}
$Pb(OH)_4^{2-}$	−178.78	Karapet'yants and Karapet'yants (1970)	K°: PbO(c, yellow)
Pb_2OH^{3+}	−59.76	Smith and Martell (1976)	K°: Pb^{2+}
$Pb_3(OH)_4^{2+}$	−211.86	Smith and Martell (1976)	K°: Pb^{2+}
$Pb_4(OH)_4^{4+}$	−221.85	Smith and Martell (1976)	K°: Pb^{2+}
$Pb_6(OH)_8^{4+}$	−429.45	Smith and Martell (1976)	K°: Pb^{2+}
$Pb(OH)_2$(c)	−108.15	Naumov et al. (1971)	K°: Pb^{2+}
PbO(c, yellow)	−45.00	Espada et al. (1970)	ΔH_f°
		King (1958b)	S°
PbO(c, red)	−45.24	Knacke and Prescher (1964)	ΔH_r°: PbO(c, yellow)
		King (1958b)	S°
PbO_2(c)	−51.51	Espada et al. (1970)	ΔH_f°
		Duisman and Giauque (1968)	S°

Pb (Lead) (*Continued*)

Species	ΔG_f°	Source	Remarks
$Pb_3O_4(c)$	−143.80	Espada et al. (1970)	ΔH_f°: PbO(c)
		Chase et al. (1974a)	S°
$PbO \cdot PbCO_3(c)$	−195.72	Marshall and Bruzs (1925)	ΔH_f°: PbO(c), $PbCO_3(c)$
		Kelley and Anderson (1935)	S°
$PbHPO_4^\circ$	−272.16	Nriagu (1972c)	K°: Pb^{2+}, HPO_4^{2-}
$PbH_2PO_4^+$	−279.80	Nriagu (1972c)	K°: Pb^{2+}, $H_2PO_4^-$
$PbP_2O_7^{2-}$	−478.55	Wagman et al. (1968)	K°: Pb^{2+}, $HP_2O_7^{3-}$
$PbHPO_4(c)$	−283.55	Nriagu (1972c)	K°: Pb^{2+}, HPO_4^{2-}
$Pb(H_2PO_4)_2(c)$	−563.04	Nriagu (1973a)	K°: Pb^{2+}, HPO_4^{2-}
$Pb_3(PO_4)_2(c)$	−568.57	Nriagu (1972c)	K°: Pb^{2+}, HPO_4^{2-}
$Pb_4O(PO_4)_2(c)$	−620.94	Nriagu (1972c)	K°: Pb^{2+}, HPO_4^{2-}
$Pb_5(PO_4)_3Br$(c, bromopyromorphite)	−896.51	Nriagu (1973b)	K°: Pb^{2+}, PO_4^{3-}, Br^-
$Pb_5(PO_4)_3Cl$(c, chloropyromorphite)	−910.59	Nriagu (1973a)	K°: Pb^{2+}, PO_4^{3-}, Cl^-
$Pb_5(PO_4)_3F$(c, fluoropyromorphite)	−930.68	Nriagu (1973b)	K°: Pb^{2+}, PO_4^{3-}, F^-
$Pb_5(PO_4)_3OH$(c, hydroxypyromorphite)	−907.39	Nriagu (1972c)	K°: Pb^{2+}, HPO_4^{2-},
$PbSO_4^\circ$	−187.42	Riet and Kolthoff (1960)	K°: Pb^{2+}, SO_4^{2-}
$Pb(SO_4)_2^{2-}$	−366.53	Riet and Kolthoff (1960)	K°: Pb^{2+}, SO_4^{2-}
PbS(c, galena)	−22.91	Smith and Martell (1976)	K°: Pb^{2+}, S^{2-}
$PbSO_4$(c, anglesite)	−194.48	Smith and Martell (1976)	K°: Pb^{2+}, SO_4^{2-}
		Gallagher et al. (1960)	S°
$PbSO_4 \cdot PbO(c)$	−246.70	Wagman et al. (1968)	ΔH_f°, S°
$PbSO_4 \cdot 2PbO(c)$	−294.00	Wagman et al. (1968)	ΔH_f°, S°
$PbSO_4 \cdot 3PbO$	−341.2	Wagman et al. (1968)	ΔH_f°, S°
$PbSiO_3(c)$	−253.77	Wagman et al. (1968)	ΔH_f°, S°
$Pb_2SiO_4(c)$	−299.30	Wagman et al. (1968)	ΔH_f°
		King (1959)	S°

S (Sulfur)

Species	ΔG_f°	Source	Remarks
S(c, rhombic)	0.0	Standard state	ΔH_f°
		Wagman et al. (1968)	S°
S^{2-} (sulfide ion)	20.51	Garrels and Naeser (1958)	K°: HS^-
S_2^{2-} (disulfide ion)	19.75	Maronny (1959)	E°: $S(rh)/S_2^{2-}$
S_3^{2-} (trisulfide ion)	17.97	Maronny (1959)	E°: $S(rh)/S_3^{2-}$
S_4^{2-} (tetrasulfide ion)	16.62	Maronny (1959)	E°: $S(rh)/S_4^{2-}$
S_5^{2-} (pentasulfide ion)	15.69	Maronny (1959)	E°: $S(rh)/S_5^{2-}$
HS(g, bisulfide)	27.58	Mackle (1963)	ΔH_f°
		Stull et al. (1971)	S°

(*Continued*)

S (Sulphur) (Continued)

Species	ΔG_f°	Source	Remarks
HS⁻ (bisulfide ion)	2.91	Smith and Martell (1976)	K°: H_2S°
H₂S(g, hydrogen sulfide)	−8.02	Kapustinskii and Kanakovskii (1958)	ΔH_f°
		Wagman et al. (1968)	S°
H₂S°(hydrogen sulfide)	−6.67	Smith and Martell (1976)	K°: $H_2S(g)$
SO(g, sulfur monoxide)	−5.12	Norrish and Oldershaw (1959)	ΔH_f°: $O_2(g)$, $S(g)$
		Stull et al. (1971)	S°
SO₂(g, sulfur dioxide)	−71.45	Evans and Wagman (1952)	ΔH_f°
		Kelley and King (1961)	S°
SO₂°(sulfur dioxide ion)	−71.87	Johnstone and Leppla (1934)	ΔG_r°: $SO_2(g)$
SO₃(g, sulfur trioxide)	−88.69	Lovejoy et al. (1962)	ΔH_f°, S°
SO₃²⁻ (sulfite ion)	−116.18	Smith and Martell (1976)	K°: HSO_3^-
HSO₃⁻ (bisulfite ion)	−125.97	Smith and Martell (1976)	K°: $H_2SO_3^\circ$
H₂SO₃°(sulfurous acid)	−128.57	Flis et al. (1965)	K°: SO_2°
SO₄²⁻ (sulfate ion)	−177.95	CODATA (1977b)	ΔH_f°, S°
HSO₄⁻ (bisulfate ion)	−180.65	Reardon (1975)	K°: SO_4^{2-}
H₂SO₄°(sulfuric acid)	−177.95	CODATA (1977b)	ΔH_f°, S°
S₂O(g)	−18.22	Stull et al. (1971)	ΔH_f°, S°
S₂O₃²⁻ (thiosulfate ion)	−123.43	Mel et al. (1956)	ΔH_r°: Br_2°, Br^-, SO_4^{2-}
		Mel et al. (1956)	S°
HS₂O₃⁻ (bithiosulfate ion)	−125.63	Smith and Martell (1976)	K°: $S_2O_3^{2-}$
H₂S₂O₃°(thiosulfuric acid)	−126.45	Page (1953)	K°: $HS_2O_3^-$
S₂O₄²⁻ (dithionite ion)	−143.5	Wagman et al. (1968)	?
HS₂O₄⁻ (bithionite ion)	−146.9	Wagman et al. (1968)	?
H₂S₂O₄°(dithionite acid)	−147.4	Wagman et al. (1968)	?

Si (Silicon)

Species	ΔG_f°	Source	Remarks
Si(c)	0.0	Standard state	ΔH_f°
		CODATA (1976)	S°
SiO₂(amorp)	−203.02	Elgawhary and Lindsay (1972)	K°: $H_4SiO_4^\circ$
SiO₂(coesite)	−203.44	Holm et al. (1967)	ΔH_r°: α-SiO_2(q)
		Holm et al. (1967)	S°
α-SiO₂(cristobalite)	−204.66	Fournier and Rowe (1962)	ΔG_r°: α-SiO_2(q)
α-SiO₂(quartz)	−204.75	Wise et al. (1963)	ΔH_f°
		CODATA (1976)	S°

Si (Silicon) (Continued)

Species	ΔG_f°	Source	Remarks
SiO_2(silica glass)	-202.98	Holm et al. (1967)	ΔH_r°: $\alpha\text{-}SiO_2(q)$
		Kelley and King (1961)	S°
SiO_2(soil)	-203.51	Elgawhary and Lindsay (1972)	K°: $H_4SiO_4^\circ$
$\alpha\text{-}SiO_2$(tridymite)	-204.42	Naumov et al. (1971)	ΔH_f°
		Kelley and King (1961)	S°
SiO_4^{4-}	-249.99	Smith and Martell (1976)	K°: $HSiO_4^{3-}$
$HSiO_4^{3-}$	-267.86	Smith and Martell (1976)	K°: $H_2SiO_4^{2-}$
$H_2SiO_4^{2-}$	-281.31	Ingri (1959)	K°: $H_3SiO_4^-$
		Langerström (1959)	
$H_3SiO_4^-$	-299.42	Blinski and Ingri (1967)	K°: $H_4SiO_4^\circ$
		Lagerström (1959)	
$H_4SiO_4^\circ$	-312.66	Morey et al (1962)	K°: $\alpha\text{-}SiO_2(q)$
$H_6Si_4O_{12}^{2-}$	-1005.72	Ingri (1959)	K°: $H_4SiO_4^\circ$
		Lagerström (1959)	

Zn (Zinc)

Species	ΔG_f°	Source	Remarks
Zn(c)	0.0	Standard state	ΔH_f°
		CODATA (1976)	S°
Zn^{2+}	-35.186	CODATA (1976)	$\Delta H_f^\circ, S^\circ$
$ZnCO_3$(smithsonite)	-175.34	Roth (1928)	ΔH_r°: $ZnO(c), CO_2(g)$
		Anderson (1934)	S°
$ZnCl^+$	-67.14	Smith and Martell (1976)	K°: Zn^{2+}, Cl^-
$ZnCl_2^\circ$	-97.93	Smith and Martell (1976)	K°: Zn^{2+}, Cl^-
$ZnCl_3^-$	-129.98	Smith and Martell (1976)	K°: Zn^{2+}, Cl^-
$ZnCl_4^{2-}$	-160.94	Smith and Martell (1976)	K°: Zn^{2+}, Cl^-
$ZnCl_2$(c)	-88.28	Bates (1939)	E°: $CuCl_2(c)/Ag/Zn/AgCl(c)$
$ZnFe_2O_4$(franklinite)	-256.54	Fischer (1966)	ΔH_r°: $ZnO(c), Fe_2O_3$ (hematite)
		Kelley and King (1961)	
$ZnMoO_4$(c)	-241.15	Naumov et al. (1971)	K°: Zn^{2+}, MoO_4^{2-}
$ZnNO_3^+$	-62.37	Smith and Martell (1976)	K°: Zn^{2+}, NO_3^-
$Zn(NO_3)_2^\circ$	-88.06	Smith and Martell (1976)	K°: Zn^{2+}, NO_3^-
$ZnOH^+$	-81.39	Gubeli and Marie (1967)	K°: Zn^{2+}
$Zn(OH)_2^\circ$	-125.64	Gubeli and Marie (1967)	K°: $Zn(OH)^+$
$Zn(OH)_3^-$	-167.49	Gubeli and Marie (1967)	K°: $Zn(OH)_2^\circ$
$Zn(OH)_4^{2-}$	-209.71	Gubeli and Marie (1967)	K°: $Zn(OH)_3^-$
ZnO(zincite)	-76.65	Berg and Vanderzee (1975)	ΔH_f°
		Kelley and King (1961)	S°
$Zn(OH)_2$(amorp)	-131.54	Schindler et al. (1964)	K°: Zn^{2+}

(Continued)

APPENDIX SELECTED STANDARD FREE ENERGIES OF FORMATION

α-Zn(OH)$_2$(c)	−131.93	Latimer (1952)	$K°$: Zn^{2+}
β-Zn(OH)$_2$(c)	−132.49	Schindler et al. (1964)	$K°$: Zn^{2+}
γ-Zn(OH)$_2$(c)	−132.55	Schindler et al. (1964)	$K°$: Zn^{2+}
ε-Zn(OH)$_2$(c)	−132.83	Schindler et al. (1964)	$K°$: Zn^{2+}
ZnHPO$_4^°$	−301.72	Nriagu (1973)	$K°$: Zn^{2+}, HPO$_4^{2-}$
ZnH$_2$PO$_4^+$	−309.22	Katayama (1976)	$K°$: Zn^{2+}, H$_2$PO$_4^-$
Zn$_3$(PO$_4$)$_2$·4H$_2$O (hopeite)	−870.82	Nriagu (1973a)	$K°$: Zn^{2+}, PO$_4^{3-}$
ZnSO$_4^°$	−216.31	Katayama (1976)	$K°$: Zn^{2+}, SO$_4^{2-}$
α-ZnS(sphalerite)	−48.37	Smith and Martell (1976)	$K°$: Zn^{2+}, S^{2-}
β-ZnS(wurtzite)	−45.37	Smith and Martell (1976)	$K°$: Zn^{2+}, S^{2-}
ZnSO$_4$(zinkosite)	−208.48	Adami and King (1965)	$\Delta H_r°$: ZnO(c), H$_2$SO$_4$; 7.06 H$_2$O
		Weller (1966)	$S°$
ZnO·2ZnSO$_4$(c)	−492.07	Wagman et al. (1968)	$\Delta H_f°, S°$
Zn(OH)$_2$·ZnSO$_4$(c)	−351.47	Wagman et al. (1968)	$K°$: Zn^{2+}, SO$_4^{2-}$
ZnSiO$_3$(c)	none	See Norvell and Lindsay (1970)	
		Latimer (1952)	$S°$
Zn$_2$SiO$_4$(willemite)	−365.09	King (1951)	$\Delta H_r°$: ZnO(c), SiO$_2$(q)
		Kelley and King (1961)	$S°$

LITERATURE CITED FOR THE APPENDIX

Abraham, M. H., J. F. C. Oliver, and J. A. Richards. 1970. J. Chem. Soc. A, 199–202.
Adami, L. H. and K. C. Conway. 1966. U.S. Bur. Mines. Rept. Inv. 6822.
Adami, L. H. and K. K. Kelley. 1963. U.S. Bur. Mines Rept. Inv. 6270.
Adami, L. H. and E. G. King. 1964. U.S. Bur. Mines Rept. Inv. 6495.
Adami, L. H. and E. G. King. 1965. U.S. Bur. Mines Rept. Inv. 6617.
Adams, F. 1971. Soil Sci. Soc. Am. Proc. 35:420–426.
Ainsworth, R. G. 1973. J. Chem. Soc., Faraday Trans. I. 69:1028–1033.
Anderson, C. T. 1931. J. Am. Chem. Soc. 53:476–483.
Anderson, C. T. 1934. J. Am. Chem. Soc. 56:849–851.
Anderson, K. P., E. A. Butler, and E. M. Woolley. 1971. J. Phys. Chem. 75:93–97.
Andrews, L. V. and D. J. Brown. 1935. J. Am. Chem. Soc. 57:254–256.
Anotonetti, P. G., G. Ferroni, and J. Galea. 1976. Bull. Soc. Chim. Fr. 747–750.
Antikainen, P. J. and D. Dryssen. 1960. Acta Chem. Scand. 14:86–101.
Arden, T. V. 1950. J. Chem. Soc. 882–885.
Arnek, R. and C. C. Patel. 1968. Acta Chem. Scand. 22:1097–1101.
Avnimelech, Y., E. C. Moreno, and W. E. Brown. 1973. Nat. Bur. Stand. J. Res. 77A:149–155.

LITERATURE CITED FOR THE APPENDIX

Bache, B. W. 1963. J. Soil Sci. 14:113-123.
Barany, R. 1959. U.S. Bur. Mines Rept. Inv. 5466.
Barany, R. 1962a. U.S. Bur. Mines Rept. Inv. 5900.
Barany, R. 1962b. U.S. Bur. Mines Rept. Inv. 6143.
Barany, R. 1965. U.S. Bur. Mines Rept. Inv. 6784.
Barany, R. and L. H. Adami. 1965. U.S. Bur. Mines Rept. Inv. 6687.
Barany, R., L. B. Pankratz, and W. W. Weller. 1964. U.S. Bur. Mines Rept. Inv. 6513.
Barton, A. F. M. and S. R. McConnel. 1974. J. Chem. Soc., Faraday Trans. I. 70:2355-2361.
Bates, R. G. 1939. J. Am. Chem. Soc. 61:1040-1044.
Bates, R. G. and G. Pinching. 1950. J. Am. Chem. Soc. 72:1393-1396.
Bates, R. G. And W. C. Vosburg. 1938. J. Am. Chem. Soc. 60:137-141.
Baumann, E. W. 1969. J. Inorg. Nucl. Chem. 31:3155-3162.
Beattie, I. R. and S. W. Bell. 1957. J. Chem. Soc. 1681-1686.
Behr, B. and H. Wendt. 1962. Z. Elektrochem. 66:223-228.
Bennett, A. C. and F. Adams. 1976. Soil Sci. Soc. Am. J. 40:39-42.
Berg, R. L. and C. E. Vanderzee. 1975. J. Chem. Thermodyn. 7:229-239.
Berne, E. and I. Leden. 1953. Z. Naturforsch. 8a:719-726.
Biedermann, G. and L. G. Sillen. 1960. Acta Chem. Scand. 14:717-725.
Biggs, A. I., H. N. Parton, and R. A. Robinson. 1955. J. Am. Chem. Soc. 77:5844-5848.
Biltz, W. and F. Weichmann. 1936. Z. Anorg. Allg. Chem. 228:268.
Bjerrum, J. 1946. Kgl danske videnskab Selskab, Mat FYS Medd. 22, No. 18.
Bjerrum, J. 1972. Acta Chem. Scand. 26:2734-3742.
Blinski, H. and N. Ingri. 1967. Acta Chem. Scand. 21:2503-2510.
Bolzan, J. A. and A. J. Arvia. 1963. Electrochem. Acta 8:375-383.
Bonner, O. D., H. Dolynuik, C. F. Jordan, and G. B. Hanson. 1962. J. Inorg. Nucl. Chem. 24:689-692.
Brewer, L. 1953. Chem. Rev. 52:1-75.
Bricker, O. 1965. Am. Miner. 50:1296-1354.
Brodsky, A. E. 1938. Z. Elektrochem. 35:833.
Brown, J. B. 1970. Can. Miner. 10:696-703.
Burnett, J. L. and M. H. Zirin. 1966. J. Inorg. Nucl. Chem. 28:902-904.
Carrington, A. and M. C. R. Symons. 1956. J. Chem. Soc. 3373-3380.
Charlu, T.V., R.C. Newton, and O.J. Kleppa. 1975. Geochim. Cosmochim. Acta 39:1487-1497.
Chase, M. W., J. L. Curnett, H. Prophet, R. A. McDonald, and A. N. Syverud. JANAF Thermochemical Tables. J. Phys. Chem. Ref. Data.
Chase, M. W., J. L. Curnett, A. T. Hu, H. Prophet, A. N. Syverud, and L. C. Walker. JANAF Thermochemical Tables, 1974b Supp. Dow Chemical Company, Midland, Mi.
Chateau, H., M. Durante, and B. Hervier. 1956a. Sci. Ind. Phot. 27:81-90.
Chateau, H., M. Durante, and B. Hervier. 1956b. Sci. Ind. Phot. 27:257-262.
Cheronosky, K. V., Jr. 1974. Am. Miner. 59:496-507.
Chien, S. H. and C. A. Black. 1976. Soil Sci. Soc. Am. J. 40:234-239.
Christ, G. L., P. B. Hostetler, and R. M. Siebert. 1973. Am. J. Sci. 273:65-83.
Chughtai, A., R. Marshall, and G. H. Nancollas. 1968. J. Phys. Chem. 72:208-211.

Cloke, P. L. 1963. Geochim. Cosmochim. Acta 27:1265-1299.
CODATA. 1971. J. Chem. Thermodyn. 4:331-333.
CODATA. 1973. CODATA Bull. 10.
CODATA. 1976. CODATA Bull. 17.
CODATA. 1977a. CODATA Bull. 21.
CODATA. 1977b. CODATA Bull. 22.
Connick, R. E. and A. D. Paul. 1961. J. Phys. Chem. 65:1216-1220.
Coughlin, J. P. 1954. U.S. Bur. Mines Bull. 542.
Covington, A. K., T. Cressey, B. G. Lever, and H. R. Thirsk. 1962. Trans. Faraday Soc. 58:1975-1988.
Daniels, F. and E. H. Johnston. 1921. J. Am. Chem. Soc. 43:53-71.
Dellien, I., F. M. Hall, and L. G. Hepler. 1976a. Chem. Rev. 76:283-310.
Dellien, I., K. G. McCurdy, and L. G. Hepler. 1976b. J. Chem. Thermodyn. 8:203-207.
DeSorbo, W. and W. W. Tyler. 1953. J. Chem. Phys. 21:1660-1663.
Devlin, J. P. and I. C. Hisatsune. 1961. Spectrochim. Acta 17:218-225.
Devyatykh, G. G. and A. S. Yushin. 1964. Zh. Fiz. Khim. 38:957-962.
Dewing, E. W. and F. D. Richardson. 1959. Trans. Faraday Soc. 55:611-615.
Dirske, T. P. 1962. J. Electrochem. Soc. 109:173-177.
Dirske, T. P. and W. M. Graven. 1959. J. Electrochem. Soc. 106:284-287.
Dodig, D. and Z. Pavlovic. 1969. J. Chim. Phys. Physicochem. Biol. 66:1213-1317.
Dolliver, M. A., T. L. Grashem, G. B. Kistiakowsky, and W. E. Vaughn. 1938. J. Am. Chem. Soc. 60:440-450.
Dubinskii, V. I. and V. M. Shul'man. 1970. Russ. J. Inorg. Chem. 15:764-766.
Duisman, J. A. and W. F. Giauque. 1968. J. Phys. Chem. 72:562-573.
Egan, E. P., Jr. and Z. T. Wakefield. 1960. J. Phys. Chem. 64:1953-1955.
Elgawhary, S. M. and W. L. Lindsay. 1972. Soil Sci. Soc. Am. Proc. 36:439-442.
Espada, L., G. Pilcher, and H. A. Skinner. 1970. J. Chem. Thermody. 2:647-653.
Evans, W. H., R. T. Munson, and D. D. Wagman. 1955. Nat. Bur. Stand. J. Res. 55:147-154.
Evans, W. F. and D. D. Wagman. 1952. Nat. Bur. Stand. J. Res. 49:141-148.
Faweett, J. J. and H. S. Yoder. 1966. Am. Miner. 51:353-380.
Feber, R. C. and C. C. Herrick. 1967. Los Almos Sci. Lab. Rept. LA-3597.
Fedorov, V. A., A. M. Robov, I. I. Shumydko, N. V. Vorontsova, and V. E. Mironov. 1974. Russ. J. Inorg. Chem. 19:950-953.
Finch, A., P. H. Gardner, K. S. Hussain, and K. K. Gupta. 1968. Chem. Commun. 872.
Fischer, M. 1966. Z. Anorg. Allg. Chem. 345:134-136.
Flis, I. E., G. P. Arkhipova, and K. P. Mishchenko. 1965. Z. Priklad. Khim. 38:1494-1500.
Frink, C. R. and M. Peech. 1963. Inorg. Chem. 2:473-478.
Frisch. M. A. 1961. Diss. Absts. 23:1940.
Fournier, R. O. and J. J. Rowe. 1962. Am. Miner. 47:897-902.
Fox, R. K., D. F. Swinehart, and A. B. Garret. 1941. J. Am. Chem. Soc. 63:1779-1782.
Furukawa, G. T., T. B. Douglas, R. E. McCoskey, and D. G. Grinnings. 1956. U.S. Nat. Bur. Stand. J. Res. 57:67-82.

LITERATURE CITED FOR THE APPENDIX

Furukawa, G. T., W. G. Saba, and M. L. Reilly. 1968. Nat. Stand. Ref. Data Ser. 18, Nat. Bur. Stand.

Gallagher, L., G. E. Brodale, and T. E. Hopkins. 1960. J. Phys. Chem. 64:687-688.

Gamsjäger, H., H. U. Stuber, and P. Schindler. 1965. Helv. Chim. Acta 48:723-729.

Gardner, A. W. and E. Glueckauf. 1970. Trans. Faraday Soc. 66:1081-1088.

Gardner, J. 1975. Water Res. 10:507-514.

Gardner, W. L., R. E. Mitchell, and J. W. Cobble. 1969. J. Phys. Chem. 73:2021-2024.

Garrels, R. M. and C. L. Christ. 1965. Harper and Row, New York.

Garrels, R. M. and C. R. Naeser. 1958. Geochim. Cosmochim. Acta 15:113-130.

Garrels, R. M. and M. E. Thompson. 1962. Am. J. Sci. 260:57-66.

Garrels, R. M., M. E. Thompson, and R. Siever. 1961. Am. J. Sci. 259:24-45.

Gaydon, A. G. 1953. 2nd Ed., Chapman and Hall, London.

Glew, D. N. and D. A. Hames. 1971. Can. J. Chem. 49:3114-3121.

Gordon, J. S. 1963. J. Chem. Eng. Data 8:294-297.

Graham, R. L. and L. G. Hepler. 1956. J. Am. Chem. Soc. 78:4846-4848.

Green, J. H. S. 1961a. J. Chem. Soc. 2241-2242.

Green, J. H. S. 1961b. Trans. Faraday Soc. 57:2132-2137.

Greenberg, S. A. and L. E. Copeland. 1960. J. Phys. Chem. 64:1057-1059.

Gregor, L. V. and K. S. Pitzer. 1962. J. Am. Chem. Soc. 84:2671-2673.

Gregory, T. M., E. C. Moreno, and W. E. Brown. 1970. Nat. Bur. Stand. J. Res. 74A:461-475.

Gronvold, F. and E. F. Westrum, Jr. 1959. J. Am. Chem. Soc. 81:1780-1783.

Gronvold, F. and E. F. Westrum, Jr. 1962. Inorg. Chem. 1:36-48.

Gronvold, F. and E. F. Westrum, Jr. 1976. J. Chem. Thermodyn. 8:1039-1048.

Grube, G. and K. Huberich. 1923. Z. Elecktrochem. 29:8-17.

Gubeli, A. O. and J. Marie. 1967. Can. J. Chem. 45:827-832.

Haar, L. 1968. Nat. Stand. Ref. Data Ser. 19.

Haas, H. and M. J. Holdaway. 1970. Am. Geophy. Union Trans. 51:437.

Harned, H. S. and R. Davis. 1943. J. Am. Chem. Soc. 65:2030-2037.

Harned, H. S. and R. W. Ehlers. 1933. J. Am. Chem. Soc. 55:625-656.

Harned, H. S. and B. B. Owen. 1958. The Physical Chemistry of Electrolytic Solutions. Reinhold, New York.

Hatton, W. E., D. L. Hildenbrand, G. C. Sinke, and D. R. Stull. 1959. J. Am. Chem. Soc. 81:5028-5030.

Heitman, S. and E. Hogfeldt. 1976. Chem. Ser. 9:24-29.

Helgeson, H. C. 1969. Am. J. Sci. 267:729-804.

Hem, J. D. and W. H. Cropper. 1959. U.S. Geol. Sur. Water Supp. Paper 1459-A.

Hem, J. D., C. E. Roberson, C. J. Lind, and W. L. Polzer. 1973. U.S. Geol. Sur. Water Supp. Paper 1827-E.

Hemingway, B. S. and R. A. Robie. 1977a. Geochim. Cosmochim. Acta 41:1402-1404.

Hemingway, B. S. and R. A. Robie. 1977. U.S. Geol. Surv. J. Res. 5:413-429.

Hemingway, B. S., R. A. Robie, J. R. Fisher, and W. H. Wilson. 1977. U.S. Geol. Surv. J. Res. 5:797-806.

Hemley, J. J., J. W. Montoya, C. L. Christ, and P. B. Hostetler. 1976a. Am. J. Sci. 277:322–351.

Hemley, J. J., J. W. Montoya, and D. R. Shaw. 1976b. Am. J. Sci. 277:353–383.

Hemley, J. J., J. W. Montoya, C. L. Christ, and P. B. Hostetler. 1977. Am. J. Sci. 277:322–335.

Hepler, L. G. and G. Olofsson. 1975. Chem. Rev. 75:585–602.

Hertz, H. G. 1960. Z. Elektrochem. 64:53–67.

Hildenbrand, D. L., W. R. Kramer, R. A. McDonald, and D. R. Stull. 1958. J. Am. Chem. Soc. 80:4129–4132.

Hodgman, C. D., Ed. 1949. Handbook of Chemistry and Physics, 31st ed., pp. 1423. CRC, Cleveland, Ohio.

Holm, J. L. and O. J. Kleppa. 1966a. J. Phys. Chem. 70:1690.

Holm, J. L. and O. J. Kleppa. 1966b. Am. Miner. 51:1608–1622.

Holm, J. L. and O. J. Kleppa. 1966c. Inorg. Chem. 5:698.

Holm, J. L., O. J. Kleppa, and E. F. Westrum, Jr. 1967. Geochim. Cosmochim. Acta 31:2289–2307.

Hopkins, H. P. and C. A. Wulff. 1965. J. Phys. Chem. 69:9–11.

Hostetler, P. B. 1963. Am. J. Sci. 261:238–258.

Huang, W. H. and W. D. Keller. 1972. Am. Miner. 57:1152–1162.

Hultgren, R., R. L. Orr, P. D. Anderson, and K. K. Kelley. 1963. Selected Values of Thermodynamic Properties of Metals and Alloys. Wiley-Interscience, New York.

Hultgren, R., R. L. Orr, and K. K. Kelley. 1970. Selected Values of Thermodynamic Properties of Metals and Alloys. Wiley-Interscience, New York.

Humphrey, G. L. 1951. J. Am. Chem. Soc. 73:1857–1858.

Humphrey, G. L., E. G. King, and K. K. Kelley. 1952. U.S. Bur. Mines Rept. Inv. 4870.

Ingri, N. 1959. Acta Chem. Scand. 13:758–775.

Irani, R. R. and C. F. Callis. 1960. J. Phys. Chem. 64:1398–1407.

Irani, A. R. and T. A. Taulli. 1966. J. Inorg. Nucl. Chem. 28:1011–1020.

Ivakin, A. A. and E. M. Voronova. 1973. Russ. J. Inorg. Chem. 18:465–468.

Izatt, R. M., D. Eatough, J. J. Christensen, and C. H. Bartholomew. 1969. J. Chem. Soc. A:47–53.

Jacobson, R. L. and D. Langmuir. 1974. Geochim. Cosmochim. Acta 38:301–718.

Jaza, R., G. Wainoff, and W. Uphoff. 1958. Z. Anorg. Chem. 296:157–163.

Jellinek, K. and H. Z. Gordon. 1924. Z. Phys. Chem. 112:207–249.

Johnson, G. K. 1977. J. Chem. Thermodyn. 9:835–841.

Johnson, G. K., P. N. Smith, and W. N. Hubbard. 1973. J. Chem. Thermodyn. 5:793–809.

Johnson, J. W., W. J. Silva, and D. Cubicciotti. 1966. J. Phys. Chem. 70:2985–2988.

Johnstone, H. F. and P. W. Leppla. 1934. J. Am. Chem. Soc. 56:2233–2238.

Jonte, J. H. and D. S. Martin. 1952. J. Am. Chem. Soc. 74:2052–2054.

Jordan, P. C. H. and H. C. Lonquet-Higgins. 1962. Molec. Phys. 5:121–138.

Juza, R., G. Wainhoff, and W. Uphoff. 1958. Z. Anorg. Allg. Chem. 296:157–163.

Kapustinskii, A. F. and R. T. Kanakovskii. 1958. Zh. Fiz. Khim. 32:2810–2816.

Karapet'yants, M. Kh. 1955. TrudyMKhTi im D.g. Mendelle Va 20:10.

Karapet'yants, M. Kh. and M. L. Karapet'yants, 1970. Thermodynamic Constants of Inorganic and Organic Compounds. Ann Arbor-Humphrey Scientific Pub. Ann Arbor, London.

LITERATURE CITED FOR THE APPENDIX

Katayama, S. 1976. J. Soln. Chem. 5:241-248.

Kelley, K. K. 1941. U.S. Bur. Mines Tech. Paper 625. pp. 17-24.

Kelley, K. K. 1943. J. Am. Chem. Soc. 65:339-341.

Kelley, K. K. 1962. U.S. Bur. Mines Rept. Inv. 5901.

Kelley, K. K. and C. T. Anderson. 1935. U.S. Bur. Mines Bull. 384.

Kelley, K. K. and E. G. King. 1961. U.S. Bur. Mines Bull. 592.

Kelley, K. K. and G. E. Moore. 1943. J. Am. Chem. Soc. 65:782-785.

Kelley, K. K., C. H. Shomate, F. E. Young, B. G. Naylor, A. E. Salo, and E. H. Huffman. 1946. U.S. Bur. Mines Tech. Paper 688.

Kelley, K. K., S. S. Todd, R. L. Orr, E. G. King, and E. R. Bonnickson. 1953. U.S. Bur. Mines Rept. Inv. 4955.

Kerr, J. A., R. C. Sekhar, and A. F. Trotman-Dickenson. 1963. J. Chem. Soc. 3217-3225.

King, E. G. 1951. J. Am. Chem. Soc. 73:656-658.

King, E. G. 1954a. J. Am. Chem. Soc. 76:3289-3291.

King, E. G. 1954. J. Am. Chem. Soc. 76:5849-5850.

King, E. G. 1957a. J. Am. Chem. Soc. 79:3639-3641.

King, E. G. 1957b. J. Am. Chem. Soc. 79:5437-5443.

King, E. G. 1958a. J. Am. Chem. Soc. 80:1799-1800.

King, E. G. 1958b. J. Am. Chem. Soc. 80:2400-2409.

King, E. G. 1959. J. Am. Chem. Soc. 81:799-802.

King, E. G., R. Barany, W. W. Weller, and L. B. Pankratz. 1967. U.S. Bur. Mines Rept. Inv. 6962.

King, E. G., M. J. Ferrante, and L. B. Pankratz. 1975. U.S. Bur. Mines Rept. Inv. 8041.

King, E. G. and W. W. Weller. 1961a. U.S. Bur. Mines Rept. Inv. 5810.

King, E. G. and W. W. Weller. 1961b. U.S. Bur. Mines Rept. Inv. 5855.

King, E. G. and W. W. Weller. 1962. U.S. Bur. Mines Rept. Inv. 6001.

King, R. C. and G. T. Armstrong. 1968. Nat. Bur. Stand. J. Res. 72A:113-131.

Kireev, V. A. 1947. Izd. An. SSSR. pp. 181.

Kittrick, J.A. 1966. Soil Sci. Soc. Am. Proc. 30:595-598.

Knacke, O. and K. E. Prescher. 1964. Z. Erzbergb. Metallhütt Wes 17:23.

Knopf, H. J. and H. Staude. 1955. Z. Phys. Chem. 204:265-275.

Koehler, M. F., R. Barany, and K. K. Kelley. 1961. U.S. Bur. Mines Rept. Inv. 5711.

Koehler, M. F. and J. P. Coughlin. 1959. J. Phys. Chem. 63:605-608.

Kostryakov, V. N. 1961. Zh. Fiz. Khim. 35:1759.

Lagerström, G. 1959. Acta Chem. Scand. 13:722-736.

Lahiri, S. C. 1965. J. Ind. Chem. Soc. 42:715-724.

Larson, J. W., P. Cerutti, H. K. Garber, and L. G. Hepler. 1968. J. Phys. Chem. 72:2902-2907.

Latimer, W. L. 1952. Oxidation Potentials. Prentice-Hall, Englewood Cliffs, N.J.

Latimer, W. M., K. S. Pitzer, and W. V. Smith. 1938. J. Am. Chem. Soc. 60:1829-1831.

Lietzke, M. H. and J. O. Hall. 1967. J. Inorg. Nucl. Chem. 29:2441-2452.

Lindsay, W. L., M. Peech, and J. S. Clark. 1959. Soil Sci. Soc. Am. Proc. 23:357-360.

Lisov, V. N. 1967. Ukr. Khim. Zh. 33:849-854.

418 APPENDIX SELECTED STANDARD FREE ENERGIES OF FORMATION

Liu, S. T. and G. H. Nancollas. 1973. Desalination 12:75–84.

Lovejoy, R. W., J. H. Colwell. D. F. Eggers, and G. D. Hasley. 1962. J. Chem. Phys. 36:612–617.

Luff, B. B. and R. B. Reed. 1978a. J. Chem. Eng. Data 23:58–60.

Luff, B. B. and R. B. Reed. 1978b. J. Chem. Eng. Data 23:60–62.

Mackle, H. 1963. Tetrahedron. 19:1159–1170.

Mah, A. D. 1957. J. Phys. Chem. 61:1572–1573.

Mah, A. D. 1960. U.S. Bur. Mines Rept. Inv. 5600.

Mah, A. D., L. B. Pankratz, W. W. Weller, and E. G. King. 1967. U.S. Bur. Mines Rept. Inv. 7026.

Makitie, O. and M. L. Savolainen. 1968. Suomen Kem. B. 41:242–245.

Malcolm, G. N., H. N. Parton, and I. D. Watson. 1961. J. Phys. Chem. 65:1900–1902.

Malyszko, J. and L. Duda. 1975. Monatsh. Chemie 106:633–642.

Markham, A. E. and K. A. Kobe. 1941. J. Am. Chem. Soc. 63:449–454.

Marino, G. M., D. M. Hendricks, G. R. Dutt, and W. H. Fuller. 1976. Soil Sci. 121:76–85.

Marland, G. 1975. Geochim. Cosmochim. Acta 39:83–92.

Maronny, G. 1959. J. Chim. Phys. 56:202–213.

Marshall, A. L. and B. Bruzs. 1925. J. Phys. Chem. 29:1184–1186.

Mattigod, S. V. and G. Sposito. 1977. Soil Sci. Soc. Am. J. 41:1092–1097.

Mattigod, S. V. and G. Sposito. 1978. Geochim. Cosmochim. Acta. 1753–1762.

Mattoo, B. N. 1959. Z. Phys. Chem. (Frankfurt) 19:157–167.

McDowell, H., W. E. Brown, and J. R. Sutter. 1971. Inorg. Chem. 10:1638–1643.

McIntyre, J. M. and E. S. Amis. 1968. J. Chem. Eng. Data 13:371–375.

Mel, H. C., Z. Z. Hugus, and W. H. Latimer. 1956. J. Am. Chem. Soc. 78:1822–1826.

Mesmer, R. E. and C. F. Base, Jr. 1974. J. Soln. Chem. 3:307–321.

Milburn, R. M. 1957. J. Am. Chem. Soc. 79:537–540.

Millar, R. W. 1928. J. Am. Chem. Soc. 50:2653–2656.

Moore, G. E. and K. K. Kelley. 1942. J. Am. Chem. Soc. 64:2949–2953.

Moreno, E. C., W. E. Brown, and G. Osborn. 1960. Soil Sci. Soc. Am. Proc. 21:99–102.

Morey, G. W., R. O. Fournier, and J. J. Rowe. 1962. Geochem. Cosmochem. Acta 26:1029–1043.

Mustafaev, F. M. and A. S. Abbasov. 1970. Izv. Akad. Nauk Azerb. SSR, Ser. Fiz-Tekh. Mat. Nauk 3:90–91.

Nair, V. S. K. and G. H. Nancollis. 1959. J. Chem. Soc. 3934–3939.

Nakayama, F. S. 1971. J. Chem. Eng. Data 16:178–181.

Näsänen, R. 1960. Suomen Kem. 33B:47–52.

Näsänen, R., P. Merilänien, and M. Hyle. 1963. Suomen Kem. 36B:73–76.

Naumov, G. B., B. N. Ryzhenko, and F. L. Khodakovsky. 1971. Atomizdat, Moscow (Russian Translation). U.S. Geol. Survey.

Nelson, F. and K. A. Kraus, 1954. J. Am. Chem. Soc. 76:5916–5920.

Neuvonen, K. 1952. J. Bull. Comm. Geol. Finlande No. 158.

Norrish, R. G. W. and F. A. Oldershaw. 1959. Proc. R. Soc. A249:498–513.

Norvell, W. A. and W. L. Lindsay. 1970. Soil Sci. Soc. Am. Proc. 34:360–361.

LITERATURE CITED FOR THE APPENDIX

Norvell, W. A. and W. L. Lindsay. 1981. Soil Sci. Soc. Am. J. 45: (in press).
Nriagu, J. O. 1972a. Am. J. Sci. 272:476–484.
Nriagu, J. O. 1972b. Geochim. Cosmochim. 36:459–470.
Nriagu, J. O. 1972c. Inorg. Chem. 11:2499–2503.
Nriagu, J. O. 1973a. Geochem. Cosmochem. Acta 37:2357–2361.
Nriagu, J. O. 1973b. Geochim. Cosmochim. Acta 37:367–377.
Nriagu, J. O. 1973c. Geochim. Cosmochim. Acta 37:1735–1743.
Nriagu, J. O. 1975. Am. Miner. 60:834–839.
O'Connor, G. A. 1975. Tech. Rept. pp. 3109–3144. New Mexico Water Resource Res. Inst., Las Cruces, NM.
Oetting, F. L. and R. A. McDonald. 1963. J. Phys. Chem. 67:2737–2743.
O'Hare, P. A. G., E. Benn, F. Y. Cheng, and G. Kuzmycz. 1970. J. Chem. Thermodyn. 2:797–804.
Openshaw, R. E., B. S. Hemingway, R. A. Robie, D. R. Waldbaum, and K. M. Krupka. 1976. U.S. Geol. Sur. J. Res. 4:195–204.
Otto, E. M. 1964. J. Electrochem. Soc. 111:88–92.
Page, F. M. 1953. J. Chem. Soc. 1719–1724.
Pankratz, L. B. 1970. U.S. Bur. Mines Rept. Inv. 7430.
Panthaleon, C. L. 1958. Thesis, University of Leiden, The Netherlands.
Parker, V. B. 1965. U.S. NBS Ref. DATA 2:1–66.
Parker, V. B., D. D. Wagman, and W. H. Evans. 1971. Nat. Bur. Stand. Tech. Note 270-6.
Partridge, J. A., R. M. Izatt, and J. J. Christensen. 1965. J. Chem. Soc. 4231.
Paterson, R., S. K. Jalota, and H. S. Dunsmore. 1971. J. Chem. Soc. A:2116–2121.
Paul, A. D. 1955. Thesis, University of California, Berkeley.
Peacock, R. D. 1959. Chem. and Ind. 904.
Perrin, D. D. 1962. J. Chem. Soc. 2197–2200.
Pitzer, K. S. and L. F. Silvester. 1976. J. Soln. Chem. 5:269–278.
Pitzer, K. S. and W. Weltner, Jr. 1949. J. Am. Chem. Soc. 71:2842–2844.
Pleskov, Y. V. and B. N. Kabanov. 1957. Z. Neorg. Khim. 2:1807–1811.
Popadopoulos, M. N. and W. F. Giauque. 1955. J. Am. Chem. Soc. 77:2740–2744.
Prosen, E. J. and F. D. Rossini. 1945. Nat. Bur. Stand. J. Res. 34:263–269.
Provost, R. H. and C. A. Wulff. 1970. J. Chem. Thermodyn. 2:655–658.
Rabinowitch, E. and W. H. Stockmayer. 1942. J. Am. Chem. Soc. 64:335–347.
Ramette, R. W. 1972. J. Chem. Ed. 49:423–424.
Randall, M. and H. M. Spencer. 1928. J. Am. Chem. Soc. 50:1572–1583.
Ray, J. D. and R. A. Ogg. 1957. J. Phys. Chem. 61:1087–1088.
Read, A. J. 1975. J. Soln. Chem. 4:53–70.
Reardon, E. J. 1975. J. Phys. Chem. 79:422–425.
Reardon, E. J., R. L. Jacobson, and D. Langmuir. 1973. Am. Geophys. Union Trans. 54:260–271.
Reardon, E. G. and D. Langmuir. 1974. Am. J. Sci. 274:599–612.
Richardson, F. D. and T. H. E. Jeffes. 1952. J. Iron Steel. Inst. 171:165–175.

Richburg, J. S. and F. Adams. 1970. Soil Sci. Soc. Am. Proc. 34:728-734.
Riet, B. Vont and I. M. Kolthoff. 1960. J. Phys. Chem. 64:1045-1047.
Robie, R. A. 1965. U.S. Geol. Sur. Prof. Paper 525D:65-72.
Robie, R. A., B. S. Hemingway, and W. H. Wilson. 1976. U.S. Geol. Surv. J. Res. 4:631-644.
Robie, R. A. and D. R. Waldbaum. 1968. Geol. Surv. Bull. 1259.
Rohwer, E. F. C. and J. J. Cruywagen. 1963. J. S. Afr. Chem. Inst. 16:26.
Rosenquist, J. 1949. J. Metals 1:451.
Rossini, F. D., K. S. Pitzer, R. L. Arnett, R. M. Braun, and G. S. Pimentel. 1953. Selected values of physical and thermodynamic properties of hydrocarbons and related compounds. Carnegie Press, Pittsburg.
Rossini, F. D., D. R. Wagman, W. H. Evans, S. Levine, and G. Jaffe. 1952. U.S. Nat. Bur. Stands. Cir. 500.
Roth, C. 1928. Z. Electrochem. 34:185.
Russell, A. S., J. D. Edward, and C. D. Taylor. 1955. Trans. Am. Inst. Min. Met. Eng. 203:1123-1128.
Sadiq, M. and W. L. Lindsay. 1979. Selection of standard free energies of formation for use in soil chemistry. Colorado State Univ. Exp. Sta. Tech. Bull. 134.
Sadiq, M., R. E. McFadden, and W. L. Lindsay. 1979. Soil Sci. Soc. Am. J. 43: (in press).
Saegusa, F. 1950. (Cited by Karapet'yants and Karapet'yants, 1970. Sci. Rept. Tohoku Univ. Ser. 1 34:1-55.
Salmon, O. N. 1961. J. Phys. Chem. 65:550-556.
Santillan-Medrano, J. and J. J. Jurinak. 1975. Soil Sci. Soc. Am. Proc. 39:851-856.
Schindler, P., H. Althaus, and W. Feitknecht. 1964. Helv. Chim. Acta 47:982-991.
Schindler, P. W., W. Michaelis, and W. Feitknecht. 1963. Helv. Chim. Acta 46:444-449.
Schindler, P., M. Reinhart, and H. Gamsjäger. 1968. Helv. Chim. Acta 51:1845-1861.
Schorsch, G. 1965. Bull. Soc. Chim. Fr. 988-995.
Schmalz, R. F. 1959. Geophys. Res. J. 64:575-579.
Schmid, G. and U. Neumann. 1967. Z. Phys. Chem. (Frankfurt) 54:150-165.
Schumm, R. H., E. J. Porsen, and D. D. Wagman. 1974. Nat. Bur. Stand. J. Res. Sec. A. 78:375-386.
Schumm, R. H., D. D. Wagman, S. Bailey, W. H. Evans, and V. B. Parker. 1973. Nat. Bur. Stand. Tech. Note 270-7.
Schupp, O. E., III, P. E. Sturrock, and J. I. Watters. 1963. Inorg. Chem. 2:106-112.
Schuylenborgh, J. V. 1973. Trans. Comm. V, VI Intern. Soc. Soil Sci. 93-102.
Scott, D. W., G. D. Oliver, M. E. Gross, W. N. Hubbard, and H. M. Haffman. 1949. J. Am. Chem. Soc. 7:2293-2297.
Shchukarev, L. S. A., L. S. Lilich, and V. A. Latysheva. 1953. Dokl. Akad. Nauk SSSR 91:273-276.
Shearer, J. A. and O. J. Kleppa. 1973. J. Inorg. Nucl. Chem. 35:1073-1078.
Shomate, C. H. 1945. J. Am. Chem. Soc. 67:765-766.
Siebert, R. M. and P. B. Hostetler. 1977. Am. J. Sci. 277:697-715.
Sigel, H., K. Becker, and D. B. McCormic. 1967. Biochim. Biophys. Acta 148:655-664.
Sillen, L. G. and A. E. Martell. 1964. Special Publ. No. 17. The Chemical Society, London.
Singer, P. C. and W. Stumm. 1970. J. Am. Water Works Assoc. 62:198-202.

LITERATURE CITED FOR THE APPENDIX

Singh, S. S. 1974. Soil Sci. Soc. Am. Proc. 38:415-417.

Smith, R. M. and A. E. Martell. 1976. Plenum Press, New York.

Southard, J. C. and C. H. Shomate. 1942. J. Am. Chem. Soc. 64:1770-1774.

Spivakovskii, V. B. and L. P. Moisa. 1964. Z. Neorg. Khim. 9:2287-2294.

Staskiewitcz, B. A., J. R. Tucker, and P. E. Snyder. 1955. J. Am. Chem. Soc. 77:2987-2989.

Stecher, P. G., Ed. 1968. The Merck Index, 8th Ed. Merck & Co., Inc.

Stehlik, B. 1963. Chem. Szesti. 17:6-13.

Stephenson, C. C. 1944. J. Am. Chem. Soc. 66:1436-1437.

Stevens, C. G. and E. T. Turkdogan. 1954. Trans. Faraday Soc. 44:370-376.

Stout, J. W. and R. A. Robie. 1963. J. Phys. Chem. 67:2248-2453.

Stull, D. R. et al. 1966. JANAF Thermochemical Tables. Dow Chemical Co., Midland. Mi.

Stull, D. R. et al. 1971. JANAF Thermochemical Tables, 2nd Ed. Dow Chemical Co., Midland, Mi.

Stull, D. R., E. F. Westrum, Jr., and G. C. Sinke. 1969. The Chemical Thermodynamics of Organic Compounds. John Wiley, New York.

Stuve, J. M., D. W. Richardson and E. G. King. 1975. U.S. Bur. Mines Rept. Inv. 8045.

Talipov, S. H. T. and O. F. Kutumova. 1956. Dokl. Akad. Nauk. Uzbek. SSR 8:23-27.

Taylor, A. H. and K. H. Christ. 1941. J. Am. Chem. Soc. 63:1377-1385.

Taylor, A. W., A. W. Frazier, E. L. Gurney, and J. P. Smith. 1963a. Trans. Faraday Soc. 59:1585-1589.

Taylor, A. W., A. W. Frazier, and E. L. Burney. 1963b. Trans. Faraday Soc. 59:1580-1584.

Taylor, W. A. and E. L. Gurney. 1961. J. Phys. Chem. 65:1613-1616.

Teterevkov, A. I. and V. V. Pechkovskii. 1974. Dokl. Akad. Nauk Belorussk. SSR 18:442-444.

Thompson, A. B. 1973. Am. Mineral. 58:277-286.

Todd, S. S. 1951. J. Am. Chem. Soc. 73:3277-3278.

Todd, S. S. and K. R. Bonnickson. 1951. J. Am. Chem. Soc. 73:3894-3895.

Torgeson, O. R. and T. G. Sahama. 1948. J. Am. Chem. Soc. 70:2156-2160.

Tur'yan, Y. I. 1962. Z. Neorg. Khim. 6:162-168.

Vanderzee, C. E. and D. L. King. 1972. J. Chem. Thermodyn. 4:675-683.

Vanderzee, C. E., M. L. N. Rodenberg, and R. L. Berg. 1974. J. Chem. Thermodyn. 6:17-23.

Vanderzee, C. E. and J. A. Swanson. 1974. J. Chem. Thermodyn. 6:827-843.

Vasil'ev, V. P. 1961. Izcest. Vuz. Khim. 4:936.

Vasil'ev, V. P. 1962. Z. Neorg. Khim. 7:1788-1794.

Volkov, A. I., V. N. Yaglov, and G. I. Noviko. 1976. Zh. Fiz. Khim. 50:797-798.

Vosburgh, W. C. and R. S. McClure. 1943. J. Am. Chem. Soc. 65:1060-1063.

Wagman, D. D., W. H. Evans, V. B. Parker, I. Halow, S. M. Bailey, and R. H. Schumm. 1968. Nat. Bur. Stand. Tech. Note 270-3.

Wagman, D. D., W. H. Evans, V. B. Parker, I. Halow, S. M. Bailey, and R. H. Schumm. 1969. Nat. Bur. Stand. Tech. Note 270-4.

Wagman, D. D. and M. V. Kilday. 1973. Nat. Bur. Stand. J. Res. 77:569-579.

Walbaum, D. R. 1966. Ph.D. Thesis, Harvard University.

Wartenberg, H. V. 1943. Z. Anorg. Chem. 252:136-143.

Watters, J. I. and R. Simonaitis. 1964. Talanta. 11:247-254.

Weaver, R. M., M. L. Jackson, and J. K. Syers. 1971. Soil Sci. Soc. Am. Proc. 35:823–830.

Weller, W. W. 1966. U.S. Bur. Mines Rept. Inv. 6782.

Weller, W. W. and K. K. Kelley. 1964. U.S. Bur. Mines Rept. Inv. 6357.

Weller. W. W. and E. G. King. 1963. U.S. Bur. Mines Rept. Inv. 6147.

Wells, C. F. and G. Davies. 1967. J. Chem. Soc. A:1858–1861.

Westerum, E. F. and J. J. McBride. 1955. Phys. Rev. 98:270.

Wilcox, D. E. and L. A. Bromley. 1963. Ind. Eng. Chem. 55:32–39.

Williams, R. J. P. 1954. J. Phys. Chem. 58:121–126.

Willix, R. L. S. 1963. Trans. Faraday Soc. 59:1325–1323.

Wilson. L. E. and N. W. Gregory. 1958. J. Phys. Chem. 62:433–437.

Wise, S. S., J. L. Margrave, H. M. Feder, and W. N. Hubbard. 1963. J. Phys. Chem. 67:815–821.

Wolhoff, J. A. and J. T. G. Overbeek. 1959. Rec. Trav. Chim. 78:759–782.

Wourtzel, E. 1919. Compt. Rend. 169:1397–1417.

Wu, C. H., R. J. Witonsky, P. George, and R. J. Rutman. 1967. J. Am. Chem. Soc. 89:1987–1991.

Yokokawa, T. and O. J. Kleppa. 1964. J. Phys. Chem. 68:3246–3249.

Young, F. E. 1945. J. Am. Chem. Soc. 67:257–261.

Zen, E-an. 1972. Am. Mineral. 57:524–553.

Zhuk, N. P. 1954. Zh. Fiz. Khim. 28:1523.

Zirino, A. and S. Yamamoto. 1972. Limonol. Oceanogr. 17:661–671.

Zordan, T. A. and L. G. Hepler. 1968. Chem. Rev. 68:737–745.

INDEX

Acetaldehyde
  oxidation state, 374-375
  stability diagram, 378, 380-381
Acetate
  dissociation constants, 377
  equilibrium reactions, 376
  reaction gradients, 378
  stability diagrams, 378, 380-381
Acetic acid metabolism
  effect of redox on, 380-381
  products of, 380-381
  reaction gradients, 380-381
Acetylene
  oxidation state, 375
  stability diagram, 378, 380-381
Acid-base reactions, 129
Activity coefficients
  Davies equation, 14
    calculated values, 18
  Debye-Hückel equations, 13-14
    calculated values, 17
    ionic parameters for, 15-16
  definition, 13
  from electrical conductivities, 16-17
  Guggenheim equation, 14
  Guntelberg equation, 14
Activity constants, defined, 12, 18-19. *See also* Equilibrium constants
Adsorbed ions
  buffering effect, 4-5
Alabanite
  cation controlling phases, 292-293
  equilibrium constants, 285
  pe + pH of formation, 291, 293-294
  effect of pH on stability, 294
  stability diagram, 291
Albite, high
  equilibrium constant, 58, 120
  stability diagram, 63, 122
  stability relationships, 122-123
Albite, low
  equilibrium constant, 58, 120
  diagram development, 63-64, 119
  stability diagram, 122-123
  stability relationships, 119, 122-123
Aluminosilicates, 57-76
  of calcium, 66-68
  effect on Al^{3+} activity, 48
  equilibrium constants, 58-60
  free energies estimated, 75
  of magnesium, 68-70
  of mixed cations, 71-73
  of potassium, 65-66

of sodium, 63-64
summary diagram, 71
unsubstituted, 57, 61-62
Aluminum
content in lithosphere, 7, 35
content in soils, 7, 35
redox equilibria, 36, 46
Aluminum activity
calculation of, 45-46, 171
controlled by gibbsite, 36-37, 73-74
controlled by kaolinite-quartz, 73-74, 87, 91-93, 169-170
controlled by montmorillonite, 73-75
equilibrium diagram, 74
summary diagram, 74
from total aluminum, 45-46
Aluminum complexes
equilibrium constants, 36
with fluoride, 41-43
with nitrate, 44
stability diagrams, 42, 44
with sulfate, 43-44
Aluminum hydrolysis
equilibrium constants, 36
equilibrium with gibbsite, 40-41
hexahydronium ion, 39
polymeric species, 41
stability diagram, 40
Aluminum hydroxides, *see* Aluminum oxides
Aluminum hydroxysulfate, hypothesized in soils, 39
Aluminum minerals
equilibrium constants, 36
solubility relationships, 35-41
Aluminum oxides
effect of pH on, 35-38
equilibrium constants, 36
stability diagram, 37
Aluminum phosphates
amorphous, 173
diagram development, 169-171
double function parameters, 171-173
effect of redox on, 177-180
equilibrium constants, 164
equilibrium with gibbsite, 171-173
equilibrium with kaolinite and silica, 172-173

stability diagrams, 170, 172
from TPS, 202-204
Aluminum redox relationships, 36, 46
Aluminum in soils
constancy of log Al^{3+} + 3pH, 47-48
exchangeable, 47-48
extracted by $CaCl_2$, 47-48
Aluminum sulfates, solubilities, 36, 38-39
Alunite
formation in soils, 38-39
solubility product, 36
Amidogen
chemical formula, 269
equilibrium reaction, 270
oxidation state, 269
stability diagram, 273
Ammonia
chemical formula, 269
equilibrium with ammonium, 270, 273
equilibrium reaction, 270
oxidation state, 269
stability diagrams, 271
Ammonium ion
chemical formula, 269
equilibrium reaction, 270
oxidation state, 269
stability diagram, 271, 273-279
Ammonium phosphates, 189
Amorphous silica, *see* Silica, amorphous
Analcime
stability constant, 58, 120
stability diagram, 63, 122
stability relationships, 122-123
Andalusite
equilibrium constant, 58
stability diagram, 61
stability relationships, 61-62
Anglesite
equilibrium constants, 330
stability diagram, 333
stability relationship, 329
Anorthite
equilibrium constant, 59, 88
stability diagram, 67, 91
stability relationships, 66-68, 91-92
Anorthite, hexagonal
equilibrium constant, 59, 88
stability diagram, 67, 91
stability relationships, 66-68, 91-92
Apatites, isomorphous substitutions, 184

INDEX

Aragonite
  equilibrium constant, 88
  stability diagram, 94
  stability relationships, 93-94
Arsenic, content in soils, 7
Azide
  chemical formula, 269
  equilibrium constant, 270
  intermediate in denitrification, 277
  not reported in soils, 277
  oxidation state, 269
  redox range of metastability, 276-277
  stability diagram, 275, 276
Azurite
  equilibrium constants, 224
  stability diagram, 223
  stability relationships, 222, 223

Barium, content in soils, 7
Beidellite
  equilibrium constant, 58, 120
  exchangeable Na^+, 123
  stability diagram, 64
Berlinite
  equilibrium constants, 164
  stability diagram, 170
  stability relationships, 170-171
  stable at high temperature, 171
Beryllium, content in soils, 7
Bicarbonate ion
  dissociation constant, 79
  effect of CO_2 on, 80
  equilibrium reactions, 80-83
Birnessite
  equilibrium constant, 151
  stability diagrams, 153, 156
  stability relationships, 153, 155-156
Bisulfate ion
  dissociation constants, 284
  effect of pH on, 289
  in soils, 287
Bisulfide
  equilibrium reactions, 284
  oxidation state, 282
  stability diagrams, 283, 288
  stability relationships, 283, 286
Bisulfide ion
  dissociation constants, 284
  effect of pH on, 287
  effect of redox on, 287

Bixbyite
  equilibrium constant, 151
  stability diagrams, 153, 156
  stability relationships, 153, 155-156
Bobierite, equilibrium constant, 165
Boltzmann equation, 47
Bones, 190
Boron, content in soils, 7
Bromide activity, effect of Ag^+, 304-306
Bromium, content in soils, 7
Bromopyromorphite
  equilibrium constants, 331
  stability diagram, 334
  stability relationship, 334-336
Brucite
  equilibrium constant, 108
  stability diagram, 114
  stability relationships, 113-114
Brushite
  equilibrium constant, 165
  from MCP, 197-204
  stability relationships, 180-184

Cadmium
  content in lithosphere, 7, 316
  content in soils, 7, 36
  environmental contaminant, 316
  redox relationships, 316, 318
  solubilization of ammonium fertilizers, 323
  use in metal plating, 316
Cadmium, elemental, redox relationships, 316, 318
Cadmium activity
  reference level for soils, 316-317
  see also Soil-Cd
Cadmium carbonate
  equilibrium constants, 318
  reference for high pH soils, 317
  stability diagram, 317
Cadmium complexes
  of ammonia, 322-323
  of bicarbonate, 323-325
  of carbonate, 324-325
  equilibrium constant, 318-319
  equilibrium with octavite, 321
  equilibrium with soil-Cd, 318-319
  of halides, 322-323
  of nitrate, 324-325
  of phosphate, 324-325

of pyrophosphate, 325-326
  stability diagrams, 322, 324
  of sulfate, 324-325
Cadmium hydrolysis
  contribution to soluble cadmium, 321
  in equilibrium with octavite, 321
  in equilibrium with soil-Cd, 321
  stability constants, 318
  stability diagram, 321
Cadmium hydroxide
  equilibrium constants, 318
  stability diagram, 317
  unstable in soils, 317, 320
Cadmium hydroxysulfate
  equilibrium constants, 318
  stability diagram, 317
  unstable in soils, 317
Cadmium minerals
  equilibrium constants, 318
  redox relationships, 316
  stability diagram, 317
  stability in soils, 316-317, 319-320
Cadmium oxides
  equilibrium constants, 318
  stability diagram, 317
  unstable in soils, 317
Cadmium phosphate
  effect of phosphate minerals on, 319-320
  equilibrium constants, 318
  stability diagram, 317
  stability in soils, 319
Cadmium silicate
  equilibrium constants, 318
  stability diagram, 317
  unstable in soils, 317
Cadmium sulfates
  equilibrium constants, 318
  unstable in soils, 320
Cadmium sulfide
  equilibrium constants, 318
  redox stability range, 320
  stability in soils, 320
Calcareous soils, pH range of, 98
Calcite
  $CaO\text{-}CO_2\text{-}H_2O$ system, 98-101
  effect of CO_2 on, 98-101
  effect of gypsum on, 283
  electroneutrality equation, 100-101
  equilibrium constant, 88

  equilibrium pH, 100-101
  equilibrium with portlandite, 93-100
  equilibrium reactions, 98-99
  log Ca^{2+} + 2pH, 97-99
  phase rule applied to, 101-102
  stability diagrams, 91, 93
  stability relationships, 92-95
  transformation to oldamite, 293-294
Calcium
  content in lithosphere, 7, 87
  content in soils, 7, 87
  redox relationships, 90, 98
Calcium activity
  reference for acid soils, 92
  reference for calcareous soil, 93
  see also Soil-Ca
Calcium aluminosilicates
  equilibrium constants, 88
  stability diagram, 91
  stability relationships, 87, 91-93
Calcium carbonates
  equilibrium constants, 88
  equilibrium equations, 98-100
  stability diagrams, 94, 100
  stability relationships, 93-95, 98-101
  see also Calcite
Calcium complexes
  of bicarbonate, 95-96
  of carbonate, 95-96
  of chloride, 95-96
  equilibrium constants, 89
  of nitrate, 95-96
  of phosphate, 97
  stability diagrams, 96-97
  of sulfate, 97
Calcium fluoride, see fluorite
Calcium glass
  equilibrium constant, 59, 88
  stability diagram, 67, 91
  stability relationships, 66-68, 91-92
Calcium hydrolysis
  stability constants, 89
  stability diagram, 97
  stability relationships, 97-98
Calcium minerals
  unstable in acid soils, 92
Calcium molybdate
  equilibrium constants, 366
  stability diagram, 368
  stability relationships, 367-372

INDEX

Calcium phosphates
  CaO–P$_2$O$_5$–H$_2$O system, 200-202
  double function plots, 182-184
  effect of calcium on, 180-184
  effect of CO$_2$ on, 180-184
  effect of fluorite on, 180-185
  effect of pH on, 180-185
  effect of redox on, 185-186
  equilibrium with calcite, 180-184
  equilibrium constants, 165
  equilibrium with soil-Ca, 180-184
  stability diagrams, 181, 183, 185
  stability relationships, 180-185
  stability relative to strengite, 180-183
  stability relative to variscite, 180-183
  transformation to vivianite, 186
Calcium pyrophosphate, convert to orthophosphate, 191-193
Calcium silicates
  equilibrium constants, 88
  as liming agents, 92
  stability diagram, 91
  stability relationships, 87, 91-93
Calculations
  activity coefficients, 13-18
  equilibrium constants, 18-22
  standard cell potentials, 26
Capacity factors, 4-5
Carbon
  content in lithosphere, 374
  content in soils, 7, 374
  oxidation states, 374-375
Carbon, elemental, see Graphite
Carbonates
  in alkaline soils, 83-84
  effect of CO$_2$ on, 81-83
  effect of pH on, 81-84
  equilibrium constants, 79
  equilibrium reactions, 79-84
  importance in soils, 79, 83-84
  mole fraction diagram, 81-82
  open systems in soils, 79
  oxidation state, 324-325
  solubility in pure water, 81-84
  unstable in acid soils, 83-84
Carbon black, 231
Carbon compounds, equilibrium reactions, 376-377
Carbonic acid
  dissociation constants, 79
  dissociation reactions, 80
  effect of CO$_2$ on, 79-80
Carbon monoxide
  equilibrium with CO$_2$ (g), 377
  oxidation state, 374-375
  stability diagram, 378, 380-381
CDTA
  formation constants, 242-243
  in hydroponic solution, 256-257
  technical name, 244
Cell potential, standard
  calculation from log K°, 26-27
  pe vs Eh, 27-28
Cemetaries, 190
Cerussite
  effect of CO$_2$ on, 329
  equilibrium constants, 330
  stability diagram, 333-334
  stability relationship, 329-337
  transformation to galena, 291-294
Cesium, content in soils, 7
Chalcocite
  controlling soil phases, 292-293
  effect of pH on, 295-297
  equilibrium constants, 294
  equilibrium diagram, 291
  pe + pH of formation, 291, 293-294
  redox stability range, 295-297
Chalcocyanite
  equilibrium constants, 224
  unstable in soils, 223
Charcoal black, adsorption of organic complexes, 219
Chelated metals
  diagrammatical representation, 245
  dissociation at roots, 258-259
  equations for, 247-248
  mole fraction diagrams, 249-256
  mole fraction equations, 249
  movements to roots, 258-259
  not absorbed by roots, 258-259
Chelates
  definition, 239
  EDTA as example, 239-241
  effect of CO$_2$ on, 248-255
  effect of pH on, 244-257
  equilibrium constants, 242-243
  to measure ion activities, 259-261
  as micronutrient fertilizers, 239

mole fraction calculations, 244-247
natural, 263-264
as soil test extractants, 261-263
Chelation, 239-263
  effect on available nutrients, 239
  effect of redox on, 252-256
  in hydroponics, 256-259
Chloride activity, effect of Ag^+, 304-306
Chlorine, content in soils, 7
Chlorite
  equilibrium constants, 59, 107
  stability diagram, 69, 112
  stability relationships, 68-70, 112-113
Chloropyromorphite
  equilibrium constants, 331
  extremely stable in soils, 335
  stability diagram, 334
  stability relationships, 334-336
Chromium, content in soils, 7
Chrysotolite
  equilibrium constants, 107
  stability diagram, 110
  stability relationships, 106, 110-111
Cinnabar
  equilibrium constants, 348
  redox stability range, 359
Cinnabar, black
  controlling soil phases, 292-293
  equilibrium constants, 285
  equilibrium diagram, 291
  pe + pH of formation, 293
Cinnabar, red
  controlling soil phases, 292-293
  equilibrium constants, 285
  equilibrium diagram, 291
  pe + pH of formation, 293
Clay minerals, see Aluminosilicates
Clinoenstatite
  equilibrium constant, 107
  stability diagram, 110
  stability relationships, 106, 110-111
Coal, 383
Cobalt, content in soils, 7
Coesite
  equilibrium constants, 51
  stability diagram, 52
  solubility relationships, 51-52
Complexes, see separate elements
Concentration constants
  definition, 12, 18

  see also Equilibrium constants
Congruent dissolution, of MCP, 198-202
Convection gradients, 4
Copper
  effect of redox on, 225, 231-235
  in lithosphere, 7, 222
  in soils, 7, 222
Copper, elemental
  formation in soils, 233
  redox reactions, 225
  stability diagram, 232
Copper activity
  equation for calculating, 234
  see also Soil-Cu
Copper carbonates
  equilibrium constants, 224
  stability diagram, 223
  unstable in soils, 222-223
Copper(I) complexes
  with chloride, 234-235
  contribution to total soluble copper, 234-235
  effect of redox on, 234-235
  in equilibrium with cuprous ferrite, 234-235
  significant species, 234-235
  stability constants, 225
Copper(II) complexes
  with bicarbonate, 225-230
  with carbonate, 225, 229-231
  with chloride, 224, 230
  with nitrate, 225, 230
  with organic matter, 231
  with phosphate, 225, 230
  with pyrophosphate, 227-228
  stability constants, 224-225
  stability diagram, 230
  with sulfate, 225, 229-231
Copper(II) hydrolysis
  stability constants, 224
  stability diagram, 229
  stability relationships, 228-229
Copper (I) hydroxides
  equilibrium constants, 225
  stability diagram, 232
  unstable in soils, 234
Copper (II) hydroxides, see Copper oxide
Copper hydroxycarbonates
  equilibrium constants, 224
  stability diagram, 226

INDEX 429

stability relationships, 222-223
Copper hydroxysulfates
  equilibrium constants, 224
  stability diagram, 226
  unstable in soils, 223
Copper(I) minerals
  effect of redox on, 231-234
  equilibrium constants, 226
  stability diagram, 232
  stability relationships, 231-234
Copper(II) minerals
  equilibrium constants, 224
  stability diagrams, 223, 226-227
  stability relationships, 222-228
Copper molybdate
  equilibrium constants, 366
  stability diagram, 368
Copper(II) orthophosphates
  available copper, 227
  available phosphate, 227
  equilibrium constants, 224
  equilibrium diagram, 226
Copper(II) oxide
  equilibrium constants, 224
  solubility diagram, 223
  stability relationships, 222
Copper oxysulfates
  equilibrium constants, 224
  stability diagram, 226
  unstable in soils, 223
Copper(II) pyrophosphate
  equilibrium constant, 224
  stability diagram, 227
  unstable in soils, 227-228
Copper sulfate
  equilibrium constants, 224
  unstable in soils, 223
Cordierite
  equilibrium constant, 59, 107
  stability diagram, 69, 112
  stability relationships, 69-70, 112-113
Covellite
  controlling soil phases, 292-293
  equilibrium constants, 284
  equilibrium diagram, 291
  pe + pH of formation, 293
Cristobalite
  equilibrium constant, 51
  solubility relationships, 51-52
Critical nutrient level, 5

Cuperite
  equilibrium constants, 225
  stability diagram, 232
  unstable in soils, 234
Cupric ferrite
  effect of iron on, 222-223
  equilibrium constants, 224
  stability diagram, 223
  stability relationships, 222-223
Cupric ion
  effect of pH on, 295-297
  effect of redox on, 295-297
Cuprous ferrite
  diagram development, 232-234
  effect of iron oxides on, 231-233
  effect of redox on, 231-234
  equilibrium constants, 225
  redox stability range, 295-296
  stability diagram, 232
  transformation to chalcocite, 291-294
Cuprous ion
  effect of chalcocite on, 295-296
  effect of pH on, 295-297
  effect of redox on, 295-297
Cyclohexane-1,2-diaminetetraacetic acid,
    see CDTA

Davies equation, 14
  calculated values, 18
DCP, see Dicalcium phosphate
DCPD, see Dicalcium phosphate dihydrate
Debye-Hückel equations, 15-16
  calculated values, 17
Dicalcium phosphate
  equilibrium in $CaO-P_2O_5-H_2O$ system,
    197-202
  equilibrium constant, 165
  stability diagram, 181
  stability relationships, 180-184
Dicalcium phosphate dihydrate
  equilibrium in $CaO-P_2O_5-H_2O$ system,
    197-202
  equilibrium constant, 165
  from MCP, 197-204
  stability relationships, 180-184
Dickite
  equilibrium constant, 58
  stability diagram, 61
  stability relationships, 61-62

INDEX

Diethylenetriaminepentaacetic acid, *see* DTPA
Diffusion gradients, 4
Dinitrogen
  chemical formula, 269
  content in atmosphere, 268
  equilibrium reaction, 270
  oxidation state, 269
  stability diagram, 271, 273-274
Dinitrogen pentoxide
  chemical formula, 269
  equilibrium diagram, 271, 273
  equilibrium reaction, 270
  oxidation state, 269
Dinitrogen tetroxide
  chemical formula, 269
  equilibrium reaction, 270
  oxidation state, 269
  stability diagram, 271, 273
Dinitrogen trioxide
  chemical formula, 269
  equilibrium reaction, 270
  oxidation state, 269
  stability diagram, 271, 273
Diopside
  equilibrium constants, 88, 107
  stability diagram, 91, 110
  stability relationships, 91-92, 110-111
Dissociation constant, definition, 11
Disulfur oxide
  equilibrium reactions, 284
  oxidation state, 282
  stability diagram, 283, 288
  unstable in soils, 283, 286
Dithionite
  equilibrium reactions, 284
  oxidation state, 282
  stability diagram, 283
  unstable in soils, 283, 286
Dithionitic acid
  dissociation constants, 284
  effect of pH on, 284
  unstable in soils, 284
Dolomite
  equilibrium constant, 88, 108
  transformation to magnesium sulfide, 293-294
Dolomite-calcite equilibria
  effect of CO_2 on, 106, 111
  effect of pH on, 106, 111
  equilibrium reactions, 111
  limit on Mg^{2+} activity, 106, 111
  stability diagrams, 110, 112, 114
DTPA
  formation constants, 242-243
  in hydroponic solution, 256-257
  mole fraction diagrams, 250-251, 256
  summation diagram for zinc, 251
  technical name, 244
DTPA soil test
  critical levels, 261
  development of, 261-263
  effect of pH on, 261-262
  effect of redox on, 262
  extracting capacity, 262-263
  procedure, 261
  success of, 261, 263
  theoretical basis, 261-263
Dynamic equilibria in soils, 2-4

EDDHA
  formation constants, 242-243
  in hydroponics, 256-257
  technical name, 244
EDTA
  acid distribution diagram, 240
  chemical of iron(II), 252-255
  chelation of manganese, 252-255
  ^{14}C-labeled, 259
  effect of redox on, 252-256
  FeEDTA structure, 241
  formation constants, 242-243
  definition of, 241
  free ligand, 240
  in hydroponics, 256-258
  mole fraction calculations, 245-257
  mole fraction diagrams, 249-254, 256
  structures, 239
  summation diagram for zinc, 251
  technical name, 244
EGTA
  formation constants, 242-243
  in hydroponic solution, 256-257
  technical name, 244
Eh
  aqueous Eh-pH diagram, 25
  cell reaction, 27-28
  conversion to pe, 25, 27-28
  definition of, 23, 27-28
  disadvantages of, 23

INDEX

range for soils, 25, 28
see also pe
Electrical conductivities, to estimate ionic strength, 16
Electron activity, *see* pe
Electroneutrality equation, for $CaO-CO_2-H_2O$ system, 100-101
Electron titration
 of iron oxides, 145-146
 mineral transformations, 145-146
 redox poises, 145
Elemental sulfur, *see* Sulfur, rhombic
Elements
 in lithosphere, 7-8
 in soils, 7-8
 in soil solution, 6-8
Enthalpies of reaction, standard, 21-22
Entropies of reaction, standard, 21-22
Equilibrium
 defined, 11, 21
 determination of, 11
 dynamic in soil, 2-4
 reaction rates, 11
 slowly attained, 3-4
Equilibrium constants
 activity constants, 18
 calculated from E^0, 26
 calculated from ΔG^0_r, 21-22
 calculated from ΔH^0_r and ΔS^0_r, 21-22
 concentration constants, 18
 definition, 11-12
 for different ionic strengths, 241
 mixed constants, 18
 from thermodynamic data, 21-22
 transformations, 18-20
Ethane
 oxidation state, 374-375
 produced from acetic acid, 381
 stability diagram, 378, 380-381
Ethanol
 oxidation state, 374-375
 stability diagram, 378, 380-381
 vapor pressure of, 377
Ethylene
 oxidation state, 375
 stability diagram, 378, 380-381
Ethylenediaminedi(*o*-hydroxyphenylacetic acid), *see* EDDHA
Ethylenediaminetetraacetic acid, *see* EDTA

Ethyleneglycolbis(ethylamine)tetraacetic acid, *see* EGTA
Exchangeable ions
 buffering action, 4-5
 exchangeable aluminum, 47-48
Exchange reactions, effect on soil solution, 2-3

Fayalite
 effect of $H_4SiO_4^0$ on, 142-144
 equilibrium constant, 131
 stability diagram, 142-143
 stability relationships, 142-144
Ferrimolybdate, hypothesized in soils, 369
Ferrite minerals, effect of iron on, 148
Ferromagnesium minerals, weather to Fe(III) oxides, 129
Ferrosic oxide
 effect on phosphate, 180
 equilibrium constant, 131
 mixed valence states, 143
 stability diagram, 144
 stability relationships, 143-144
Fertilizers
 effect on soil solution, 3
 reaction products of, 197-204
Flashes of light, 190
Fluorapatite
 effect of fluorite on, 180-185
 effect of fluorphlogopite on, 184-185
 equilibrium constants, 165
 stability diagram, 181, 183, 185
 stability relationships, 180-185
Fluorine, content in soils, 7
Fluorite
 control of F^- activity, 42-43, 95, 335
 effect on phosphate, 180-185
 equilibrium constants, 36, 42, 89
 equilibrium with soil-Ca, 95
Fluoropyromorphite
 equilibrium constants, 331
 stability diagram, 334
Fluorphlogopite
 effect of F^- on, 43, 70
 effect on phosphate, 184-185
 equilibrium constant, 59
 stability diagram, 69
 stability relationships, 69-70
Formate
 oxidation state, 374-375

stability diagram, 378, 380-381
Formation constants, definition, 11
Forsterite
  equilibrium constants, 107
  stability diagram, 110
  stability relationships, 106, 110-111
Franklinite
  effect of iron on, 214-215
  equilibrium constant, 212
  stability diagram, 214
  stability relationships, 213-214
Free energy
  estimated for clay minerals, 75
  minimum at equilibrium, 21
  see also Standard free energy

Galena
  controlling soil phases, 292-293, 336-337
  equilibrium constants, 285, 331
  equilibrium diagram, 291
  pe + pH of formation, 291, 293-294, 336-337
Gallium, contents in soils, 7
Germanium, content in soils, 7
Gibbsite
  effect on aluminum solubility, 35-38, 73-74
  stability vs. aluminum oxides, 35-38
    vs. alunite, 38-39
  stability diagram, 37, 63, 66, 71
Glucose
  equilibrium reactions with other carbon species, 376
  oxidation state, 374-375
  product of photosynthesis, 375
  reference level, 378
  stability diagrams, 378, 380-381
Glucose metabolism
  effect of redox on, 378-379
  oxidation products of, 377
  reaction gradients, 378
  reduction products of, 379
Goethite
  equilibrium constants, 130
  most stable Fe(III) mineral, 132-133
  stability diagram, 132
  stability relationships, 129, 132-133, 143-145
Graphite
  effect of CO_2 on, 383
  equilibrium constants, 377-378

formation in soils, 382-383
  oxidation state, 374-375
  redox stability range, 383
Greennokite
  controlling soil phases, 292-293
  effect of otavite on, 293
  equilibrium constants, 284, 318
  equilibrium diagram, 291
  pe + pH of formation, 291, 293-294
  redox stability range, 320
Guggenheim equation, 14
Guntelberg equation, 14
Gypsum
  in arid soils, 94
  coexistent with calcite, 102-103
  effect on sulfate, 44, 283
  equilibrium constants, 36, 98, 285
  equilibrium with soil-Ca, 283
  in polder soils, 94
  stability diagram, 94

Halide activities, reference levels, 335
Halloysite
  equilibrium constant, 58
  stability diagram, 61
  stability relationships, 61-62
Hauerite
  controlling soil phases, 292-293
  effect of pH on stability, 294
  equilibrium constants, 285
  pe + pH of formation, 291, 293
  stability diagram, 291
Hausmannite
  equilibrium constants, 151
  stability diagrams, 153, 156
  stability relationships, 153-156
HEEDTA
  formation constants, 242-243
  in hydroponic solution, 256-257
  technical name, 244
Hematite
  equilibrium constants, 130
  stability diagram, 132
  stability relationships, 129, 132-133
Hoagland nutrient solution
  chelation of iron, 256-257
  composition of, 257
  zinc activity in, 258
Hopeite
  equilibrium constants, 212

INDEX

stability diagram, 215
stability relationships, 216
stability relative to soil-Zn, 215-216
Hydrazine
  chemical formula, 269
  equilibrium reaction, 270
  oxidation state, 269
  protonated, 269-270
  stability diagram, 273
Hydrogen half cell, 26-28
Hydrogen sulfide
  dissociation constants, 284
  effect of pH on, 287-290
  effect of redox on, 287-290
  in equilibrium with rhombic sulfur, 289-290
  partial pressure of, 287
  solubility relationships, 287
  stability in soils, 287
Hydrolysis, *see separate elements*
Hydroxyapatite
  effect on Pb^{2+} solubility, 334
  effect of redox on, 186
  equilibrium constants, 165
  stability diagram, 181, 183, 185
  stability relationships, 180-185
  stability relative to manganese phosphates, 180-189
Hydroxyethylethylenediaminetriacetic acid, *see* HEEDTA
Hydroxylamine
  chemical formula, 269
  equilibrium reaction, 270
  oxidation state, 269
  stability diagram, 271-273
Hydroxypyromorphite
  equilibrium constants, 331
  stability diagram, 334
  stability relationships, 334-336

Ikaite
  equilibrium constants, 88
  stability diagram, 94
  stability relationships, 93-94
Illite
  equilibrium constant, 60, 107, 120
  stability diagrams, 71, 112, 124
  stability relationships, 71-72, 112-113, 124-125
Imidogen
  chemical formula, 269
  equilibrium reaction, 270
  oxidation state, 269
  stability diagram, 273
Intensity factors, 45
Iodide activity
  effect on Ag^+, 304-306
  effect on Hg^{2+}, 345, 348
Iodine, content in soils, 7
Ionic strength
  for chelate calculations, 241
  definition, 12-13
  equation, 12-13
Iron
  content in lithosphere, 7, 129
  content in soils, 7, 129
  redox relationships, 131, 139-148
Iron activity, *see* Soil-Fe
Iron(II) complexes
  of bromide, 147
  of chloride, 147
  contribution to soluble iron, 146-148
  effect of redox on, 146-148
  equilibrium constants, 131
  equilibrium with magnetite, 146-147
  equilibrium with soil-Fe, 146-148
  of phosphate, 146-148
  stability diagram, 147
  of sulfate, 146-148
Iron(III) complexes
  with bromide, 136, 137
  with chloride, 136, 137
  contribution to total iron, 138-139
  in equilibrium with soil-Fe, 136-138
  with fluoride, 136, 137
  with hydroxyl, 138
  with nitrate, 137-138
  with phosphate, 137-138
Iron(II) hydrolysis
  contribution to soluble iron, 146
  effect of pH on, 146-148
  effect of redox on, 146-148
  equilibrium constants, 131
  equilibrium with magnetite-soil-Fe, 146-148
  stability diagram, 147
Iron(III) hydrolysis
  contribution to total iron, 136
  equilibrium constants, 130
  equilibrium with other Fe(III) oxides, 136

equilibrium with soil-Fe, 135-136
stability diagram, 135
stability relationships, 134-136
Iron (II) hydroxide
  equilibrium constants, 131
  instability in soils, 142
  stability diagram, 142
  stability relationships, 142-144
Iron(III) hydroxides, *see* Iron(III) oxides
Iron minerals
  effect of redox on, 141-146
  solubility relationships, 143-144
  stability of, 141-146
Iron molybdate
  equilibrium constants, 366
  stability diagram, 371
  stability relationships, 369
Iron(II) oxide
  equilibrium constant, 131
  stability diagram, 142
  stability relationships, 142-143
Iron(III) oxides
  effect on Al^{3+}, 73-75
  effect of pH on, 129, 132
  equilibrium constants, 130
  stability diagram, 132
  stability relationships, 129, 132-133
Iron phosphates
  double function parameters, 174-176
  effect of iron oxides on, 174-177
  effect of redox on, 177-180
  equilibrium constants, 164
  equilibrium with soil-Fe, 174-177
  precipitation from TPS, 202-204
  stability diagrams, 175-177
  stability relationships, 174-177
Iron pyrophosphate
  effect of iron oxides on, 193-194
  effect of pH on, 193-194
  effect of redox on, 193-194
  equilibrium constants, 167
  stability diagram, 194
  stability relative to calcium pyrophosphate, 193-194
  unlikely in soils, 194
Iron(II) silicates
  equilibrium constants, 131
  stability diagrams, 142-143
  stability relationships, 142-144

Iron solubility
  controlled by Fe(III) oxides, 129
  critical redox, 148
  effect of redox on, 148
Iron(II) solubility
  effect of Fe(III) oxides on, 140
  effect of pH on, 139-140
  effect of redox on, 139-140
  in equilibrium with soil-Fe, 139-140
  stability diagram, 140
Iron(II) sulfates
  equilibrium constants, 131
  stability relationships, 146
Iron(III) sulfates
  equilibrium constants, 130
  stability relationships, 133
Iron sulfides
  controlling soil phases, 291, 292-294
  equilibrium constants, 284-285
  relative stabilities, 294
  stability diagram, 291
Islemannite, hypothesized in soils, 369
Isomorphous substitution
  alters free energy, 75
  effect of redox on, 75

Jadeite
  equilibrium constants, 58, 120
  stability diagram, 63, 122
  stability relationships, 122-123
Jarosite
  in acid sulfate soils, 133
  equilibrium constant, 130
  from pyrite, 133
  stability diagram, 134
  stability relationships, 133-134

Kaliophilite
  equilibrium constants, 59, 120
  stability diagram, 66, 124
  stability relationships, 65-66, 124
Kaolinite
  effect on Al^{3+} activity, 73-75
  equilibrium constant, 58
  stability diagrams, 61, 71
  stability in soils, 57, 61-62, 71-75
Kyanite
  equilibrium constants, 58
  stability diagram, 61
  stability relationships, 61-62

INDEX

Lansfordite
  equilibrium constant, 108
  stability diagram, 114
  stability relationships, 113-114
Lanthanum, content in soils, 7
Larnite
  equilibrium constants, 88
  stability diagram, 91
  stability relationships, 91-92
Lawrencite
  equilibrium constant, 131
  equilibrium relationships, 146
Lawsonite
  equilibrium constant, 59, 88
  stability diagram, 67, 91
  stability relationships, 66-68, 91-93
Lead
  contaminant in environment, 7, 329
  content in lithosphere, 7, 329
  content in soils, 7, 329
  effect on molybdenum solubility, 336
  redox relationships, 329, 330, 331, 337
Lead, elemental
  possibility in soils, 337
  redox reaction, 331
Lead activity, see Soil-Pb
Lead bromophosphates
  equilibrium constants, 331
  stability diagram, 334
  stability relationship, 335
Lead carbonates
  effect of CO_2 on, 329
  effect on molybdenum, 368
  equilibrium constants, 330
  stability diagram, 333-334, 368
  stability relationships, 329-337
Lead chlorocarbonates
  equilibrium constants, 330
  stability diagram, 333
  stability relationship, 329
Lead chlorophosphate
  equilibrium constants, 331
  stability diagram, 334
Lead complexes
  equilibrium constants, 331-332
  with halides, 339-340
  with nitrate, 340-341
  with phosphate, 340-341
  with pyrophosphate, 340-341
  stability diagrams, 339-340
  stability relationships, 339-341
  with sulfate, 340-341
Lead fluorophosphate
  equilibrium constants, 331
  stability diagram, 334
  stability relationship, 335
Lead hydrolysis
  formation constants, 341
  stability diagram, 338
  stability relationships, 338-339
Lead hydroxide
  equilibrium constants, 330
  stability diagram, 333
  stability relationships, 329
Lead hydroxycarbonates
  equilibrium constants, 330
  stability diagram, 333
  stability relationships, 329
Lead minerals
  equilibrium constants, 330-331
  stability diagrams, 333-334
  stability relationships, 329-337
Lead molybdate, see Wulfenite
Lead oxides
  effect of redox on, 329
  equilibrium constants, 330-331
  oxidation states, 329
  stability diagram, 333
  stability relationships, 329
Lead oxycarbonates
  equilibrium constants, 330
  stability diagram, 333
  stability relationships, 329
Lead oxyphosphate
  equilibrium constants, 331
  stability diagram, 334
  stability relationship, 334-335
Lead oxysulfates
  equilibrium constants, 330
  stability diagram, 333
  stability relationships, 329
Lead phosphates
  control of Pb^{2+}, 334
  equilibrium constants, 330-331
  stability diagram, 334
  stability relationships, 334-336
Lead silicates
  effect of silica on, 333
  equilibrium constants, 330
  stability diagram, 334

stability relationship, 333
Lead sulfates
  equilibrium constants, 330
  stability diagram, 333
  stability relationship, 329
Lead sulfide
  development of equation, 326-337
  equilibrium constants, 331
  redox for precipitation, 337
  stability relationship, 336-337
Leonhardite
  equilibrium constants, 59, 88
  stability diagram, 67, 91
  stability relationships, 66-68, 91-93
Lepidocrocite
  equilibrium constant, 130
  stability diagram, 132
  stability relationships, 129, 132
Leucite
  stability constant, 59, 120
  stability diagram, 65, 124
  stability relationships, 65-66, 124
Lime
  equilibrium constant, 89
  reactions with soil, 92
  stability relationships, 93-94
Limonite, definition, 133
Lithium, content in soils, 7
Lithosphere, elements in, 6-8

Maghemite
  equilibrium constants, 130
  stability diagram, 132
  stability relationships, 129, 132-133
Magnesite
  equilibrium constants, 108
  stability diagram, 114
  stability relationships, 113-114
Magnesium
  content in lithosphere, 7, 106
  content in soils, 7, 106
  redox relationships, 108, 116
  removed by leaching, 111
Magnesium activity
  for acid soils, 111
  for calcareous soils, 111
  effect on aluminum, 73-75
  see also Soil-Mg
Magnesium aluminosilicates
  effect of silica on, 113

equilibrium constants, 107
  stability diagram, 112
  stability relationships, 112-113
Magnesium carbonates
  equilibrium constants, 108
  stability diagram, 114
  stability relationships, 113-114
Magnesium complexes
  with bicarbonate, 115-116
  with carbonate, 115-116
  with chloride, 115-116
  equilibrium constants, 108
  with nitrate, 115-116
  with phosphate, 115
  stability diagrams, 115
  stability relationships, 114-116
  with sulfate, 115
Magnesium fluorapatite, 186
Magnesium hydrolysis
  stability constants, 108
  stability diagram, 115
  stability relationships, 115-116
Magnesium hydroxides
  equilibrium constants, 108
  stability diagram, 114
  stability relationships, 113-114
Magnesium hydroxyapatite, 186
Magnesium molybdate, equilibrium constants, 366
Magnesium oxides
  equilibrium constants, 108
  stability diagram, 114
  stability relationships, 113-114
Magnesium oxyapatite, 186
Magnesium phosphates
  equilibrium constants, 165
  as fertilizers, 186-187
  reaction products in soils, 186
  stability diagram, 187
  stability relationships, 186-187
  stability relative to calcium phosphates, 186-187
Magnesium silicates
  equilibrium constant, 107
  soluble in acid soils, 110-111
  stability diagram, 110
  stability relationships, 106, 110-111
Magnesium sulfates
  equilibrium constants, 108

INDEX

stability diagram, 114
stability relationships, 113-114
Magnesium sulfide
  controlling soil phases, 292-293
  equilibrium constants, 285
  pe + pH of formation, 293-294
  unstable in soils, 292, 294
Magnetic minerals, maghemite, 133
Magnetite
  effect of CO_2 on stability, 141-142
  effect on phosphates, 177-180
  equilibrium with siderite, 141-142
  equilibrium with soil-Fe, 141-142
  redox stability range, 141-142
  stability diagram, 141-142
  stability relationships, 141-145
  transformation, to ferric sulfide, 291-293
    to markasite, 291-293
    to pyrite, 291-294
    to pyrrhotite, 291-294
Malachite
  equilibrium constants, 224
  stability diagram, 223
  stability relationships, 222-223
Manganese
  content in lithosphere, 7, 151
  content in soils, 7, 151
  in flooded soils, 151
  maximum solubility, 156-157
  redox relationships, 151-160
Manganese activity, solid phase controls, 154-157
Manganese(II) complexes
  with bicarbonate, 152, 157, 159
  with carbonate, 152, 157, 159
  with chloride, 152, 157, 159
  equilibrium constants, 152
  stability diagram, 159
  with sulfate, 152, 157, 159
Manganese(II) hydrolysis species
  effect of redox on, 157-158
  equilibrium constants, 152
  equilibrium with magnetite and pyrolusite, 157-160
  stability diagram, 158
Manganese hydroxide, see Manganese oxide
Manganese minerals
  effect of CO_2 on, 153-160

effect of pH on, 153-160
effect on phosphate, 188-189
effect of redox on, 153-160
effect of silica on, 153-154
equilibrium constants, 151, 152
mixed valency states, 151, 155
stability diagrams, 153, 156
stability relationships, 153-160
Manganese molybdate
  equilibrium constants, 366
  stability diagram, 371
  unstable in soils, 369
Manganese oxides
  equilibrium constants, 151
  stability diagrams, 153, 156
  stability relationships, 153-156
Manganese phosphates
  effect of calcite on, 188-189
  effect of CO_2 on, 188-189
  effect of manganese minerals on, 188-189
  effect of Mn/P ratio, 189
  effect of pH on, 188-189
  effect of redox on, 187-189
  equilibrium constants, 165
  stability, relative to hydroxyapatite, 188
    to strengite, 188
  stability diagram, 188
Manganese sulfates
  equilibrium constants, 152
  stability relationships, 157
Manganese sulfide
  controlling soil phases, 292-293
  equilibrium constants, 285
  pe + pH of formation, 291, 293-294
  stability diagram, 291
  unlikely in soils, 295
Manganese toxicities, in highly reduced soils, 156
Manganite
  coexistence with pyrolusite, 153-154
  coexistence with rhodochrosite, 155, 159-160
  equilibrium constant, 151
  stability diagrams, 153, 156
  stability relationships, 153-160
Manganosite
  equilibrium constant, 151
  stability diagrams, 153, 156
  stability relationships, 153-154
Mardin silt loam, log Al^{3+} + 3pH in, 47-48

Markasite
  controlling soil phases, 292-293
  equilibrium constants, 285
  pe + pH of formation, 291, 293
  stability diagram, 291
Maximum solubility in soils, determination of, 6, 8
MCP, *see* Monocalcium phosphate monohydrate
Mercury
  chalcophilic element, 344
  content in lithosphere, 7, 344
  content in soils, 7, 344
  cycled by rain, 344
  effect of halides on, 360-361
  effect of redox on, 359-361
  effect of total mercury on distribution, 360-361
  liquid, transformation to $Hg_2S(c)$, 291-293
  ore deposits, 344
  organic, 362
  oxidation states, 344
  summary redox diagram, 360
  ubiquitous element, 344
Mercury, elemental
  aqueous mercury, 355
  equilibrium constants, 347
  oxidation to Hg^{2+}, 355
  oxidation to Hg_2^{2+}, 355
  solubility in water, 355
  vapor pressure, 355
Mercury(I) complexes
  with ammonia, 352-353
  equilibrium constants, 347
  with halides, 354
  with pyrophosphates, 354-355
  unstable in soils, 354
Mercury(II) complexes
  with ammonia, 352-353
  effect of total Hg, 348-349
  equilibrium constants, 346-348
  equilibrium with Hg_2I_2 (c), 357-358
  equilibrium with HgI_2 (c), 357-358
  equilibrium with $Hg(\ell)$, 357-358
  with halides, 348-351
  mole fraction in solution, 357-358
  stability diagram, 349-352
  stability relationships, 348-353, 360-361
Mercury(I) halides
  effect of halide activities on, 356
  effect of redox on, 356
  stabilities relative to $Hg(\ell)$, 356
Mercury (II) halides
  equilibrium constants, 346
  stability diagrams, 345, 349-350
  stability relationships, 344-345
Mercury(II) hydrolysis
  effect of total mercury on, 353
  stability constants, 346
  stability diagram, 351
  stability relationships, 352-353
Mercury(II) hydroxide
  stability constants, 346
  stability diagram, 345
  stability relationships, 345
Mercury(I) minerals
  carbonates, 353
  Hg_2I_2 (c) most stable, 353, 355
  hydroxide, 353
  phosphates, 353
  stability constants, 347
  stabilities in soil, 353-355
  sulfates, 353
Mercury(II) minerals
  equilibrium constants, 346
  stability diagrams, 345, 349, 350, 351, 352
  stability in soils, 344-355
Mercury(II) oxides
  equilibrium constants, 346
  red, yellow, etc., 345-346
  stability diagram, 345
  stability relationship, 344, 351-353
Mercury(II) sulfates
  equilibrium constants, 346
  unstable in soils, 345
Mercury(I) sulfide
  equilibrium constants, 285, 348
  example calculations, 291-292
  redox stability range, 291, 288-294, 358-359
  stability diagram, 291-359
  stability relationships, 358-359
  transformation to cinnabar, 292
Mercury(II) sulfide
  equilibrium constants, 348
  stability diagram, 359
  stability relationship, 358-359
Metal chelation, *see* Chelates

INDEX

Metal solubilities
  for copper, 295-297
  effect of sulfides on, 295-297
Metal sulfides
  effect of redox on formation, 290-295
  equilibrium constants for, 284-285
  example calculation of, 290-292
  redox for precipitation of, 291-294
  soil cation reactions, 285
  stability diagram of, 291
  unstable in soils, 287
Methane
  equilibrium reaction with, 377
  formation from anaerobic digestion, 382
  oxidation state, 374-375
  solubility constant, 377
  stability diagram, 378, 380-381
Methanol
  oxidation state, 375
  stability diagram, 378, 380-381
Microcline
  equilibrium constant, 59, 120
  stability diagram, 66, 124
  stability relationships, 65-66, 124
Minerals, control soil solution, 4-5
Mixed constants
  definition, 18, 241
  for metal chelates, 242-243
Molecular ratios in soils, importance of, 6
Molybdenite
  controlling soil phases, 292-293
  equilibrium constants, 285, 366
  equilibrium diagram, 291, 371
  pe + pH of formation, 291, 293-294
  stability relationships, 370
Molybdenum
  availability to plants in highly reduced soils, 371-372
  constants limited, 365
  content of lithosphere, 7, 365
  content of soils, 7, 365
  environmental contaminant, 365
  essential to plants and animals, 365
  not control Pb^{2+}, 336
  oxidation states, 368
  polymerizes at high concentration, 365
  reactions with iron oxides, 369
Molybdenum, elemental
  redox reactions, 368
  stability diagram, 371
  unstable in soils, 369
Molybdenum activity, reference level for soils, 369. See also Soil-Mo
Molybdenum hydrolysis
  stability constants, 366
  stability diagram, 367
  stability relationships, 365-367
Molybdenum minerals
  stability constants, 366
  stability diagram, 368-371
  stability relationships, 367-371
Molybdenum oxides
  equilibrium constants, 366
  stability diagram, 368
  stability relationships, 368
Molybdenum sulfites, see Molybdenite
Molybdic acid
  dissociation constants, 366
  stability diagram, 367
  stability relationships, 365-367, 368
Molybdite
  equilibrium constants, 366
  stability diagram, 368
  stability relationships, 368
Molysite
  equilibrium constant, 130
  stability relationships, 133
Monetite
  equilibrium constants, 165
  from MCP, 197-204
  stability relationships, 180-184
Monocalcium phosphate monohydrate
  congruent dissolution, 198-202
  dissolution of iron and aluminum, 202-204
  dissolving granule, 197-204
  equilibrium constant, 165
  formation of MTPS, 198-202
  formation of TPS, 198-202
  reaction products of, 200-204
  reactions in $CaO-P_2O_5-H_2O$ system, 197-202
  in superphosphate, 197
Monteponite
  equilibrium constants, 318
  stability diagram, 317
  unstable in soils, 317
Montmorillonite, magnesium
  effect on aluminum solubility, 73-75
  effect of silica on, 113

equilibrium constants, 107
stability diagram, 112
stability relationships, 112-113
summary diagram, 71
MTPS
  in $CaO-P_2O_5-H_2O$ system, 197-202
  composition of, 199
  definition, 198-199
  formed from MCP, 198, 200
  pH of, 199
  reaction with soil, 202-204
Muscovite
  equilibrium constant, 59, 120
  stability diagram, 66, 124
  stability relationships, 65-66, 72-76, 124
  summary diagram, 71
MXL
  definition of, 247
  measurement of, 248-249

Natural chelates, 263-264
Nepheline
  equilibrium constant, 58, 120
  stability diagram, 63, 122
  stability relationships, 122-123
Nernst equation, 27
Nesquehonite
  equilibrium constant, 108
  stability diagram, 114
  stability relationships, 113-114
Newberryite, equilibrium constants, 165
Nickel, content in soils, 7
Nitrate ion
  chemical formula, 269
  energy-free fertilizer, 271-272
  equilibrium reaction, 270
  oxidation state, 269
  reference level for soils, 272
  stability diagrams, 271, 273-279
  stable under atmospheric conditions, 271-272
Nitric oxide
  chemical formula, 269
  equilibrium reaction, 270
  oxidation state, 269
  possible intermediate in denitrification, 277-279
  redox stability range, 277-279
  stability diagrams, 271, 273, 275-279
Nitrogen

compounds of, 269
content in atmosphere, 268
content in soils, 7, 268
equilibrium constants, 270
as N_2 (g) in atmosphere, 269
nonequilibrium, 269-279
oxidation states, 268-269
reference levels for soils, 272, 274, 275, 277
Nitrogen dioxide
  chemical formula, 269
  equilibrium reaction, 270
  oxidation state, 269
  stability diagram, 271-273
Nitrogen equilibria
  effect of redox on, 272-279
  excluding N_2, 275-276
  excluding N_2 and N_2O, 276-277
  excluding N_2, N_2O, and N^-, 277-279
  with N_2 and O_2 of atmosphere, 269-272
  stability diagrams, 271, 278-279
Nitrogen trioxide
  chemical formula, 269
  equilibrium diagram, 271
  equilibrium reaction, 270
  oxidation state, 269
Nitrous oxide
  chemical formula, 269
  content in atmosphere, 276
  equilibrium reactions, 270
  oxidation state, 269
  stability diagrams, 271, 273, 275
Nsutite
  equilibrium constant, 151
  stability diagrams, 153, 156
  stability relationships, 153, 155-156

OCP, *see* Octacalcium phosphate
Octacalcium phosphate
  effect of redox on, 186
  equilibrium constant, 165
  stability diagram, 181, 183
  stability relationships, 180-184
Octavite
  equilibrium constants, 318
  stability diagram, 317
Oil, 383
Oldhamite
  controlling soil phases, 292-293
  equilibrium constants, 284

INDEX

pe + pH of formation, 293-294
  unstable in soils, 294
Olivine
  equilibrium constant, 107
  stability diagram, 110
  stability relationships, 106, 110-111
Olivine, calcium
  equilibrium constants, 88
  stability diagram, 91
  stability relationships, 91-92
Organic acids
  dissociation constants, 377
  equilibrium relationships, 379
Organic compounds
  effect of redox on, 381-382
  oxidation to CO_2 and H_2O, 381-382
  reduction to methane, 382
Organic matter
  average content in soils, 374
  exchange with soil solution, 3
  unstable in soils, 374
Organic mercury, *see* Mercury, organic
Organic reactions, catalyzed, 379
Organisms, source of electron, 30
Orthophosphoric acid
  activity equation, 169
  dimer of, 164, 169
  dissociation constants, 164
  effect of pH on, 168
  mole fraction development, 163, 168
  mole fraction diagram, 168
  stable in soils, 163
Otavite, transformation to greennokite, 291-294
Oxalate
  dissociation constants, 377
  oxidation state, 374-375
  stability diagrams, 378, 380-381
Oxidation, *see* Redox
Oxygen
  content in soils, 7
  near roots, 30
  nonequilibrium conditions, 30
  utilized in soils, 4

Paragonite
  equilibrium constant, 58, 120
  stability diagram, 122
  stability relationships, 122-123

pe
  absolute activity, 25-26
  advantages of, 23, 25
  aqueous pe-pH diagram, 25
  conversion to Eh, 25, 27-28
  definition, 23, 25
  see also pe + pH
pe + pH
  advantages, 25, 28-30
  aqueous redox limits, 23-25
  constancy in soils, 28-29
  definition of, 23-25
  equilibrium with atmospheric O_2, 24-25
  equilibrium with $H_2(g)$, 23-25
  equilibrium with $O_2(g)$, 24-25
  measurement in soils, 28-30
  partition of hydrogen ions, 30
  range for soils, 25
  in waterlogged soils, 30
Periclase
  equilibrium constant, 108
  stability diagram, 114
  stability relationships, 113-114
Phase rule
  application to
    $CaO-CO_2-H_2O-H_2SO_4$ system, 102-103
    $CaO-CO_2-H_2O$ system, 102
    CO_2-H_2O system, 102
  definition, 101
Phosgenite
  equilibrium constants, 330
  stability diagram, 333
  stability relationship, 329
Phosphate
  effect of liming, 181-182
  effect of redox on, 177-180, 185-186
  initial reaction products, 204
  pH of maximum solubility, 181-182
Phosphate activity, calculation of, 169, 173
Phosphate complexes
  with calcium, 167, 195-197
  contribution to total phosphorus, 195-197
  effect of pH on, 195-197
  effect of redox on, 196-197
  formation constants, 167
  importance of, 197
  with iron, 167, 195-197
  of magnesium, 167, 195-197
  mole fraction diagram, 195
Phosphate fertilizer
  concentrated superphosphate, 197-198

reactions in soils, 197-204
Phosphine
  effect of redox on, 190
  equilibrium constant, 166
  unstable in soils, 190
Phosphoric acid, *see* Orthophosphoric acid
Phosphorus
  content in soils, 7, 163
  in lithosphere, 17, 163
  oxidation states, 189-190
Phosphorus, elemental, not stable in soils, 190
Phosphorous acid
  dissociation constants, 166
  effect of redox on, 190
  oxidation state, 190
  transformation to orthophosphate, 190
Platinum electrode, reliability, 29-30
Podzolization, role of chelates in, 239
Polyphosphates
  availability to plants, 194
  as fertilizer, 190-194
  inclusion of cations, 191
  unstable in soils, 193
Polysulfides
  equilibrium reactions, 284
  oxidation states, 282
  stability diagrams, 283
  stability relationships, 283, 286
Portlandite
  coexistence with calcite, 93, 100
  equilibrium constant, 89
  stability diagram, 94
  stability relationships, 93-94
Potassium
  content in soils, 7, 119
  in lithosphere, 7, 119
  redox relationships, 121, 126-127
  solubility relative to sodium, 125
Potassium activity, *see* Soil-K
Potassium aluminosilicates
  effect of silica on, 124
  equilibrium constant, 120
  stability diagram, 124
  stability relationships, 124-125
Potassium complexes
  with carbonate, 125-126
  with chloride, 125-126
  equilibrium constants, 121
  stability diagram, 126
  with sulfate, 125-126
Potassium glass
  equilibrium constant, 59, 120
  stability diagram, 66, 124
  stability relationships, 65-66, 124
Potassium hydrolysis
  stability constant, 121
  stability diagram, 126
  stability relationships, 125-126
Potassium ions, fixed by clay minerals, 125
Potassium minerals
  effect of silica on, 124
  equilibrium constants, 120
  stability diagram, 124
  stability relationships, 124-125
Potassium phosphates, equilibrium constants, 166
Primary minerals, 57
Pseudowollastonite
  equilibrium constant, 88
  stability diagram, 91
  stability relationships, 91-92
Pyrite
  controlling soil phases, 292-293
  equilibrium constants, 285
  pe + pH of formation, 291, 293-294
  stability diagram, 291
Pyrochroite
  equilibrium constant, 151
  stability diagrams, 153, 156
  stability relationships, 153, 155-156
Pyrolusite
  coexistence with manganite, 153-154
  equilibrium constant, 151
  stability diagrams, 153, 156
  stability relationships, 153-160
Pyrophosphate
  of calcium, 167, 191
  conversion to orthophosphates, 190-194
  equilibrium constants, 167
  hydrolysis, 190-193
  of iron, 167, 193-194
  of magnesium, 167, 191
  unstable relative to orthophosphates, 190-194
Pyrophosphate complexes
  stability constants, 167
  stability diagram, 192
  stability relationships, 192-193

INDEX 443

Pyrophosphoric acid
  dimer of, 166
  dissociation constants, 166
  effect of water on composition, 190-192
  in equilibrium with calcium pyrophosphate, 191-193
  formed by heating, 191
  mole fraction diagrams, 191
Pyrophyllite
  equilibrium constant, 58
  stability diagram, 61, 71
  stability relationships, 61-62, 71-72
Pyroxene
  equilibrium constant, 59, 88
  lime equivalent, 92
  stability diagram, 67, 91
  stability relationships, 66-68
Pyrrhotite
  controlling soil phases, 292-293
  equilibrium constants, 284
  pe + pH of formation, 291, 293
  stability diagram, 291

Quartz
  equilibrium constant, 51
  stability diagram, 52
  stability in soils, 51-52

Rainfall, dilution of soil solution, 3
Redox measurements in soils, 28-30
Redox relationships, 23-30. *See also* pe + pH
Rhodochrosite
  equilibrium constant, 152
  equilibrium with manganite, 154-155, 159-160
  stability diagrams, 153, 156
  stability relationships, 154-157, 159-160
Rhodonite
  equilibrium constant, 152
  stability diagram, 142
  stability relationships, 153-154
Rhombic sulfur, *see* Sulfur, rhombic
Rice plants
  phosphate availability, 179-180
  redox relationships near roots, 179-180
Rodochrosite
  transformation to alabanite, 291-294
  transformation to hauerite, 291-294
  transformation to manganese sulfide, 291-294
Rubidium, content in soils, 7

Sanidine, high
  stability constant, 59, 120
  stability diagram, 66, 124
  stability relationships, 65-66, 124
Scandium, content in soils, 7
Secondary minerals, stability in soils, 57
Selenium, content in soils, 8
Sepolite
  equilibrium constant, 107
  stability diagram, 110
  stability relationships, 106, 110-111
Serpentine
  equilibrium constant, 107
  stability diagram, 110
  stability relationships, 106-111
Siderite
  effect of CO_2 on, 141-145
  effect of redox on, 141-145
  equilibrium constant, 131
  stability diagram, 142, 144-145
  stability relationships, 141-145
Silica
  effect on aluminum solubility, 73-75
  equilibrium constants, 52
  forms in soils, 51-52
Silica, amorphous
  equilibrium constant, 51
  flocculated by $0.02\,M$ $CaCl_2$, 54
  solubility relationships, 51-52
  stability diagram, 52
Silica minerals, equilibrium constants, 51
Silicate ions
  dissociation constants, 51
  effect of pH on, 53-54
  major species, 51-53
  stability diagram, 53
Silicic acid
  dissociation constants, 51
  hydrated, 52
  major species, 51-53
Silicon
  content in lithosphere, 7
  content in soils, 8, 51
Sillimanite
  equilibrium constant, 58
  stability diagram, 61
  stability relationships, 61-62

Silver
  in lithosphere, 8, 300
  metallic, see Silver, elemental
  oxidation states, 300, 303-304, 308, 310
  reference levels for soils, 309
  in soils, 7, 300
Silver, elemental
  redox stability diagrams, 301, 305
  stability constant, 302
  stability range in soils, 300-301, 304-305, 312
  transformation to sulfides, 291-294
Silver activity, reference level, 309
Silver bromide
  equilibrium constants, 302
  stability diagram, 305
  stability relationships, 304-306, 312
Silver carbonate
  equilibrium constants, 302
  stability diagram, 307
  stability in soils, 306, 312
Silver chloride
  equilibrium constants, 302
  stability diagram, 305
  stability relationships, 304-306, 312
Silver complexes
  with ammonia, 311-312
  contribution to soluble silver, 308-309
  equilibrium constants, 302-303
  with halides, 308-309
  with nitrate, 310-312
  stability diagram, 308
  with sulfate, 311-312
  with sulfite, 311-313
Silver fluoride
  equilibrium constants, 302
  unstable in soils, 304, 312
Silver halides
  equilibrium constants, 302
  example soils, 304-306
  stability diagram, 305
  stability in soils, 304-306, 312
Silver hydrolysis
  insignificant in soils, 310
  stability constants, 303
  stability diagram, 310
Silver hydroxide, see Silver oxides
Silver iodide
  for cloud seeding, 300
  equilibrium constant, 302
  equilibrium diagram, 305
  stability relationships, 304-306, 312
Silver minerals
  effect of redox on, 300-301, 303-306
  equilibrium constants, 302
  stability diagrams, 301, 305, 307
Silver molybdate
  equilibrium constants, 302, 366
  stability diagram, 307
  stability in soils, 308
Silver nitrate
  equilibrium constants, 302
  stability diagram, 307
  unstable in soils, 306, 312
Silver oxides
  equilibrium constants, 302
  redox stability diagram, 301
  unstable in soils, 300-301, 312
Silver phosphate
  equilibrium constants, 302
  stability diagram, 307
  stability in soils, 306, 312
Silver sulfides
  α, β forms, 302, 303
  α, β forms coexist, 294
  controlling soil phases, 292-293
  diagram development, 301
  effect of pH on, 301
  effect of redox on, 301, 303, 306
  equilibrium constants, 284, 302
  equilibrium diagrams, 291
  pe + pH of formation, 291, 293-294
  redox stability diagram, 301, 305
  stability relationships, 301, 303, 305-306, 312
  transformation to elemental silver, 301-302
Silver sulfite
  equilibrium constants, 302
  stability diagram, 307
  unstable in soils, 307, 312
Smithsonite
  equilibrium constants, 212
  stability diagrams, 214, 215
  stability relationships, 213-216
Sodium
  accumulation in soils, 119, 123
  content in soils, 7, 119
  in lithosphere, 7, 119

INDEX

445

movement to oceans, 64
redox relationships, 121, 126-127
Sodium activity, *see* Soil-Na
Sodium aluminosilicates
  effect of silica on, 122-123
  equilibrium constants, 120
  equilibrium diagram, 122
  equilibrium relationships, 119, 122-123
  unstable in soils, 123
Sodium complexes
  with bicarbonate, 125-126
  with carbonate, 125-126
  with chloride, 125-126
  equilibrium constants, 121
  stability diagram, 125
  with sulfate, 125-126
Sodium glass
  equilibrium constant, 58, 120
  stability diagram, 63, 122
  stability relationships, 122-123
Sodium hydrolysis
  stability constant, 121
  stability diagram, 125-126
Sodium minerals
  effect of silica on, 122-123
  equilibrium constants, 120
  stability diagram, 122
  stability relationships, 119, 122-123
  unstable in soils, 119, 123
Sodium phosphates, 189
Soil
  definition of, 2
  dynamic equilibria in, 2-4
  elements in, 6-8
  multiphase system, 2-4
Soil air, effect on soil solution, 3-4
Soil-Ca
  definition, 92
  reference level in soils, 92
  stability diagrams, 91, 94
  transformation to oldhamite, 293-294
Soil-Cd
  in acid soils, 316-317
  need for further study of, 326
  reference state, 316-317
Soil-Cu
  definition, 222
  determination of, 261
  equilibrium constants, 222, 224
  redox stability range, 295-296

reference level for soils, 222
solubility diagrams, 223, 226-227, 229, 230, 232, 235
stability relationships, 222-235
Soil-Fe
  definition, 129, 132
  determination of, 259-261
  effect of redox on, 141-146
  equilibrium constant, 130
  stability relationships, 129, 132-133
Soil-K
  defined, 124
  reference level for soils, 124
  stability diagrams, 124, 126
Soil-Mg
  definition, 111
  equilibrium constants, 108, 111
  stability diagrams, 110, 112, 114
Soil-Mo
  equilibrium constants, 366, 371
  experimental basis, 369
  near wulfenite, 369
  stability relationship, 370
  transformation to molybdenite, 291-294, 370-371
Soil-Na
  definition, 123
  stability diagram, 122
Soil-Pb
  basis of selection of, 335
  effect on molybdenum, 368
  reference level, 335
  stability diagram, 368
Soil-Si
  definition, 52
  reference solubility, 51-52
  silica-rich, 119
  solubility, 51-52
Soil solution
  controlled by minerals, 4-5
  effect on adsorbed ions, 4-5
  factors affecting composition, 2-4
  maximum content at 10% moisture, 6-9
Soil tests
  chelates useful for, 261-263
  use of DTPA, 261-263
Soil-Zn
  definition, 211
  determination of, 261
  equilibrium constants, 212

possibility of franklinite, 214
stability diagrams, 214, 216, 217-218
stability relationships, 211-216
Solid phases in soils
control of cations, 244-245
equilibrium constants for, 244-248
Solubility in soils, 8
Sphalerite
controlling soil phases, 292-293
equilibrium constants, 285
pe + pH of formation, 291, 293-294
stability diagram, 291
Standard cell potential, see Cell potential, standard
Standard free energies of formation, given in appendix
Standard free energy of reaction, 21-22
Standard hydrogen half cell, 26-28
Stishovite, stability, 51-52
Strengite
double function parameters, 174-176
effect of redox on, 177-180
equilibrium constant, 164
equilibrium with soil-Fe, 174-177
reaction product in soils, 174-177
stability diagrams, 175-177
stability relative to
calcium phosphates, 180-183
magnesium phosphates, 188
variscite, 175-177
vivianite, 178-180
Strontium, content in soils, 8
Struvite, equilibrium constant, 165
Submerged soils, phosphates in, 179-180
Sulfate
effect of calcite on, 283
equilibrium diagrams, 283, 288, 296
in equilibrium with gypsum, 283
equilibrium reactions, 284
oxidation states, 282
range in soils, 39
reference levels in soils, 283
stable in most soils, 282, 286
Sulfide
equilibrium reactions, 284
example calculation of, 283, 286
of metal, see Metal sulfides
oxidation state, 282
stability diagrams, 283, 288
stability relationships in soils, 283, 286

Sulfite
equilibrium reactions, 284
metal, see Metal sulfides
oxidation states, 282
stability diagram, 283
unstable in soils, 283, 286
Sulfur
chemical species of, 282
content in soils, 7
effect on metal solubilities, 282, 295-297
effect of redox on, 282-283, 286-287
equilibrium constants for, 284-285
essential for all organisms, 282
in lithosphere, 7, 282
oxidation states, 282
reference level for soils, 283
in soils, 282
stability relationships, 282-297
in submerged soils, 282
Sulfur, elemental, see Sulfur, rhombic
Sulfur, rhombic
conditions for formation in soils, 288-290
definition, 286
effect on other sulfur species, 283, 286-288
effect of pH on, 283, 287-288, 289-290
effect of redox on, 283, 287-288, 289-290, 291, 293-294
effect of total sulfur on, 283, 289-290
equilibrium reaction, 284
$H_2S(g)$ pressure of, 289-290
oxidation state, 282
stability diagrams, 283, 289, 291
stability relationships, 283, 286
unlikely in soils, 291
Sulfur acids, dissociation constants, 284
Sulfur dioxide
equilibrium reactions, 284
oxidation state, 282
stability diagram, 283
unstable in soils, 283, 286
Sulfuric acid
dissociated in soils, 287
dissociation constants of, 284
effect of pH on, 287
Sulfur monoxide
equilibrium reactions, 284
oxidation state, 282
stability diagram, 283
unstable in soils, 283, 286

INDEX

Sulfurous acid
 dissociation constants of, 284
 effect of pH on, 287
 unstable in soils, 287
Sulfur trioxide
 equilibrium reactions, 284
 oxidation state, 282
 stability diagram, 283
 unstable in soils, 283, 286
Superphosphoric acid
 chain length, 190-192
 effect of water on, 190-192
Supersaturation, 4-5

Talc
 equilibrium constant, 107
 stability diagram, 110
 stability relationships, 106, 110-111
Taranakites
 effect of K^+ and NH_4^+ on, 171
 equilibrium constants, 164
 stability diagram, 170
 stability relationships, 170-171
Tenorite
 equilibrium constants, 224
 stability diagram, 223
 stability relationships, 222-223
Tephroite
 equilibrium constant, 152
 stability diagrams, 153
 stability relationships, 153-154
Thiosulfate
 equilibrium reactions, 284
 oxidation state, 282
 stability diagram, 283
 unstable in soils, 283-286
Thiosulfuric acid
 dissociation constants, 284
 effect of pH on, 284
 unstable in soils, 284
Tin, content in soils, 8
Titanium, content in soils, 8
TPS
 in $CaO-P_2O_5-H_2O$ system, 197-202
 composition of, 199
 definition, 198-199
 dissolution of iron and aluminum, 202-204
 formed from MCP, 198-200
 pH of, 199
 reactions with soil, 202-204
 vapor pressure of, 198-200
Trace element phosphates, 189
Transforming equilibrium constants, 18-21
Tricalcium phosphate, α, β
 effect on lead solubility, 334
 equilibrium constant, 165
 stability diagram, 181, 183
 stability relationships, 180-184
Tridymite
 equilibrium constant, 51
 stability diagram, 52
Troilite
 controlling soil phase, 292-293
 equilibrium constants, 285
 pe + pH of formation, 291, 293
 stability diagram, 291

Vanadium, content in soils, 8
Variscite
 diagram development, 169-171
 effect of redox on, 177-180
 equilibrium constant, 164
 reaction product in soils, 172-173
 stability diagrams, 170, 172
 stability relative to calcium phosphate, 180-183
 stability relative to strengite, 175-177
 transformation to vivianite, 179-180
Vermiculite
 equilibrium constant, 60, 107
 stability diagrams, 69, 110-112
 stability relationships, 69-70, 106, 110-111
Vivianite
 effect of iron oxides on, 178-180
 effect of redox on, 178-180
 equilibrium constant, 164
 formed from calcium phosphates, 186
 stability diagram, 178

Wagnerite, 186
Wairakite
 equilibrium constant, 59, 88
 stability diagram, 67, 91
 stability relationships, 66-68, 87, 91-92
Water, redox limits, 23-25
Weathering of soils
 factors affecting, 57

formation
  of aluminosilicates, 35
  of aluminum oxides, 32-38
  of iron oxides, 38
loss
  of aluminosilicates, 37-38
  of calcium minerals, 87
  of carbonates, 87
  of silica, 35
  of sodium, 119
release of potassium, 119
Willemite
  equilibrium constants, 212
  stability diagram, 214
  stability relationships, 213-215
Wollastonite
  equilibrium constant, 88
  stability diagram, 91
  stability relationships, 91-92
Wulfenite
  equilibrium constants, 331, 366
  possibility in soils, 369
  stability diagram, 368
  stability relationships, 336, 368
Wurtzite
  controlling soil phases, 292-293
  equilibrium constants, 285
  pe + pH of formation, 291, 293
  stability diagram, 291
Wustite
  equilibrium constant, 131
  stability diagram, 142
  stability relationships, 142-143

Yttrium, content in soils, 8

Zinc
  content in lithosphere, 8, 211
  content in soils, 8, 211
  oxidation states, 211
Zinc, elemental
  equilibrium constants, 212
  unstable in soils, 211
Zinc activity
  calculation of, 217-219
  critical level for plants, 258-259
  diffusion gradients, 258-259
  effect of DTPA on, 258
  equation for, 217-219
  see also Soil-Zn
Zinc chlorides
  stability constants, 212
  unstable in soils, 215
Zinc complexes
  beneficial in fertilizers, 216
  of chloride, 213, 216, 218
  in equilibrium with soil-Zn, 216-218
  importance in soils, 216-218
  mole fraction of total zinc, 216-218
  of nitrate, 213, 216, 218
  with organic matter, 219
  of phosphate, 213, 216, 218
  stability constants, 213
  stability diagram, 218
  of sulfates, 213, 216, 218
Zinc deficiency, not attributable to zinc sulfide, 294-295
Zinc hydrolysis
  in equilibrium with soil-Zn, 216-217
  stability constants, 213
  stability diagram, 217
  stability relationships, 216-217
Zinc hydroxide
  amorphous, 212
  α-, 212-215
  β-, 212-215
  γ-, 212-215
  ϵ-, 212-215
  as fertilizer, 215
  stability diagram, 214
  stability relationships, 213-215
Zincite
  equilibrium constants, 212
  as fertilizer, 215
  stability diagram, 214
  stability relationships, 213-215
Zinc minerals
  equilibrium constants, 212
  equilibrium relationships, 211-216
  stability diagrams, 214-215
Zinc molybdate
  equilibrium constants, 366
  stability diagram, 268
Zinc oxysulfates
  equilibrium constants, 212
  unstable in soils, 215-216
Zinc phosphate
  equilibrium constants, 212
  effect of phosphates on, 216
  stability diagram, 215
  stability relationships, 216

INDEX

Zinc silicates
  equilibrium constants, 212
  stability diagram, 214
  stability relationships, 213-215
  $ZnSiO_3(c)$ not in soils, 214
Zinc sulfates
  equilibrium constants, 212
  unstable in soils, 215
Zinkosite
  equilibrium constants, 212
  unstable in soils, 215
Zirconium, content in soils, 8

CPSIA information can be obtained at www.ICGtesting.com
Printed in the USA
BVOW030420121112

305265BV00001B/11/A